T0261699

THE GREAT FORMAL
MACHINERY WORKS

THE GREAT FORMAL MACHINERY WORKS

Theories of Deduction and Computation at the Origins of the Digital Age

Jan von Plato

PRINCETON UNIVERSITY PRESS
PRINCETON AND OXFORD

Published by Princeton University Press,
41 William Street, Princeton, New Jersey 08540
In the United Kingdom: Princeton University Press,
6 Oxford Street, Woodstock, Oxfordshire OX20 1TR

press.princeton.edu

Jacket photograph: Charlie Chaplin in *Modern Times* © Roy Export S.A.S. Scan courtesy of
Cinetica di Bologna

ISBN 978-0-691-17417-4

Library of Congress Control Number 20179386675

British Library Cataloging-in-Publication Data is available

This book has been composed in Times New Roman and Univers LT Std

Printed on acid-free paper ∞

Typeset by Nova Techset Pvt Ltd, Bangalore, India
Printed in the United States of America

1 3 5 7 9 10 8 6 4 2

"The great cogwheels of the formal machinery works turn slowly but securely. Servicemen are needed as much as those who invented the machines; nuts and bolts here and there need to be tightened to keep the wheels going. Here is the chief serviceman with his manual..."

CONTENTS

PREFACE

The idea of a formal proof in mathematics emerged gradually in the second half of the nineteenth century, hand in hand with the notion of a formal process of computation. These developments were brought to perfection by the end of the 1930s, with precise theories of formal languages and formal deduction, and parallel theories of algorithmic computability. Soon the idea of a programmable computing machine was formulated. It took another fifty years from these pioneering times for the computer to become an everyday object, and for the present information society to emerge; a brainchild of the mostly unknown but great chapter of science reported in this book.

The final form of the individual chapters stems mainly from notes for my lectures and the scholarly colleague would do well to keep this origin in mind: there is no elaborate discussion of secondary literature, but my aim has instead been to develop in the student the habit of reading original texts. The original works on which I draw were written in English, German, French, Italian, Norwegian, Finnish, Dutch, Latin, and Greek. I have mainly used my own readings and translations, with expert help in the two ancient languages.

Jan von Plato

THE GREAT FORMAL
MACHINERY WORKS

PROLOGUE

LOGICAL ROOTS OF THE DIGITAL AGE

It may sound paradoxical, but if around 1930 Kurt Gödel had not thought very deeply about the foundations of mathematics, there would be no information society in the form in which we have it today. Gödel's solitary work was the single most important factor in the development of precise theories of formal languages, ones that through the coding he invented could be handled by a machine. Likewise, his work led to precise notions of algorithmic computability from which a direct path led to the first theoretical ideas of a computer, in the work of Alan Turing in 1936 and John von Neumann some years later.

The development of logic and foundational study until about 1930 is known in broad outline through the pioneering work of Jean van Heijenoort. In 1967, he published a collection of original papers under the title *From Frege to Gödel: The Development of Mathematical Logic 1879–1931*. No attempt at organized work of a historical nature with the sources was made, however. Such detailed work is still in its beginnings, and crucial sources for the development of logic remain unexplored.

Deep intellectual history cannot be achieved if only published work is considered. A comparison to a parallel and indeed even partly overlapping development will be illuminating, namely that of quantum mechanics: logic and foundations of mathematics, as well as quantum mechanics, received a decisive impetus around the year 1900, with Planck's discovery of the quantum of action and with the discovery of Russell's paradox and the idea of formalization as a way out of the foundational crisis.

Both of these developments had led to a well-established theoretical account by the early 1930s, right before the Nazi takeover in Germany: quantum mechanics in the second part of the 1920s in the work of the Göttingers Heisenberg and Born in the first place, and then logic and foundations with the full development of first-order logic by the early 1930s by the Göttingers Hilbert, Bernays, Ackermann, and Gentzen, assisted mainly by Gödel's striking discoveries. Historical work on the development of quantum mechanics began in the 1960s, in a massive effort in which every conceivable source was unearthed and made available in several hundred microfilm reels, the Archive for the History of Quantum Physics.

Sadly, nothing of this kind was ever done with logic and foundations. This goes even for the work of Gödel, whose incompleteness theorems of 1931 spurred the great development that led to truly formal systems of logic, to the more general idea of formal languages, and to precise notions of algorithmic computability. Formal languages and computability are part of the conceptual apparatus that by the 1950s had led to programming languages and computers. Thus, our present information society owes part of its existence to a well-hidden but essential line of development, the theoretical study of logic and foundations of mathematics boosted by Gödel's discovery.

From a broader perspective, I hope this book will contribute to the understanding of a chapter of European science. One could initially maintain that the development of logic and foundations could not be compared to, say, quantum mechanics that has had an enormous effect on science and technology. On the other hand, a significant part of the computerized society was initiated by the study of formal systems of logic and of theories of computability, the former by Frege, Peano, Russell, Bernays, Gödel, and Gentzen in the first place, the latter by Peano, Skolem, Ackermann, Bernays, Gödel, Politzer, Church, Turing, Kleene, and von Neumann, among others.

The concept of computability has its roots in the foundational research in arithmetic of the nineteeth century. An important milestone was *recursive arithmetic*, ascribed to Thoralf Skolem in 1923, with anticipations deep in the earlier parts of the preceding century. It gives an exact theory of finitary, bounded computations by what are called primitive recursive functions.

In the 1920s, the frontiers of computability were extended by Ackermann's discovery of computable functions that are not primitive recursive. In the early 1930s, recursion theory was developed in the hands of Bernays, and especially Rosa Politzer (later Rozsa Péter) and Laszlo Kalmár.

In 1934, Gödel gave lectures on his incompleteness theorem in Princeton, with emphasis on the definition of computable functions. This event led directly to the remarkable development of the connection between recursive and λ-definable functions by Church and by Kleene and Rosser. Also in 1934, Max Newman gave a course on Gödel's results in Cambridge, attended by Alan Turing. It took Turing one year to find yet another equivalent definition of computable functions, in terms of Turing machines, and directly establish the undecidability of predicate logic through a negative solution of what is now called *the halting problem*.

Beyond Gödel's fundamental result and its immediate impact, the aftermath of the discovery of incompleteness was a series of events mostly known through a few key publications. The difficult circumstances of the 1930s, with the Nazi takeover of German universities and with the Second World War, precluded a normal exchange of results and views. Thus, here we have nothing comparable to some of the great scientific-intellectual debates of the 1920s, like the one between Hilbert and Brouwer on the nature of mathematics, the one between Einstein and Bohr on the nature of microphysical reality, or those on the nature of probability and chance. The actors were there, but the debate lay implicit in what remained mostly unpublished notes.

The year 1936 was a turning point in the developments that led to our contemporary theories of deduction and computation. Gentzen had shown how the proof systems of natural deduction and sequent calculus he created could be applied to the foundational problems of mathematics and indicated a continuation of this line of research within structural proof theory and ordinal proof theory. In other quarters, the beginnings of the information age were being set by Turing, von Neumann, and those who followed them. These latter developments have been the focus of a lot of attention recently, in connection with the Turing centenary of 2012. One collection of studies is the book *Computability: Turing, Gödel, Church, and Beyond* edited by Jack Copeland and others. The editors write in their introduction (Copeland et al. 2013, p. viii):

> In his strikingly original paper of 1936, Turing characterized the intuitive idea of computability in terms of the activity of an abstract automatic computing machine. This machine, now called simply the "Turing machine," figures in modern theoretical computer science as the most fundamental model of computation. . . .

Within a dozen years or so, Turing's universal machine had been realized electronically—the world's first stored-program electronic digital computers had arrived.

There is no denying the fact that computers were a by-product of the research in logic and foundations of mathematics, yet no one could have foreseen that development, not even in the mid-1930s. Bernays, Gödel, Gentzen, Church, and Turing, to mention the five foremost names, did their research in the 1930s in a purely academic environment, with exclusively intellectual interests. Yet the societal impact of their research, to use a term in vogue, was enormous. Turing used the programmable electronic computer in his successful attempt at deciphering the ENIGMA machine, an encryption device used by the German submarines that attacked the Allied convoys of cargo ships to Britain. On the other side of the Atlantic, von Neumann used the electronic computer to numerically calculate the effect of a nuclear explosion.

1

AN ANCIENT TRADITION

1.1. REDUCTION TO THE EVIDENT

The original idea of mathematical proofs was to make immediately evident that which is somehow hidden and beyond what can be directly seen. The immediately evident was given in the form of axioms that formed the basis of proofs of theorems. Examples from ancient Greek geometry are easily found: say, the sum of angles of any quadrangle makes a full circle. To make this evident, first check that any quadrangle can be divided into two triangles. For a convex quadrangle, just take any two opposite vertices and join them by a line. The same cutting into two triangles works also if the quadrangle has a cavity on one side:

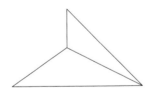 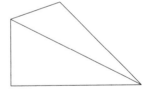

Now take one of the triangles and draw a perpendicular through one of the vertices. This gives you two smaller right-angled triangles. Take one of these and draw another congruent triangle with the same hypothenuse, and you have a rectangle the angles of which add up to a circle:

The angles of each of the two triangles have to add up to half a circle. Now do the same with the bigger triangle that was one half of the original quadrangle. Here, with some details filled in, we could add the famous *QED, quod erat demonstrandum* (as was to be proved).

Ancient Greek geometry was an attempt at determining explicitly what it is that makes proofs like the preceding one evident, namely the axioms of geometry and their combination step after step, as codified in Euclid's book *The Elements*. The preceding proof contains certain "auxiliary constructions" with specific properties, all detailed in Euclid.

1. First, the *connecting line* between two vertices of the quadrangle is drawn. *The Elements* contains a *construction postulate* by which the geometer assumes to have the capacity to produce lines that pass, with infinite precision one could say, through two given points. Therefore, it is justified to conclude that two triangles are produced.

2. In the second diagram, the auxiliary construction is the *perpendicular* to the base of the triangle, drawn through the opposite vertex. This could be produced by another construction postulate, but the organization in Euclid is different. He poses the construction of such a perpendicular or orthogonal line as a *problem*. Namely, his results are divided into theorems in which it is required to prove a result, and into problems in which it is required to produce a geometric object with some specific properties. The chosen conceptual order of things now requires a *definition* of perpendicularity that Euclid gives through the requirement that the two adjacent angles be equal. When this is the case, the angles are by definition *right*.

3. A third auxiliary construction is the triangle that has the same hypothenuse as the one produced by the perpendicular line. Principles of proof are required from which it follows that the two triangles have the same angles, and then the final result follows by some summing up of conclusions and angles.

In the end, the system of geometry becomes complicated, and one needs training in it. Mere mastery of the principles is not sufficient, for it cannot be determined in advance what auxiliary constructions are needed to solve a problem; their discovery comes from the ingenuity of the geometer.

Geometric proofs have been a constant part of higher education for more than two thousand years. Another part has been Aristotle's theory of logic, his syllogistic inferences. Here the basis is given by abstract forms of language such as *Every A is B* and *Some A is B*. Consider the two inference patterns

Assumption α:	*Every A is B*	*Some A is B*
Assumption β:	*Some B is C*	*Every B is C*
Conclusion:	*Some A is C*	*Some A is C*

Strangely enough, the first suggested inference is incorrect, but the second is correct. Where does the latter come from? Aristotle gives an explanation of the quantifiers *Every* and *Some* and thereby justifies the second inference. For the first, we can give a *counterexample*; say, let *A*, *B*, and *C* be, respectively, *number divisible by four, number divisible by two, a prime number.* Now we get, for assumption α, *Every number divisible by four is a number divisible by two*, and for assumption β, *Some number divisible by two is a prime number*. Both are obviously correct, the latter because 2 is divisible by 2. The conclusion, however, is *Some number divisible by four is a prime number*, which is a falsity. In producing the example, we had to make a little adjustment, by adding of the indefinite article in the first assumption.

The discovery of non-Euclidean geometries in the nineteenth century changed the traditional picture of axioms as evident truths: If triangles are drawn on the surface of the Earth so that each side is a part of a great circle (one that passes through two opposite points of the globe), the geometry is elliptic, and the sum of the angles of triangles is greater than that of two right angles. Axioms are now just some postulates that we choose as a basis.

For some reason, today's logic did not first follow the lead of geometry, as a theory of hypothetical reasoning from axioms, but was formulated as a theory of logical truth on which even truth in mathematics was to be based. Here, then, is the essential tension of the following account: truth or proof?

1.2. ARISTOTLE'S DEDUCTIVE LOGIC

Aristotle's system of deductive logic, also known as the "theory of syllogisms," has been interpreted in various ways in the long time since it was conceived. The situation is no different from the reading of other chapters of the formal sciences of antiquity, such as Euclid's geometry and the works of Archimedes. When Frege invented predicate logic, he finished the presentation proudly with a reconstruction of the Aristotelian forms of propositions, such as *Every A is B* that is interpreted as $\forall x(A(x) \supset B(x))$, with a universal quantification over some domain and the predicates A and B. Frege similarly reproduced Aristotelian inferences, such as the conclusion *Every A is C* obtained from the premises *Every A is B* and *Every B is C*, in the way shown in Section 4.2. Frege's interpretation has become the most common one, but we can also consider Aristotle's logic in itself, without such interpretations, and see that it works to perfection.

(A) **THE FORMS OF PROPOSITIONS.** Aristotle's system of deductive logic is presented in his book *Prior Analytics*. It begins with four forms of propositions with a *subject A* and a *predicate B*, as shown in table 1.1.

Table 1.1. The Aristotelian forms of propositions

Universal affirmative:	*Every A is B.*
Universal negative:	*No A is B.*
Particular affirmative:	*Some A is B.*
Particular negative:	*Some A is not-B.*

There is also an *indefinite* form of proposition, *A is B*, not usually present in the rules of inference, even though it can so be (*cf.* 26a28).

Subjects and predicates together are *terms*. The indefinite form *A is B* has various other readings: *subject A has the predicate B, predicate B belongs to subject A, B belongs to A,* etc. The last one is preferred by Aristotle, and he writes the other forms similarly:

B belongs to every A, B belongs to no A, B belongs to some A, B does not belong to some A.

Here the copula is written as one connected expression between the predicate and the subject, which underlines the formal character of the sentence construction.

A second reading is given to *B does not belong to some A*, namely *B does not belong to all A* (in 24a17). Another useful way of expressing the Aristotelian propositions is:

Every A is B, No A is B, Some A is B, Not every A is B.

Now the indefinite form *A is B* is a constant part of the propositions, and the varying quantifier structure is singled out. It is seen clearly that the first and last are opposites, and that the second and third are likewise opposites.

The main principle in the formation of propositions is that subjects and predicates are treated symmetrically in the universal and particular propositions: Whenever *Every A is B* is a proposition, also *Every B is A* is one, and similarly for the universal negative and the particular forms. A formal structure is imposed that is not a natural feature of natural language, as in *Some man is wise*, the *converse* of which, *Some wise is a man*, would not be a natural expression but would have to be paraphrased, as in *Some wise being is a man.*

The *universal quantifier* of the Aristotelian form *Every A is B* is explained as follows in the *Prior Analytics* (24b28):

A thing is said of all of another when there is nothing to be taken of which the other could not be said.

Aristotle is saying that universality means the lack of a counterexample, a common idea in logic ever since. The other forms of quantification are explained similarly and are used in a justification of the Aristotelian rules. However, we do not need to go into the details, because it will turn out to be sufficient to treat the four Aristotelian forms as just atomic formulas with two terms but no further internal structure.

(B) THE PRINCIPLE OF INDIRECT PROOF. The two pairs *Every A is B, Some A is not-B* and *No A is B, Some A is B* form between themselves *contradictory opposites*. Furthermore, because from *No A is B* the weaker *Some A is not-B* follows, also *Every A is B* and *No A is B* together lead to a contradictory pair. We indicate the contradictory opposite of a proposition P by the orthogonality symbol, $P^\perp$. (Note that $P^{\perp\perp}$ is identical to P.) In general, if an assumption P has led to contradictory consequences Q and $Q^\perp$, $P^\perp$ can be concluded and the assumption P *closed*. The rule of indirect proof thus takes on the schematic form in table 1.2, with an *inference line* that separates the two premisses above it from the conclusion below.

Table 1.2. The scheme of indirect proof

$$
\frac{
\begin{array}{cc}
\overset{1}{P^m} & \overset{1}{P^n} \\
\vdots & \vdots \\
Q & Q^{\perp}
\end{array}
}{P^{\perp}} \; RAA,1
$$

This schematic proof figure is to be understood as follows. The assumption may appear among those that were used in the derivations of Q and $Q^{\perp}$, respectively. Any numbers $m, n \geqslant 0$ of occurrences of P in the two *subderivations* can be closed at the inference. The closed ones are indicated by a suitable label, such as a number, so that each instance of rule *RAA* (for *reductio ad absurdum*) clearly shows which occurrences of P are closed at the inference. It is typical of Aristotle's proofs that an assumption closed in indirect proof occurs just once: either $m = 1, n = 0$ or $m = 0, n = 1$. We then have one of:

Table 1.3. Aristotelian special cases of indirect proof

$$
\frac{
\begin{array}{cc}
\overset{1}{P} & \\
\vdots & \\
Q & Q^{\perp}
\end{array}
}{P^{\perp}} \; RAA,1
\qquad\qquad
\frac{
\begin{array}{cc}
& \overset{1}{P} \\
& \vdots \\
Q & Q^{\perp}
\end{array}
}{P^{\perp}} \; RAA,1
$$

As mentioned, Aristotle's derivations have at most one instance of indirect proof, as a last rule. A rule of indirect proof in which the premisses of *RAA* are *Every A is B* and its *contrary No A is B* can be derived from the second of the following conversion rules.

(C) THE RULES OF CONVERSION AND SYLLOGISM. Aristotle's system of deductive logic begins properly with his *rules of conversion*:

$$
\frac{No\ A\ is\ B}{No\ B\ is\ A} \; No\text{-}Conv
\qquad
\frac{Every\ A\ is\ B}{Some\ B\ is\ A} \; Every\text{-}Conv
\qquad
\frac{Some\ A\ is\ B}{Some\ B\ is\ A} \; Some\text{-}Conv
$$

The third rule of conversion is a *derivable rule*. Its conclusion is derivable from its premiss by the first conversion rule and the rule of indirect proof. Aristotle, in fact, notes the same (24a22): Given the premiss *Some A is B*, assume the contrary of the conclusion of *Some-Conv*, namely *No B is A*. Then, by *No-Conv*, *No A is B*, a contradiction, so that *Some B is A* follows.

Two more rules enter into Aristotle's deductive logic, the proper *syllogisms* as this word has been understood for a long time. Its meaning in Aristotle vacillates between a single syllogism and what today is called a deduction. The major part of the *Prior Analytics* deals with derivations that consist of a single syllogism, conversions, and a single step of indirect inference. The two syllogistic rules are (25b38–26a2):

Table 1.4. Aristotle's formulation of the syllogistic rules

When A of every B and B of every C, it is necessary that A is said of every C. For we have explained above what we mean by every.

Correspondingly also when A of no B, B instead of every C, then A will not belong to any C.

The added clause hints at a justification of the rule in terms of the meaning given to universal quantification, as discussed in detail in von Plato (2016).

We write the preceding two rules as

$$\frac{Every\ A\ is\ B \quad Every\ B\ is\ C}{Every\ A\ is\ C}\ Every\text{-}Syll \qquad \frac{Every\ A\ is\ B \quad No\ B\ is\ C}{No\ A\ is\ C}\ No\text{-}Syll$$

The order of the premises, from left to right, is the reverse of that in Aristotle's proof texts. At some stage, it became customary to read the propositions with the subject first, so, to have the middle term in the middle, the order of the premises was changed.

When one reads Aristotle's examples of syllogistic inference, the real deductive structure is somewhat hidden behind the convention of a linear sentence structure. Here is an example from *Prior Analytics* (27a10):

If M belongs to every N and to no X, then neither will N belong to any X. For if M belongs to no X, then neither does X belong to any M; but M belonged to every N; therefore, X will belong to no N (for the first figure has come about). And since the privative converts, neither will N belong to any X.

Let us number the sentences of this text in the succession in which they appear, rewritten so that each single purely syllogistic sentence is identified (i.e., with the connectives and rhetorical expressions eliminated):

1. *M belongs to every N.*
2. *M belongs to no X.*
3. *N belongs to no X.*
4. *M belongs to no X.*
5. *X belongs to no M.*
6. *M belongs to every N.*
7. *X belongs to no N.*
8. *N belongs to no X.*

The *assumptions* in the syllogistic proof are 1 and 2. Line 3 states the *conclusion* of the proof. Line 4 *repeats* assumption 2, and line 5 gives the result of applying of a conversion rule to the premiss given by line 4. Line 6 repeats the assumption from line 1. Line 7 gives the conclusion of a syllogistic rule from the premisses given by lines 5 and 6. Line 8 gives the conclusion of a conversion rule applied to the premiss given by line 7. It is at the same time the sought-for conclusion expressed on line 3.

The formal nature of Aristotle's proof text is revealed by the repetition, twice, of assumptions or previous conclusions, on lines 4 and 6. These repetitions are made so that the application of a rule of inference in the proof text can follow a certain pattern: A one-premiss rule such as conversion is applied to a sentence in a way such that the conclusion immediately follows the sentence. A two-premiss rule such as a syllogism is applied to two premisses, given in succession in a predetermined order, so that the conclusion immediately follows the premisses. Let us collect these observations into a proof in which every step is justified in detail. The beginning of the proof is at the place where the word *For* occurs:

1. *M belongs to no X.*   assumption
2. *X belongs to no M.*   from 1 by *No-Conv*
3. *M belongs to every N.*   assumption
4. *X belongs to no N.*   from 2 and 3 by *No-Syll*
5. *N belongs to no X.*   from 4 by *No-Conv*

(D) THE DEDUCTIVE STRUCTURE OF SYLLOGISTIC PROOFS. The linearity of Aristotle's proofs texts hides a part of their true deductive structure. In the example, the assumptions are *independent* of each other: neither is derivable from the other by the rules. Leaving out the tentative statement of the conclusion from line 3, the *deductive dependences* on assumptions in Aristotle's proof are that line 4 depends on assumption 2, line 5 likewise on 2 through 4, line 6 on 1, line 7 on 1 and 2, and line 8 on 1 and 2.

Aristotle's linear derivations can be translated into a tree form by using the following two clauses:

1. *Take the last sentence and draw a line above it, with the name of the rule that was used to conclude it next to the line. Write the sentences that correspond to the lines of the premises of the last rule above the inference line.*

2. *Repeat the procedure until assumptions are arrived at.*

Here is what we get when the translation algorithm is applied to lines 1–5 of the preceding example:

$$\dfrac{\dfrac{1.\ M\ belongs\ to\ no\ X}{2.\ X\ belongs\ to\ no\ M}\ ^{No\text{-}Conv}\qquad 3.\ M\ belongs\ to\ every\ N}{\dfrac{4.\ X\ belongs\ to\ no\ N}{5.\ N\ belongs\ to\ no\ X}\ ^{No\text{-}Conv}}\ ^{No\text{-}Syll}$$

The *deductive dependences* in Aristotle's proof are now univocally determined, as can be seen from the derivation tree. They are as follows:

As the next example, consider the following indirect syllogistic derivation (from *Prior Analytics*, 28b17):

If R belongs to every S but P does not belong to some, then it is necessary for P not to belong to some R. For if P belongs to every R and R to every S, then P will also belong to every S; But it did not belong.

The translation into tree form gives

$$\dfrac{\overset{3}{Every\ R\ is\ P} \quad Every\ S\ is\ R}{Every\ S\ is\ P}\ {\scriptstyle Every\text{-}Syll}$$

$$\dfrac{Every\ S\ is\ P \qquad\qquad\qquad Some\ S\ is\ not\text{-}P}{Some\ R\ is\ not\text{-}P}\ {\scriptstyle RAA,3}$$

To finish this brief tour of Aristotle's logic, we show that his practice of making at most one last step of indirect inference is justified. The proof system consists of the five rules *Every-Conv, No-Conv, Every-Syll, No-Syll*, and *RAA*, and we have the following theorem.

Theorem. Normal form for derivations. *All derivations in Aristotle's deductive logic can be so transformed that the rule of indirect proof is applied at most once as a last rule.*

Proof. Consider an uppermost instance of *RAA* in a derivation. If it is followed by another instance of *RAA*, we have a part of derivation such as

$$\dfrac{\dfrac{\overset{1}{P}\quad\ \vdots}{\dfrac{Q\quad Q^{\perp}}{P^{\perp}}\ {\scriptstyle RAA,1}}\qquad \overset{2}{R}\ \vdots\ P}{R^{\perp}}\ {\scriptstyle RAA,2}$$

This derivation is transformed into

$$\dfrac{\overset{1}{R}\ \vdots\ P\ \vdots\ \dfrac{Q\quad Q^{\perp}}{}}{R^{\perp}}\ {\scriptstyle RAA,1}$$

Two derivations have been *combined* in the transformation, when the derivation of P from R has been continued by the derivation of Q from P.

The preceding transformation is repeated until there is just one instance of *RAA*. If the conclusion $R^{\perp}$ is existential, it cannot be a premiss in any rule, and the claim of the theorem follows. If the conclusion is universal, we have one of the following:

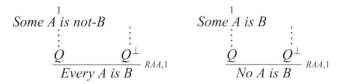

There are by assumption no instances of *RAA* other than the ones shown. Therefore, because existential formulas can be premisses only in the instances of *RAA* shown, the derivations of the left premisses of *RAA* are degenerate, with *Some A is not-B* ≡ *Q* and *Some A is B* ≡ *Q*, respectively. The derivations are therefore

$$
\dfrac{\overset{1}{\textit{Some A is not-B}} \quad \overset{\vdots}{\textit{Every A is B}}}{\textit{Every A is B}} \, {\scriptstyle RAA,1}
\qquad
\dfrac{\overset{1}{\textit{Some A is B}} \quad \overset{\vdots}{\textit{No A is B}}}{\textit{No A is B}} \, {\scriptstyle RAA,1}
$$

The conclusion is equal to the right premiss in both, and therefore the instances of *RAA* can be removed.   QED.

The proof depends crucially on a subtle property of Aristotle's minimal deductive system of five rules, namely that existential formulas are not premisses in rules other than *RAA*.

    The formulas of Aristotle's deductive logic are atomic formulas in today's terminology, and his rules act only on such atomic formulas. The question of the derivability of an atomic formula from given atomic formulas used as assumptions is known as the *word problem*. The terminology stems from algebra where the word problem concerns the derivability of an equality from given equalities. The solution of this problem in Aristotle's deductive logic, i.e., the decidability of derivability by his rules, follows at once by the preceding result on normal form. It is sufficient to show that the terms in a derivation of P from the assumptions Γ can be restricted to those included in the assumptions and the conclusion. If the proof is direct, this is so because terms in any formula in the derivation can be traced up to assumptions. Otherwise, the last step is indirect, but the closed assumption is $P^{\perp}$, so that the terms in a derivation are as in the direct case. With a bounded number of terms, there are a bounded number of distinct formulas. The number of possible consecutive steps of inference in a loop-free derivation, i.e., the height of a branch in a derivation tree, is bounded by the number of distinct formulas and we have the following theorem.

Theorem. Termination of proof search in Aristotle's deductive logic. *The derivability of a formula P from given formulas Γ used as assumptions is decidable.*

The only thing that has been added to Aristotle's proofs as he wrote them is the tree form that keeps track of what depends on what within a proof.

1.3. INFINITY AND INCOMMENSURABILITY

I shall give two ancient proofs about numbers. The first is Euclid's famous proof of the infinity of prime numbers, and the second is the Pythagorean proof of the irrationality of the square root of 2, as given by Aristotle. Both proofs have been the source of infinite misunderstandings. In particular, both have been described as *indirect proofs* which they are not. Furthermore, the first one is a clear proof by induction, in contrast to repeated claims that such proofs surfaced only with the work of Pascal in the seventeenth century. One's natural question is: How could there have been an ancient arithmetic without this crucial principle of proof? There wasn't, but a mere reading of the ancient texts may not be sufficient for seeing it.

(A) THE INFINITY OF PRIMES. Euclid's *Elements* contains 13 "books" as they were called, with books VII to IX being dedicated to arithmetic instead of geometry. The last of these includes Euclid's famous proof of the infinitude of prime numbers, theorem 20:

> *The prime numbers are more than any proposed multiplicity of prime numbers.*

The proof goes as follows, in my translation from Fabio Acerbi's formidable 2007 Greek-Italian edition of all of Euclid in one 2720-page volume! It can be read with some effort exactly as it was written some time around 200 B.C.—no explanations and the considerable delight of figuring from the context the meaning of Euclid's notation, with just the modest hint that the letters as they appear come from their alphabetical order in Greek:

Let the prime numbers proposed be A, B, Γ. I say that there are more prime numbers than A, B, Γ.

Assume the least number to have been taken that is divisible by A, B, Γ, and let it be ΔE, and let a unit ΔZ have been summed to ΔE. Then EZ is in any case either prime or not. Let it be in the first place prime. Then more prime numbers than A, B, Γ have been found, namely A, B, Γ, EZ.

But now, let EZ not be prime, so then it is divisible by a certain prime number. Let it be divisible by prime H. I say that H is not the same one of any of the A, B, Γ. If that should be possible, let it so be. And A, B, Γ divide ΔE, so then even H divides ΔE. And it divides also EZ; it divides also the unit ΔZ that remains because H is a number, which is absurd. So there is not the case that H is the same as just one of the A, B, Γ. And it was assumed to be prime. It then results that more prime numbers have been found, A, B, Γ, H, than the proposed multiplicity A, B, Γ, which was to be proved.

The same proof in today's terminology is: Let n prime numbers $p_1, \ldots p_n$ be given. Form the product $p_1 \times \cdots \times p_n$, add one to it, and denote this number by p. If p is prime, we have $n + 1$ primes. If p instead is divisible, it has at least one prime divisor q. This divisor is distinct from each of the previous p_i, because each of them leaves a rest if used to divide p. So we have $n + 1$ primes.

If we add to this the fact that 2 is a prime number, we have the base case:

2 is a prime number.

The step case is

If there are n prime numbers, there are n + 1 prime numbers.

Euclid's conclusion is that *the prime numbers are more than any proposed multiplicity of prime numbers.* It should be kept in mind how the Ancients saw the concept of infinity, as something that can be extended beyond any bound rather than as a finished totality. Euclid uses precisely this notion in his proof: He does not write that *all* numbers have a property but instead gives a procedure of extension beyond any given number. In this light, it seems rather foolhardy to maintain that Euclid did not give an inductive proof, especially as his result cannot be proved without induction or equivalent.

The notion of potential infinity applies even in the geometrical parts of the *Elements*. One of the construction postulates is *to extend a given line segment indefinitely.* Euclid's lines are not actually infinite in both directions but just rather extensible beyond any point.

Once it is accepted that Euclid gives an inductive proof expressed in the language of his notion of potential infinity, other results in Euclid's arithmetic should be found as evidence for the permanence of his proof pattern. Namely:

1. It is known that there is at least one number a such that some property $P(a)$ holds for a.

2. Assuming that there are n such numbers, there are more than n.

3. Conclusion: Such numbers "are more than any proposed multiplicity."

In Euclid's proof, the inductive step is produced by a construction process from a given n to $n + 1$. Many inductive proofs are so simple that we step silently over them, perhaps with the added word "obviously." However, "obviously" is no recognized step in a proof that stands on its own feet.

(B) THE IRRATIONALITY OF THE SQUARE ROOT OF 2. Aristotle reports in §23 of the *Prior Analytics* the following proof of the Pythagorean theorem about the "incommensurability of the side and diagonal of the square." The proof, in modern notation and reading, is as follows. Assume $\sqrt{2}$ is rational. Then there are n, m such that $\sqrt{2} = n/m$, and let these be relatively prime, i.e., assume that n and m have no common divisors. With both sides squared, we get $2 = n^2/m^2$, so $n^2 = 2m^2$. But now, if n^2 is even, n itself has to be even, so there is some k such that $n = 2k$. By $n^2 = 2m^2$, we get $2^2k^2 = 2m^2$, and dividing both sides by 2, we get $2n^2 = m^2$, by which m also is even, so 2 divides both n and m against the assumption that n and m are relatively prime. Therefore, the assumption that $\sqrt{2}$ is rational led to a contradiction, by which it is irrational.

There is no limit to the number of books, from textbooks for schoolchildren to advanced books written by otherwise competent logicians, that declare the preceding proof to be a *proof by contradiction, a proof by reductio ad absurdum, an indirect proof.* The logically minded reader who keeps things in a conceptual order can only despair and repeat the basics: What is it that we are proving? We are proving the claim that $\sqrt{2}$ is an irrational number. What does it mean to be an irrational number? It means that there do *not* exist two integers n, m such that n/m is equal to the number. At this point, says our logician, the *direct* way to prove a negative statement such as $\neg A$ is to assume A, to derive a contradiction, and to conclude $\neg A$. In an indirect proof, a negative assumption such as $\neg A$ is made, a contradiction is derived, and a positive claim A is concluded. The most typical example is: Assume there does not exist a number x such that $P(x)$, derive a contradiction, and conclude indirectly that there exists an x such that $P(x)$. The pattern of proof is different from the earlier Pythagorean proof of a negative proposition.

The logical notation and rules of proof of formal logic were invented for the purpose of making steps in mathematical proofs explicit. Let us use some of this notation and steps of proof to see that there is no step of indirect inference in the proof of irrationality of $\sqrt{2}$. We start with implication $A \supset B$. The principle of proof is that if A is *assumed* and B *derived under assumption A*, then $A \supset B$ can be concluded with no assumption A left. The neatest way to treat negation $\neg A$ is to use a constant *false proposition* $\bot$ and to define $\neg A$ as $A \supset \bot$. A proof of a negation is obtained if $\bot$ is derived from the assumption A. This is a *direct* proof of a negative proposition. In arithmetic, we can use $0 = 1$ for $\bot$.

In the Pythagorean proof, the task is to prove $\neg \exists x \exists y (\sqrt{2} = x/y)$. To prove this, assume $\exists x \exists y (x/y = \sqrt{2})$, just a formal notation for the previous "there are n, m such that $\sqrt{2} = n/m$," and try to derive a contradiction. The existential quantifiers of an assumption are treated through *instantiation by eigenvariables*; that is, fresh symbols n and m in place of x and y and the assumption $n/m = \sqrt{2}$, with the further property that n and m are relatively prime. None of the simple arithmetic steps up to the contradiction is indirect, though one step is in itself quite interesting: *If n^2 is even, n itself has to be even.* The definition in logical notation is $Even(x) \equiv \exists y (x = 2 \cdot y)$, a *finitary* definition despite the quantifier, because, given x, there is a bounded number of possible values for y. As "obviously" is not allowed for a step of proof, an attempt to cover the "interesting step" will lead to the realization that the only way is the inductive one.

Indirect inference can be formulated as the following rule: from $\neg\neg A$ to conclude A. If our logic is classical, these two are equated. Therefore, our principles of proof have to be formulated so that such equating does not creep in unwanted. We can put the matter overall as:

> *In the Pythagorean proof, the distinction between being rational and not being irrational is precisely what is needed to have a separate notion of indirect proof.*

Observations similar to those about the Pythagorean proof apply to Euclid's proof. A proof could begin with the assumption that there is only a finite number of prime numbers, and let this number be n. Now a contradiction is derived. Thus, it is not the case that the number of primes is finite, and we have a direct proof of a negative claim. Euclid's claim, though, is formulated in positive terms, and his proof gives an algorithm by which, given any number of primes, one more can be effectively produced. Euclid's algorithm does not necessarily produce a prime greater than the given ones, and it leaves enormous gaps. If we start from 2, Euclid's method gives three primes in succession, first $2 + 1 = 3$, then $2 \times 3 + 1 = 7$, then $2 \times 3 \times 7 + 1 = 43$, until $2 \times 3 \times 7 \times 43 + 1 = 1807$ is reached. This number has the prime 13 as a factor, so now we get $2 \times 3 \times 7 \times 13 \times 43 + 1 = 23479$, with the prime 53 as a factor. The algorithm keeps producing composite numbers, with the prime divisors 5 and 89, so there are at least four in succession after the first four primes produced.

Let us look a bit more closely at the logic of Euclid's proof. The first essential step in it is the phrase *Then EZ is in any case either prime or not,* an instance of the law of excluded middle $A \vee \neg A$. This would in general qualify for a genuinely classical, indirect step: Assume the premiss of an indirect inference, namely $\neg\neg A$, and the second of the possibilities in excluded middle is contradicted, with just the first one left, namely A as concluded from $\neg\neg A$. However, the property of being a prime number is *decidable* in a bounded number of steps, and the instance of excluded middle does not have the effect of producing genuinely classical, i.e., undecidable case distinctions. More generally, it can be shown that all results about the natural numbers that don't require the use of quantifiers (*for all* and *there exists*, or ∀ and ∃) for their expression have finitary proofs, appearances to the contrary.

The structure of a mathematical proof must in principle be visible from the way the proof is written. When pressed enough, a mathematician should be able to produce a proof composed in a transparent way of such simple parts that no one competent in mathematics can doubt it. When something in a proof is assumed to be false, a level of understanding is imported that has no place there. Such aspects belong to the background knowledge that guides what we try to prove. An assumption is just an assumption; it can turn out false at some point in the sense that a contradiction is reached, and then its negation can be concluded. In fact, in the presence of more than one assumption, it is the assumptions together that lead to a contradiction. The negation of any one of the assumptions can be concluded with the rest kept in place. Similarly, if a negative assumption is made, indirect proof leads to the positive conclusion.

Explicit rules of proof leave no place for a second level of consideration of truth inside a proof, beyond the level of formal inference. At most, we can say: *If* the assumptions are true, *if* the proof is correct, *then* even the conclusion is true.

1.4. DEDUCTIVE AND MARGINAL NOTIONS OF TRUTH

(A) LOGICAL TRUTH. Today's logic had its beginnings in the work of Frege, Peano, and Russell. In Russell's work, logic had taken the same form as traditional geometry: There were as starting points logical axioms, and just two rules of logical inference. Frege and Russell advocated a doctrine, called *logicism*, by which the axioms express the most basic logical truths, and the rules just make evident other such truths. The scheme even included that first arithmetic, and then the rest of mathematics, be reduced to logic. It is somewhat odd that logic followed the old geometric tradition, for by this time geometric axioms were no longer viewed as geometric truths but rather as postulates that may hold in one situation and fail in another.

Frege was the first to detail the principles of reasoning with *generality*, with the idea that if we can prove a property $A(x)$ for an *arbitrary* x, then we are allowed to conclude the universal claim $\forall x\, A(x)$. Arbitrariness means simply that the *eigenvariable* x of the step of inference does not occur free anywhere in any assumptions on which the premiss of inference to generality $A(x)$ may depend.

Frege took the crucial step into today's logic. His notation was impossible, but Bertrand Russell adapted Peano's notation to Frege's logic, in the great synthetic work *Principia Mathematica* that he wrote together with Alfred Whitehead. The *Principia* contained a correct account of universality, with the instantiation axiom $\forall x\, A(x) \supset A(a)$ for any chosen object a, and the converse rule of generalization.

Existence is a kind of dual to universality. Thus, it can be taken as an axiom that an instance implies existence; that is, $A(a) \supset \exists x\, A(x)$. There is a rule dual to generalization that tells how to deal with an existential assumption $\exists x\, A(x)$: Take an arbitrary instance $A(y)$, and if some claim C follows from $A(y)$, independently of the choice of y, it follows from $\exists x\, A(x)$. This rule, with y an eigenvariable, is used informally all through the history of logic and mathematics, but it did not get its first explicit formulation until the late 1920s.

Frege's rule and axiom give a clear sense to generality: To infer generality, prove an arbitrary instance. In the other direction, if a generality is at hand, either assumed or proved, instances can be taken. But is inference enough? How should that which can be inferred relate to truth?

Philosophical realism dictates an absolute notion of truth. Somewhere in a big imaginary book is a list of all truths. To be a truth just means to be in the list. As the list is infinite, we human beings don't have direct access to it but have to proceed on the basis of evidence for truth. In the best of cases, such evidence amounts to a proof, or inference in the sense of logic and mathematics. The list is not affected by what we happen to have proved; it's at most worth a marginal remark: This truth was even proved by human beings.

By the 1920s, especially under the influence of Ludwig Wittgenstein, the view of logical truths as *tautologies* emerged. A tautology is a logical sentence that is true under all possible circumstances, or true by virtue of its form. Such truths can be found out by analyzing this form:

1. $A \& B$ is true when both A and B are true, otherwise it is false.
2. $A \vee B$ is true when at least one of A and B is true, otherwise it is false.
3. $A \supset B$ is true if B is true as soon as A is true, otherwise it is false.
4. $\neg A$ is true when A is false, otherwise it is false.

Clause 3 can be put also as: $A \supset B$ is true when A is false or B is true, otherwise it is false.

Given a formula such as $(P \supset Q) \vee (Q \supset R)$, built out of three atomic formulas P, Q, and R that do not have any logical structure, there are altogether $2 \times 2 \times 2 = 8$ combinations of truth values to P, Q, and R. If Q is true, $P \supset Q$ is true by clause 3 and the disjunction $(P \supset Q) \vee (Q \supset R)$ as well by clause 2. If Q is false, $Q \supset R$ is true and so is $(P \supset Q) \vee (Q \supset R)$. Whichever way P, Q, and R are, $(P \supset Q) \vee (Q \supset R)$ turns out true, with no consideration of P and R. Such a tautology does not exclude any possible state of affairs, and therefore it is empty: It does not state anything. Combined with the logicist thesis, even mathematical truths are mere empty tautologies. Moreover, any relation they might bear to each other is just apparent, as in the preceding example. Say, with any two mathematical claims A and B, we get from that example, with A in place of P and R and B in place of Q, that either $A \supset B$ or $B \supset A$, a rather startling result by which any mathematical claim implies any other, or the other way around!

Wittgenstein listed in his little book *Tractatus Logico-Philosophicus*, originally just the German "Logisch-Philosophische Abhandlung," a number of logical maxims, from 1 (the world is all that is the case) to 7 (what one cannot speak about, on that one must be silent). One maxim is that tautologies stand on their own feet, so to say; there is no need for a concept of inference from other tautologies or assumptions (maxim 6.1265):

> One can always conceive logic in such a way that every theorem is its own proof.

Axiomatic propositional logic has a few axioms and just the rule of inference: If $A \supset B$ and A have been proved, B can be inferred.

The notion of a tautology can be used in place of the axioms and inferences from them; inference is indeed completely absent from the *Tractatus*.

One would normally think that if a philosophy leads to consequences such as that mathematical theorems have no content but just form, and that there is never any relation between one theorem and another as each theorem is its own proof, there is something wrong with it, but this was not the view taken in the 1920s or 1930s. Georg Kreisel, who was even a student of Wittgenstein's, told me in a discussion about this matter in the summer of 2010: "A position was taken to its extreme to show its absurdity."

Proponents of the tautology view, such as Wittgenstein and the logical empiricists who followed him, did not recognize that there is no account of quantificational logic without rules of inference. We shall see later how desperately Wittgenstein struggled with the quantifiers, but no amount of inward-bound philosophical reflection could replace the command over quantificational inferences reached in other quarters in the latter part of the 1920s, Göttingen in the first place. With his characteristic blindness to what others had accomplished, Wittgenstein's teaching had a devastating effect on some of his students, who never understood quantificational logic under his guidance. The first victim of this attitude was Frank Ramsey whose 1926 essay on mathematical logic makes for very sad reading. As to the logical empiricists, Rudolf Carnap, perhaps their main proponent, published in 1929 a short "outline" of Russell's *Principia*, the *Abriss der Logistik*. One searches in vain for the rule of universal generalization in this booklet: Carnap's logic is propositional, augmented with an axiom of universal instantiation.

Ignorant of the crucial role of a rule of generalization, the proponents of the tautology view extended, or rather pretended to extend, the truth conditions of classical propositional logic to quantificational formulas by two additional clauses:

5. $\forall x\, A(x)$ is true in a given domain $\mathcal{D}$ of objects whenever $A(a)$ is true for all a in $\mathcal{D}$, otherwise it is false.
6. $\exists x\, A(x)$ is true in a given domain $\mathcal{D}$ of objects whenever $A(a)$ is true for at least one a in $\mathcal{D}$, otherwise it is false.

What happens in the case where the domain $\mathcal{D}$ is infinite? With $a_1, a_2, a_3 \ldots$ the objects, from 5 and 6 we then have the following:

$\forall x\, A(x)A$ is true in $\mathcal{D}$ if $A(a_1)$ is true in $\mathcal{D}$ and $A(a_2)$ is true in $\mathcal{D}$ and $A(a_3)$ is true in $\mathcal{D} \ldots$

$\exists x\, A(x)A$ is true in $\mathcal{D}$ if $A(a_1)$ is true in $\mathcal{D}$ or $A(a_2)$ is true in $\mathcal{D}$ or $A(a_3)$ is true in $\mathcal{D} \ldots$

How should we put this? Perhaps it would do to say that *there is no well-founded explanation of truth* under the tautology notion extended to infinite domains. It would not be sufficient to find, next to a claim, an oracular marginal remark to the effect that the claim is true.

In Hilbert's Göttingen, it was realized by 1920 that the logical steps in mathematical proofs could be represented formally as steps in propositional and quantificational logic. Hilbert's earlier work, such as that on foundations of geometry in 1899, had left the logical part on an intuitive basis. Some twenty-odd years later, as a first step under the leadership of Hilbert's assistant Paul Bernays, pure predicate logic got a clear and concise formulation. The next step was to apply it to mathematical axiom systems. They would appear as premisses in formal logical derivations, and thus not as given truths but as hypotheses that could hold in one situation, fail in another, such as the Euclidean axiom of parallels. A crucial component here was to show that a logical derivation of a theorem P from given hypotheses H could be turned into a derivation of the implication $H \supset P$ in pure logic. Step by step, the view of logic and mathematics as a collection of tautologies gave way to the idea of logic as the deductive machinery of mathematics.

(B) UNDECIDABILITY. The tautology view of logic and mathematics, and the *verificationist* philosophy of truth more generally, derives from the *decidability* of classical propositional logic: There is an algorithm for deciding if a proposition is a logical truth. Quantificational logic instead is not decidable, but that result was not confirmed until 1936. Now the choice becomes one between the tautology view and quantificational logic.

There is another notion of decidability, quite different from the preceding one about logical truth, namely the decidability of the atomic formulas of a given mathematical theory. The arithmetic of natural numbers is the best example here. It can be formulated so that the equality of two numbers, $n = m$, is its only basic notion. Then, for any two numerical terms n and m, $n = m$ is either a true or a false numerical equation, something that can be decided by computing the values on both sides, as in $7 + 5 = 3 \times 4$. As suggested in Section 1.3, the way to express this matter inside a logical calculus is to change the underlying pure logic so that the law of excluded middle is not applied as a general logical principle but just for those basic relations that are meant to be decidable. Thus, in arithmetic we pose $n = m \vee \neg n = m$ for any numerical terms n, m. It follows that all quantifier-free formulas A are decidable; that is, that $A \vee \neg A$ is provable for any such formula. Here we have a way of expressing decidability that is lost in classical logic. The law of excluded middle is an example of a tautology, for if A is true, so is $A \vee \neg A$, and if A is false, $\neg A$ is true and thereby also $A \vee \neg A$. It does not follow, in general, that either A or $\neg A$ would be a tautology. If instead arithmetic is formulated constructively, one of A and $\neg A$ is provable whenever $A \vee \neg A$ is.

The possibility of undecidable basic relations took a long time to become understood. One clear expression of such awareness is found in Hilbert's 1894 lectures on geometry, with the idea that the intersection point of two lines escapes to infinity as the lines approach parallelism (Hilbert 2004, p. 75):

> It escapes our experience whether one can always *find a point of intersection* for 2 lines. *We leave the matter undecided for the time being and state only*:
>
> 2 lines on a plane have either one or no points in common.

In Emile Borel's version of constructivism from the first decade of the twentieth century, there is a rather clear recognition that the equality of real numbers cannot be a decidable relation. For example, there is a way of computing what is known as Riemann's constant C, and the computation has so far given $0.4999\ldots$. If all successive decimals are 9's, we have $C = 0.5$, otherwise $\neg C = 0.5$ holds, but the answer is unknown.

The same insight as in Borel's work got a more forceful expression in L. Brouwer's ideas about real numbers in the 1920s. He replaced equality of real numbers as a basic notion with the *apartness* of two reals that we can write as $a \neq b$ (Brouwer 1924a). That a and b are in this way distinct requires that there be a positive lower bound for their difference. Thus, a finite determination of values will verify apartness, though not falsify it, the precise contrary to the case of equality. The latter notion can now be defined as the negation of apartness, and its transitivity comes from the contraposition of Brouwer's *apartness axiom* $a \neq b \supset a \neq c \vee b \neq c$ that is justified as follows. Let a and b be apart, and let it be infinitely difficult to decide if $a \neq c$. Then c must make a positive distance to b so that $b \neq c$.

The point is how to reason with infinitely fine, ideal objects and concepts such as real numbers and their properties and relations. If we follow the lead of the logical positivists, there should be no undecidable basic relations. The doctrine of verificationism requires a method for deciding truth, otherwise a notion is not meaningful. All of traditional synthetic plane geometry, for example, would be declared meaningless metaphysical speculation about infinitely fine points on infinitely thin lines that remain forever unobservable. At the other extreme, there is the unlimited acceptance of classical logic, with its law of excluded middle. Say, begin a proof by $C = 0.5 \vee \neg C = 0.5$ with C Riemann's constant. There are two cases, with two different consequences. As long as the value of C remains undecided, nothing concrete follows from the cases. If instead the classical law is not allowed to enter, computability is maintained. Constructive logical reasoning will never lead from assumptions with a finitary meaning into something infinitistic. That is the main point in the use of a constructive or intuitionistic logic.

The standard view in the 1930s was that finitism and constructivism contain the requirement that all basic relations be decidable and all functions be computable. The requirement on relations is an erroneous view to which one, however, could easily be led if one considered only intuitionistic arithmetic that has a decidable equality as the only primitive relation, instead of also considering the intuitionistic theory of real numbers that cannot be based on a decidable equality, or the first intuitionistic axiomatization ever, Arend Heyting's 1925 system of intuitionistic projective geometry in which decidability of the basic apartness relations cannot be assumed. One of the first outside the intuitionist camp to realize the difference was Kurt Gödel, whose lectures on intuitionism in Princeton in 1941 have been preserved in manuscript form. The extant text begins with two pages of improvements for the lectures, written in his Gabelsberger shorthand. One of the improvements is: "The belief is put aside that a system of axioms has an intuitionistic sense only if the basic concepts are decidable."

From the actual formal work of the intuitionists, Brouwer and his student Heyting in the first place, it can be seen that they required all functions to be computable but not all basic relations to be decidable. So, why was there such a belief or requirement? One reason lies in the possibility to emulate operations through added basic relations. For example, one could substitute the operation of sum in arithmetic by a three-place relation written, say, $\Sigma(a,b,c)$, with the intended meaning that c is the sum of a and b. One result of this move from functions to relations can be seen in Gerhard Gentzen's thesis of 1933, where a theory, elementary arithmetic in this case, gets formulated in pure predicate logic so that general results for the latter can be applied, whereas this would not necessarily be so in a formulation with functions.

(C) **TRUTH VS. PROVABILITY.** With the insight that axioms can be taken as assumptions in a given situation, logic and mathematics returned to an old Aristotelian form of *hypothetical reasoning*, exemplified by syllogistic inferences that have some assumptions or previously proved statements as premisses. The task of logic is just to guarantee that the steps from assumptions to a conclusion are correct. The latter, in turn, means that whenever the assumptions hold, even the conclusion has to hold. An example from mathematics will be instructive. Let a theorem have the form *For all x and y such that $A(x)$ and $B(x,y)$, there is a z such that $C(z)$*. Let x, y, and z be real numbers so that the conditions $A(x)$ and $B(x,y)$ express some properties and relations between real numbers and $C(z)$ likewise some property of real numbers. A sufficiently detailed proof of such a theorem will do the following: Given any two real numbers a and b in place of x and y, given numerical verifications of the properties and relations $A(a)$ and $B(a,b)$, the proof will produce a real number c and a numerical verification of the property $C(c)$.

2

THE EMERGENCE OF
FOUNDATIONAL STUDY

Modern foundational study has twofold mathematical roots. One is the discovery of non-Euclidean geometries, especially the proof of independence of the parallel postulate by Eugenio Beltrami in 1868, in his *Saggio di interpretazione della geometria non-euclidea* (Treatise on the interpretation of non-Euclidean geometry). The other root is arithmetical, retraceable through Peano and others to the 1861 book *Lehrbuch der Arithmetik für höhere Lehranstalten* by the high school teacher Hermann Grassmann (Arithmetic for higher institutions of learning). Grassmann sets out to *prove* some of the basic properties of arithmetic operations and, remarkably, finds that to prove the commutativity of addition, $a + b = b + a$, one first has to prove its associativity, $a + (b + c) = (a + b) + c$. The inductive proof for $a + b = b + a$ just doesn't go through if you don't do it in this order. The inductive proofs are rendered possible because Grassmann had found the correct definitions of the arithmetic operations. The source of these definitions seems to be in the development of combinatorics that provided recursively defined algorithms for the manipulation of formal expressions. Properties of arithmetic operations had been postulated axiomatically in earlier literature, say the 1790 book *Anfangsgründe der reinen Mathesis* (Basics of pure mathesis) by the Königsberg mathematician Johann Schultz.

In each of these two cases, one has to set things straight: To prove independence in geometry, one has to ask what the axioms are, and maybe even the principles of proof. In arithmetic, you have to define your operations, which requirement leads to the idea of *recursive definition* of functions that goes hand in hand with the principle of inductive proof. Grassmann's formal approach to elementary arithmetic was a first step toward a theory of algorithmic computability.

One of David Hilbert's earliest foundational essays, the *Über den Zahlbegriff* of 1900, contrasts arithmetic and geometry on a general level. The former is an example of a *genetic* approach and method in which the objects of mathematics are built up gradually from the rockbottom of natural numbers, in the succession of naturals, integers, rationals, reals, complex numbers, and on a second level functions of these various objects. The main proponent of this approach was Leopold Kronecker. The other approach is *axiomatic* in which abstract axioms are set up that implicitly define a domain of objects with just those properties that follow from the axioms. Clearly, this second approach makes sense only if the axioms are accompanied by a precisely defined collection of principles of proof. In the essay, Hilbert wants to carry through the axiomatic approach even in the arithmetic of real numbers, with a grouping of axioms analogous to that of his *Grundlagen der Geometrie* one year earlier.

Hilbert's *Grundlagen* set the agenda of foundational problems of mathematical theories that we can summarize in the following four specific questions:

1. To determine a collection of precise principles of proof by which any question of provability and unprovability can be definitively answered.
2. To study the consistency of the principles of proof.
3. As suggested by question 1, to study the completeness and independence of the principles of proof.
4. The question of a *decision method* to determine if a claim is a theorem, i.e., the *Entscheidungsproblem*.

The formalization of the language of a mathematical theory and of mathematical proof grew out of the first requirement. Answers to these questions were obtained in turn for the propositional logic of the connectives, for quantificational logic, and for elementary arithmetic, and some of these answers were unexpected.

2.1. IN SEARCH OF THE ROOTS OF FORMAL COMPUTATION

(A) FOUNDATIONAL QUESTIONS IN ARITHMETIC. The basic laws of arithmetic have been an object of study at least since the time of Leibniz. He thought he could prove the transitivity of equality as follows (cf. Martin 1938, p. 32):

> If a and b are equal, and if c is equal to a, I say c must be equal to b. If $a = c$ as well as $a = b$, then b can be substituted with quantity maintained, which makes $b = c$, qu. e.d.

A similar argument gives $a + b = l + m$ from the assumptions $a = l$ and $b = m$. The proof begins with a reflexivity that is taken to hold in general, $a + b = a + b$, and then l is substituted for a and m for b at the right, which gives $a + b = l + m$.

These proofs use Leibniz' definition of equality (Martin, pp. 31–32): "Those that can be mutually substituted *salva veritate* are equal." Christian Wolff changed this into: "Those that can be substituted *salva quantitate* are equal."

The standard explanation of Leibniz' treatment is that the equality of a and b is defined by the requirement that they share the same properties. The substitution principle for properties P and functions f is a corollary to the definition:

$$\text{If } a = b, \text{ if } P(a), \text{ then } P(b).$$
$$\text{If } a = b, \text{ then } f(a) = f(b).$$

In Leibniz' marvelous rendering of his dialogues with the British empiricist philosopher John Locke, the *Nouveaux essais sur l'entendement humain* of 1704 but not published until 1765, the question is raised by Leibniz' adversary as to "what principle is needed to prove that two plus two is four." The proof Leibniz offers is as follows (p. 363):

Definitions: 1) *Two* is one and one.
2) *Three* is Two and one.
3) *Four* is Three and one.

Axiom, putting equals in place, equality is kept.

Proof: 2 and 2 is 2 and 1 and 1 (by def. 1)

2 and 1 and 1 is 3 and 1 (by def. 2)

3 and 1 is 4 (by def. 3) . . . . . .

Therefore (by the axiom)

$$2 + 2$$
$$2 + \overbrace{1 + 1}$$
$$\underbrace{3 + 1}$$
$$4$$

2 and 2 is 4. As was to be proved.

Leibniz adds that he could have written instead of "2 and 2 is 2 and 1 and 1" as well "2 and 2 is equal to 2 and 1 and 1." In other words, he suggests that one could as well have used the equality of numbers with the writing $2 + 2 = 4$, etc.

The horizontal curly brackets in the Leibnizian proof text are a rather curious thing; let's see what they could amount to.

1. The first, upward pointing bracket indicates that the second occurrence of 2 in the sum $2 + 2$ is by the axiom substituted by $1 + 1$ that by definition (1) is equal to it, with $1 + 1$ taken as a single expression. A linear notation with parentheses would be $2 + 2 = 2 + (1 + 1)$.
2. Secondly, the change of the upward pointing bracket into the downward pointing one indicates that one applies the step by which the result, in linear parenthesis notation $2 + (1 + 1)$, can be transformed into $(2 + 1) + 1$.
3. The second, downward pointing bracket indicates further that by definition (2), $2 + 1$ can be substituted by 3, so one obtains $3 + 1$ on line three of the proof.
4. The third, downward pointing bracket indicates that one applies definition (3) to obtain 4 as equal to expression $3 + 1$ on the preceding line.

Here are the very same steps written in a standard parenthesis notation:

$$2 + 2 = 2 + (1 + 1) = (2 + 1) + 1 = 3 + 1 = 4$$

It is of some interest to see how the sum $2 + 3$ would have been treated in the Leibnizian approach. In an abbreviated notation, we would have the sequence of equalities

$$2 + 3 = 2 + (2 + 1) = (2 + 2) + 1 = (2 + (1 + 1)) + 1 =$$
$$((2 + 1) + 1) + 1 = (3 + 1) + 1 = 4 + 1 = 5$$

One sees that the *associativity of sum*, $(a + b) + c = a + (b + c)$ in a general notation, is applied only in the case of $c = 1$:

$$a + (b + 1) = (a + b) + 1$$

Today this equality is counted not as a special case of associativity of sum, but rather as the proper *recursive definition of sum* for the *successor case*, in the second argument of a sum, from which the *general case* follows by an inductive proof on c. From the way Leibniz' example is chosen, we don't see directly how he intended it: with just $c = 1$ or in general. In the former case, we could say that the recursive definition of sum had been discovered by Leibniz by 1704!

Leibniz thought he could prove the commutativity of addition. In the *Anfangsgründe* of Johann Schultz mentioned earlier, the commutativity and associativity of addition appear as axioms. In analogy to Euclid's geometry with its axioms and construction postulates, there are in addition to these axioms two postulates for "the science of quantity." They express capacities of sorts that are *required* as things that can be done, as revealed by the German synonym given, *Forderungssatz*. The two postulates are (Schultz 1790, p. 32 and p. 40):

Postulates

Postulate 1. To transform several given quantities of the same kind into a quantity, i.e., a whole, through taking them successively together.

Postulate 2. To increase or decrease in thought every given quantity without end.

These are followed by two axioms (p. 41):

Axioms

General science of quantity:

1. The quantity of a sum is the same, should we add to the first given quantity the second or to the second the first, i.e., once and for all $a + b = b + a$, e.g., $5 + 3 = 3 + 5$.
2. The quantity of a sum is the same, should we add to the first given quantity another either once as a whole or each of its parts one after the other, i.e., once and for all $c + (a + b) = (c + a) + b = c + a + b$.

These axioms and postulates were, in Schultz' mind, formulations of principles found in Kant's philosophy of arithmetic, as explained in Schultz' two-volume *Prüfung der Kantischen Critik der reinen Vernunft* (Examination of the Kantian critique of pure reason, 1789 and 1792) that went hand in hand with the mathematics of the *Anfangsgründe* of 1790. The axioms and postulates were repeated by many authors, such as Michael Ohm (1816). Ohm tries to improve on Schultz' postulate 1 by defining the sum $a + b$ as "having so many units as the numbers a and b taken together" (1816, p. 7), but his presentation has nowhere near the sophistication found in the Schultz volumes.

Gottfried Martin studied these developments in his dissertation, published in 1938, in which he writes that the idea of axioms in arithmetic started to erode with Schultz. He refers to the question posed by Schultz (p. 57):

> How do I know, then, that this willful procedure [of changing the order of a sum] that lies in no way in the concept of addition itself, provokes no difference in the sum $7 + 5$?

This passage is from Schultz' *Prüfung* (vol. 1, p. 220). There he gives a proof of the equation $7 + 5 = 12$, perhaps the only formula in Kant's *Kritik*, a "smoking gun" of sorts in later writers. Kant's most detailed formulation is the following (p. B16):

> *I take the number 7 to begin with and, using for the concept 5 the fingers of my hand as an intuitive help, I put the units that I earlier took together to make the number 5, now in that image of mine one by one to the number 7, and thus I see the number 12 emerge.*

Schultz in turn gives first the axioms of commutativity and associativity of addition, then states that they are "indispensable for arithmetic" (p. 219), for without them, one would get only that $7 + 5 = 7 + (4 + 1) = 7 + (1 + 1 + 1 + 1 + 1)$. With commutativity and associativity, one gets instead (p. 220):

> Instead of all of 5, or its units taken together and added at once to 7, I must take them instead successively one by one, and instead of $7 + (4 + 1)$ first set $7 + (1 + 4)$, and in place of it $(7 + 1) + 4$, so then I get thanks to the concept of the number 8, namely that it is $7 + 1$,

first $8 + 4$ i.e. $8 + (3 + 1)$. Instead of this I have to set again $8 + (1 + 3)$, and $(8 + 1) + 3$ in place of it, so I get $9 + 3$, i.e., $9 + (2 + 1)$. Setting for this $9 + (1 + 2)$, and therefore $(9 + 1) + 2$, gives $10 + 2$, i.e., $10 + (1 + 1)$, and setting for this $(10 + 1) + 1$ gives finally $11 + 1$, i.e. 12. Every arithmetician knows that this is the only way through which we can come to the insight that the theorem is correct.

Here is a clear sign of awareness of the recursive definition of sum. It can be seen that Schultz missed by a hair's breadth the correct recursion equation, here $7 + 5 = 7 + (4 + 1) = (7 + 4) + 1$. He thus missed the inductive proofs of commutativity and associativity, and could do not better than claim that any attempted proof of the commutativity of addition would be circular (p. 221). In the *Anfangsgründe*, Schultz had declared the commutativity of addition to be "immediately evident" (p. 42).

Martin suggests that there is an intimation of deductive dependences in one of Schultz' discourses, namely that the commutativity of product follows from the special case of $1 \times n = n \times 1$ and the distributive law. There is a grain of truth to this, as we shall see in a while.

Schultz' proof of $n \times r = r \times n$ is as follows. First, the case $1 \times n = n \times 1$ is proved, and then come three lines with a first step on each line, Corollary 6, from the distributive law as in $n \times 2 = n \times (1 + 1) = n \times 1 + n \times 1$, and similarly for the last step, Corollary 4. The "by proof" clause (p. dem.) in between always refers to the case of one less on a previous line (*Anfangsgründe*, p. 64):

A product of two integers is the same, should one multiply the first factor by the second, or the second by the first, i.e., $n \times r = r \times n$.

For since $n = 1$ times n (§ 36. Cor. 3) $= n$ times 1 (§ 36. Cor. 2); then $n \times 1 = 1 \times n$.

$n \times 2 = n \times 1 + n \times 1$ (Cor. 6) $= 1 \times n + 1 \times n$ (p. dem.)
$$= 2 \times n \text{ (Cor. 4)}$$

$n \times 3 = n \times 2 + n \times 1$ (Cor. 6) $= 2 \times n + 1 \times n$ (p. dem.)
$$= 3 \times n \text{ (Cor. 4)}$$

$n \times 4 = n \times 3 + n \times 1$ (Cor. 6) $= 3 \times n + 1 \times n$ (p. dem.)
$$= 4 \times n \text{ (Cor. 4) etc.}$$

So we have in general $n \times r = r \times n$.

Martin (1938, p. 61) calls this "an inference from n to $n + 1$," but it certainly is not the kind of induction in which we have a base case, here $r = 1$, and a step case from an assumed value r to $r + 1$. There is instead a clear pattern indicated by the "etc" and the repetitive layout by which the proof of commutativity can be continued to any given number r. Let us single out the pattern by some little formal notation.

The expression $n \times r$ can be considered, for each value of n, a function of r in the old-fashioned sense of an expression with a free variable r for which values can be substituted as arguments of the function, denoted $f(r)$. Similarly, the expression $r \times n$ can be considered, for each value of n, a function of r, denoted $g(r)$. Schultz' first line of proof gives the result

$$f(1) = g(1)$$

His general case, even if not written down with a variable, gives

$$f(r + 1) = f(r) + f(1)$$

$$g(r + 1) = g(r) + g(1)$$

Now we see the hidden inductive step:

If $f(r) = g(r)$, then $f(r + 1) = f(r) + f(1) = g(r) + g(1) = g(r + 1)$.

The base value for $r = 1$ and the recursion clause from $r + 1$ to r in Schultz come from his two corollaries, 2 and 3 in § 36 and 6 in § 42, $n = 1 \times n = n \times 1$ and $n \times (r + 1) = n \times r + n \times 1$. These are what we today take as the recursion equations for product.

There is an even simpler way to look at Schultz' proof. We have two expressions, $n \times r$ and $r \times n$. They agree in value for $r = 1$, and when r grows by one, each grows by the value $f(1) = g(1)$. Therefore, $f(r)$ and $g(r)$ have the same value for all r.

The first to have formulated Schultz' proof method explicitly seems to have been Paul Bernays, in a talk he gave on 21 February 1928 in Göttingen, titled *Die Rekursion als Grundlage der Zahlentheorie* (Recursion as a foundation of number theory). Bernays, in a letter to be discussed in Section 5.2(C), mentions "the possibility of taking instead of the complete induction the rule of equalizing recursive terms satisfying the same recursive equations." An observation in much simpler words but to the same effect was made by Ludwig Wittgenstein some years after Bernays' talk, as reported in Waismann (1936, p. 99):

> The induction proof can also be thought of as a direction for the formation of proofs of individual numerical equations, as the general terms of a series of proofs. Indeed, the induction proof could very well be written in the form of a series of equations, with individual numbers, as a part of a series with an "etc.," and it would thereby lose none of its power.

Seven years after the *Anfangsgründe*, Schultz produced an abridged version with the title *Kurzer Lehrbegriff der Mathematik* (Short course in mathematics) that has the theorem of commutativity of multiplication with an identical wording, and the following proof (p. 36):

> For let one pose that the theorem be true for whatever multiplier m, namely let $n \times m = m \times n$; then $n \times (m + 1) = n \times m + n \times 1$ (Cor. 5) $= m \times n + 1 \times n$ (§ 21, Cor. 3) $= (m + 1) \times n$ (Cor. 3), so the theorem is in this case true also for the successive multiplier $(m + 1)$. The theorem is true for the multiplier 1, because $n \times 1 = 1 \times n$ (§ 21, Cor. 3.) so also for the multiplier 2, consequently also for the multiplier 3 etc, thereby for each multiplier r whatever; i.e., in general $n \times r = r \times n$.

What made Schultz change the proof into an explicit induction? I have found no such proof in the *Anfangsgründe*, but it does have at least two proofs that use another variant of induction, remarkably both of them in the form of an indirect existence proof. Theorem 17 is (p. 84):

> § 58. Each composite finite number m has a prime number as a divisor.
>
> **Proof.** Since m has a divisor n (§ 57), we have $m = n + n + \ldots$ (§ 53. Cor. 1). Let then n be composite; then it again has a divisor r and we have again $n = r + r + \ldots$, consequently $m = r + r + \ldots + r + r + \ldots$. Were r again composite; then r would again have a divisor u, and we would have $r = u + u + \ldots$, consequently $m = u + u + \ldots + u + u + \ldots + u + u + \ldots$. Then, had m no prime number as a divisor, then each new divisor would have each time again a

divisor without end, consequently *m* would be a sum of infinitely many whole numbers, whereby it would be infinitely great (§ 15). Because this is against the assumption, it must have a prime number as a divisor.

The theorem is followed by another, by which "every composite finite number is a product of prime numbers throughout" (p. 85). The proof uses the same principle: The contrary of the theorem would lead to an infinite product.

Both proofs are based on the principle that has been traditionally formulated as

There is no infinite descending chain of natural numbers.

In the *Lehrbegriff* of 1797, this principle is made explicit in the proof of the prime divisor theorem. Regarding the numbers in the sequence of divisors of a composite number, Schultz writes that "they cannot become always smaller without end" (p. 56).

Overall, we have that there is no trace of induction in the form of the step from *n* to *n* + 1 in the *Anfangsgründe* but just one equational proof and two by the impossibility of an infinite sum and product; altogether unusual arguments. In the 1797 *Lehrbegriff*, the former has been changed into a canonical induction from *n* to *n* + 1, and the latter has been changed into the well-known infinite descent argument. Did Schultz realize the equivalence of all these formulations? I don't think an answer would make such a big difference in the way we look at his results. *We* know they are based on one formulation or other of a principle today called induction. All four forms of induction found in Schultz have been used indiscriminately in arithmetic and elsewhere, with varying degrees of awareness of their equivalence.

(B) RECURRENCE IN COMBINATORICS. The first explicit references to a "recurrent procedure" (rekurrierendes Verfahren) seem to stem from a development quite different from the foundational concerns of Schultz, namely the combinatorics in the early nineteenth century. Andreas von Ettingshausen's (1796–1878) book *Die combinatorische Analysis als Vorbereitungslehre zum Studium der theoretischen höhern Mathematik*, published in 1826, is one such (Combinatorial analysis as a preparatory for the study of theoretical higher mathematics). Combinatorial formulas are there described as "recurrent determinations" (recurrirende Bestimmungen), in the following general terms (p. 83):

> ...this procedure is the *recurrent* one, i.e., one in which a successive member of the result aimed at is obtained only when all preceding members have already been calculated.

An engaging book by Friedrich Wilhelm Spehr (1799–1833), in its second edition of 1840, has the exhaustive title *Vollständiger Lehrbegriff der reinen Combinationslehre mit Anwendungen derselben auf Analysis und Wahrscheinlichkeitsrechnung* (Complete course for learning the pure calculus of combinatorics with applications thereof to analysis and the calculus of probability). The preface and first edition are dated 1824, but considering its talented author's death at the young age of thirty-four seven years before the second edition, it was most likely a simple reprint. In combinatorics, the nature of the things plays no role, but one considers just the ways in which they can be put together, writes Spehr (p. 1). *Each* topic is treated separately by the "independent" and "recurrent" methods, explained as follows (p. 9):

> Whenever quantities or just things whatever are at hand that are connected together successively according to laws that remain always the same, so that these connections are members of a *progression*, the members have also *among themselves* a fixed link; a rule can be given each time by which one can find another thing from some given ones. It is so overall in analysis where the lawful sequence comes out as a result of an operation, and it is so also in combinatorics.

> One can produce each member of a result in itself and *independent* of every other member, from the quantities or things the connections of which shall bring it forth; (*independent procedure, independent determination*) but one can also derive a successive one from members of the result already formed earlier. (*Recurrent procedure, recurrent determination.*)

In § 15, the recurrent method is explained in detail, with combinatorial formulas that show how the computation with a parameter value k is reduced to one with a parameter value $k - 1$. Such experience with complicated situations of combinatorics had led to the idea of a general pattern of reduction, in which in the end the value 1 is reached.

Then, turning the procedure around, the value of an expression for k could be determined step by step from the value for 1.

2.2. GRASSMANN'S FORMALIZATION OF CALCULATION

(A) GRASSMANN'S DISCOVERY OF DEFINITION BY RECURSION. In his *Ausdehnungslehre* of 1844, Hermann Grassmann wrote (p. xix):

> Proof in the formal sciences does not go into another sphere, beyond thinking itself, but resides purely in the combination of the different acts of thought. Therefore the formal sciences must not begin with axioms unlike the real ones; their foundation is made up of definitions. If axioms have been introduced into the formal sciences, such as arithmetic, this is to be seen as a misuse that can be explained only through the corresponding treatment in geometry.

The last sentence of the quotation is from an added footnote. Fifteen years after the preceding passage was published, Grassmann put his credo about avoiding axioms in arithmetic to full effect through the application of the "recurrent procedure" to the most elementary parts of arithmetic, namely the basic arithmetic operations. His 1861 *Lehrbuch der Arithmetik für höhere Lehranstalten* contains the first explicit *recursive definitions* of arithmetic operations, ones that go hand in hand with *inductive proofs* of properties of the recursively defined operations. The base case of induction is 0, or, in Grassmann's time, usually 1, and the step case is that of n to $n + 1$, the *successor* of n (Nachfolger). Grassmann's definition of sum is for an arbitrary "basic sequence" with a unit e and a succession of units as in $e, e + e, e + e + e, \ldots$, with the addition of a unit always intended at the right, in the sense of $(e + e) + e$. He actually constructs a doubly infinite series of integers with a positive and a negative unit, but the following is for brevity the part on natural numbers that begin with the positive unit denoted e. Here is Grassmann's explanation of the positive integers (1861, p. 2):

> **7.** *Explanation.* Let a sequence of quantities be built out of a quantity e [Grösse] through the following procedure: One sets e as *one* member of the sequence, $e + e$ (to be read e plus e) as the successive [nächstfolgende] member of the sequence, and one continues in this way, by deriving from

the member that is each time last the successive one through the joining of $+e$ . . . .

$$\vdots$$

When one assumes each member of this sequence to be different from all the other members of the sequence, one calls this sequence the *basic sequence* [Grundreihe], e the *positive* unit.

It is remarkable that Grassmann does not use the word "addition" in this characterization of the number sequence. The addition of the unit e is explained as follows (p. 3):

8–9. *Explanation.* If a is any member of the basic sequence, one means by $a + e$ . . . the member of the sequence that follows next to a . . . that is, if b is the member of the sequence next to a, we have

$$(8)\ b = a + e$$

One calls this operation [Verknüpfung] the addition of a unit.

Addition of a unit is a clearly defined separate operation explained as the taking of a successor in the basic series. Grassmann now sets out to prove that "the members of the sequence that follow e are sums of positive units." The proof goes as follows:

Proof. The members of the basic series that follow e have (by **7**) resulted from e by a progressive addition of positive units, are therefore sums of positive units.

There is a clear awareness in Grassmann that the sequence of natural numbers is generated from the unit through repeated application of the successor operation. The only fault is that there is no separate notation for the successor; it is denoted by $+e$. Grassmann's definition of addition should be read with the overloading of the symbol $+$ in mind (p. 4):

15. *Explanation.* If a and b are arbitrary members of the basic sequence, one understands with the sum $a + b$ that member of the basic sequence for which the formula

$$a + (b + e) = a + b + e$$

holds.

In Grassmann's terms, this equation gives a procedure for reducing arbitrary sums into members of the "basic sequence" that by the preceding proof are "sums of positive units."

The recursive definition of sum is put into use in Grassmann's "inductory" (inductorisch) proofs of the basic properties of addition, such as associativity and commutativity. Anyone who tries to prove the commutativity of addition, $a + b = b + a$, will notice that the proof requires as a preliminary a proof of associativity, $(a + b) + c = a + (b + c)$. This must have happened to Grassmann, who offers the following proof (p. 8):

22. $\qquad\qquad a + (b + c) = a + b + c.$

"Instead of adding a sum one can add the summands step by step," or "instead of adding two quantities step by step, one can add their sums."

Proof. (inductorily in relation to c). Assume formula 22 to hold for whatever value of c, then we have

$$a + [b + (c + e)] = a + [b + c + e] \quad \text{(by 15)}.$$

$$= a + (b + c) + e \quad \text{(by 15)}.$$

$$= a + b + c + e \quad \text{(by assumption)}.$$

$$= a + b + (c + e) \quad \text{(by 15)}.$$

Therefore, if formula 22 holds for whatever value c, it holds even for the one following next, thereby for all values that follow.

The use of parentheses could have made the steps a bit clearer. Next, as result 23, Grassmann proves the commutativity of addition.

Grassmann conceived of the natural numbers in a completely abstract way, as a special case of his "basic sequences," as evidenced by the occurrence of the number symbols in his treatise. Zero appears early on in his system, as the notation 0 for the sum of the positive and negative units (p. 3):

$$e + -e = 0$$

Zero is not yet a natural number but belongs to any basic sequence. The symbols 1, 2, and 3 appear much later, after the treatment of sum and its properties has been completed in § 3, in the section on *Multiplication* (§ 4, p. 17):

52. *Expl.* By $a.1$ (read a times one or a multiplied by one) one intends the quantity a itself, i.e.,

$$(52). \ a.1 = a.$$

"To multiply by one changes nothing."

53. *Expl.* A basic sequence the unit of which is equal to one is called a number sequence, its members numbers, the number $1 + 1$ is denoted by 2, the number $2 + 1$ by 3, etc.

Grassmann ends this explanation with the remark that since the number sequence is a basic sequence, the previously established laws of addition and subtraction apply to it.

Multiplication with "the rest of the numbers (beyond 1)" is defined by the recursion equation $a \cdot (\beta + 1) = a\beta + a$, where "$\beta$ is a positive number" (p. 18). Grassmann takes $a \cdot 0 = 0$ to be part of the definitional equations of a product. Much later, it became clear that it is instead derivable from the recursion equation and one of the Peano axioms.

Finally, we note Schultz' reduction of the proof of commutativity of multiplication to the special case $1 \cdot n = n \cdot 1$ and the distributive law. The first step in the proof was $n \cdot 2 = n \cdot (1 + 1) = n \cdot 1 + n \cdot 1$, now an instance of Grassmann's recursive definition of product by the equation $a \cdot (b + 1) = a \cdot b + a$, together with equation 52. Both equations are present in Schultz' work, but as "corollaries" instead of definitions.

Today, one would formulate the successor operation as a separately written function, with the notation $s(a)$ or $succ(a)$ so as not to mix sum and successor, with 0 included and the definition of sum and product as

1. $a + 0 = a \quad a \cdot 1 = a$
2. $a + s(b) = s(a + b) \quad a \cdot s(b) = a \cdot b + a$

Here it is seen how recursion clause 2 makes the second summand diminish step by step until it is 0 and vanishes.

There is in the case of sum a clear conceptual advantage in having the natural numbers begin with 0 instead of 1. The apparent circularity of the definition of sum, signaled as a defect by Frege and others in the nineteeth century, is seen in the Grassmann-style recursion equation

$$a + (b + 1) = (a + b) + 1$$

With clauses 1 and 2, recursion produces a well-founded sequence through the application of clause 2, until the sum operation is eliminated by clause 1. In Grassmann's work, the end of recursive return is instead signaled by the appearance of $+e$.

Grassmann's motivations for developing formal arithmetic are remarkably clearly stated in the introduction to his book. He writes that it "claims to be the first strictly scientific elaboration of the discipline" (p. v):

> These claims contain even the charge against earlier elaborations of lacking scientific rigour and consequential following. Whether they are justified must be documented by the work itself, because a polemical or apologetical justification of these claims is in contradiction with the special aim of this work.

He then writes about a planned further work to "elaborate the method for a scientifically trained reader" by singling out the leading ideas and by "showing in detail the necessity of the method followed." No such work ever appeared, and Grassmann got rather frustrated at the lack of interest in his work and turned to linguistics, where he became one of the leading authorities on the Sanskrit language.

Apparently, Grassmann was one of those Frege criticized in his article *Über formale Theorien der Arithmetik* in 1885 as taking numbers to be mere signs that don't signify anything, even if Frege gives no names. However, as concerns Grassmann, in his *Lehrbuch der Arithmetik* he warns against that idea in the short introduction (p. vi):

> Mathematics in its most strict form, in its uncompromising consequentiality, is alone in the position to guard the student from the fashionable domination of spirited phrases and to train him in logically consequent thinking. This aim would not, though, be reached if one wanted to just put formula after formula in sequence, without conceptual development. Both development in formulas and conceptual development must always go hand in hand.

In his instructions, he states that the student must present a proof in words each time (p. vi):

> In this way, the whole presentation proceeds through a conceptual development, during which the formula written down represents each time symbolically the conceptual step ahead.

Grassmann gives one example of how this should be done (p. 5):

17. $a + (b + -e) = a + b + -e$

Proof (progressive).

$$a + (b + -e) = a + (b + -e) + e + -e \quad \text{(by 13)}.$$
$$= a + (b + -e + e) + -e \quad \text{(by 15)}.$$
$$= a + b + -e \qquad\qquad \text{(by 14)}.$$
$$\vdots$$

Proof in words. We begin with the left side of the equation to be proved, i.e., from

$$a + (b + -e)$$

One can bring this expression to a form in which it ends as on the right side, with $+ - e$, because to add progressively a positive and a negative unit changes nothing. Then the above expression becomes

$$= a + (b + -e) + e + -e$$

Instead of adding to the sum $a + (b + -e)$ a positive unit, one can add it to the second summand, so the expression becomes

$$= a + (b + -e + e) + -e$$

To add progressively a negative and a positive unit changes nothing; this applied to the expression in parentheses gives the above expression

$$= a + b + -e$$

Therefore $a + (b + -e) = a + b + -e$, i.e.: Instead of adding a negative unit to the second summand, one can add it to the sum.

The justifications in numbers in the equational derivation are replaced by verbal explanations, and the substitutions are explained.

Grassmann ends his presentation of the properties of addition with a result that is a clear proof of a conclusion from an assumption: The "hypothesis" is $a + b = a + c$, the "thesis" $b = c$. The same proof pattern appears often when he proceeds to define and prove the basic properties of subtraction, multiplication, the order relations between natural numbers, and so on.

(B) THE RECEPTION OF GRASSMANN'S IDEA. Grassmann's approach to the foundations of arithmetic is explained in detail in the first volume of Hermann Hankel's two-volume treatise *Vorlesungen über die complexen Zahlen und ihre Functionen* of 1867. The full title of the first volume is *Theorie der complexen Zahlensysteme insbesondere der gemeinen imaginären Zahlen und der Hamiltonschen Quaternionen nebst ihren geometrischen Darstellung* (Theory of complex number systems especially of the common imaginary numbers and of the Hamiltonian quaternions together with their geometrical representation). He begins with a discourse about the natural numbers (p. 1):

> What it means to think or pose an object 1 time, 2 times, 3 times … cannot be defined because of the fundamental simplicity of the concept of posing. An absolute, entire number 1, 2, 3 … expresses that an object has to be posed 1, 2, 3 … times, and it means $1e, 2e, 3e...$, the result of repeated posing of e.

Next, addition is explained as the result of posing the numerical unit e first a times, then b times. Associativity and commutativity are mentioned as "the main laws." Further, addition is an operation with a unique result and the property that if one summand is changed and the other remains constant, the result changes also (p. 2):

> The properties of addition given here are sufficient for the derivation of all further consequences on the building of sums, without the need to remind oneself each time about the real meaning of addition. In this sense they are the conditions that are necessary and sufficient to *formally* define the operation.

The topic of § 9 is "positive entire numbers," generated from the unit 1 by setting $1 + 1 = 2, 2 + 1 = 3, 3 + 1 = 4, ...$. The sum $(A + B)$ of two numbers is defined as in Grassmann's recursion equation (Hankel, p. 37):

$$A + (B + 1) = (A + B) + 1.$$

Hankel now states that "this equation determines every sum," and he shows how it goes: By setting $B = 1$ in the equation one has $A + 2 = A + (1 + 1) = (A + 1) + 1$, and with $B = 2$ one has $A + 3 = A + (2 + 1) = (A + 2) + 1$ so that $A + 2$ and $A + 3$ are numbers in the sequence of integers (ibid.):

> In this way one finds through a recurrent procedure, one that goes on purely mechanically without any intuition, unequivocally every sum of two numbers.[1]

This proclamation is followed by a mechanical computation of the Kantian formula $7 + 5 = 12$ through the writing of 17 equations that begins with $7 + 5 = 7 + (4 + 1) = (7 + 4) + 1$ and ends with $7 + 5 = 11 + 1 = 12$. The example was very likely inspired by the proof in Schultz' *Prüfung* discussed earlier. Next, Hankel proves associativity, then the lemma $1 + A = A + 1$, and finally commutativity exactly as done by Grassmann, and then proceeds to the recursive definition of multiplication. The presentation of positive integers ends as follows (p. 40):

> The idea to derive the rules of addition and multiplication as done here owes itself in its essence to Grassmann (Lehrb. d. Arithmetik).

Grassmann's approach is next described by Ernst Schröder in his 1873 *Lehrbuch der Arithmetik und Algebra*. The book contains an introduction and chapters on arithmetic operations, with a presentation that follows Hankel's divisions directly. The integers are explained through the process of counting, with the paradigm that *"a natural number is a sum of ones"* (p. 5). The addition of one is kept strictly separate from the concept of an arbitrary sum; the symbol + is used only because otherwise, for example, writing 111 for three, one would

---

[1] The German is too beautiful to be left just in translation: Auf diese Weise findet man durch ein recurrirendes Verfahren, welches ohne alle Anschauung, rein mechanisch vor sich geht, unzweideutig jede Summe zweier Zahlen. (Auf diese Weise findet man durch ein recurrirendes Verfahren, welches ohne alle Anschauung, rein mechanisch vor sich geht, unzweideutig jede Summe zweier Zahlen.)

take that as a hundred and eleven. A number is further "independent of the order in which the units are built with the unit stroke" (p. 16).

Each of the arithmetic operations is treated twice, just as in the much earlier book of combinatorics by Spehr: first by the *independent treatment* (independente Behandlungsweise) and then by the *recurrent* one (Schröder, p. 51). The meaning of the former is not clearly explained, but it is more or less the way one learns basic arithmetic at school by becoming acquainted with counting, numbers, sums, and so on. Spehr's explanation was that a solution can be determined in itself for a given case, independent of other possible cases. The second approach aims at greater rigor and "a simplification of the conditions taken as a point of departure." Schröder explains Grassmann's recurrent mode of counting sums through detailed examples (pp. 63–64):

$$(5) \quad 2 = 1 + 1, \, 3 = 2 + 1, \, 4 = 3 + 1, \, 5 = 4 + 1, \text{etc.},$$

The natural numbers are hereby defined *recurrently*. Namely, to give in a complete way the meaning of a number, i.e., to express it through the unity, one has to go back from it to the previous number and to run through backwards (recurrere) the whole sequence.

The presentation is copied from Hankel, up to the 17 equations that lead from $7 + 5 = 7 + (4 + 1)$ to $7 + 5 = 11 + 1 = 12$ (p. 65):

One can find such a sum [of two numbers] in this way unequivocally, through a recurrent procedure that goes on purely mechanically.

Well, this goes over to the side of direct copying of Hankel. In counterbalance, Schröder's presentation of how the natural numbers are generated by the $+1$-operation is described through a notational novelty (p. 64):

If a is a number from our sequence: 1, 2, 3, 4, 5, ... then even $a + 1$ is one, namely

$$a' = a + 1$$

is the general form of equations (5).

This seems to be the first place in which the successor operation obtains a separate notation, one that later became the standard one. It was a conceptually important step, and the notation was taken into use by Dedekind in 1888.

In conclusion, we can say that with the two textbooks by Hankel and Schröder, Grassmann's recursive foundation of arithmetic became known and generally appreciated.

One who added to the knowledge of Grassmann's achievement was Hermann von Helmholtz, through his widely read essay *Zählen und Messen, erkenntnistheoretisch betrachtet* (Counting and measuring, epistemologically considered) of 1887. Grassmann's approach is explained right from the beginning: von Helmholtz writes that one had until then posed axioms in arithmetic that include the associativity and commutativity of sum, whereas *Grassmann's axiom*, as he calls the recursion equation for sum, leads to inductive proofs of the mentioned axioms. "Thereby, as we hope to show in what follows, the right foundation has been gained for the doctrine of addition of pure numbers." What this foundation is and how the basic laws of addition and product follow from Grassmann's axiom are presented with admirable clarity in von Helmholtz, mostly in reference to Schröder's *Lehrbuch*. The axiom itself is described as an equation that captures the process of counting from a number a on, with transparent inductive proofs of commutativity and associativity.

Richard Dedekind's widely read booklet *Was sind und was sollen die Zahlen?* of 1888 (What are numbers and what are they for?) mentions on the first page as the first source Schröder's "exemplary *Lehrbuch*" as well as von Helmholtz' essay; one then finds in it the use of the successor function as a primitive, with the Schröderian notation $p = n'$ for "the successive number" (p. 27). Addition is defined by the recursive clauses (p. 36)

II. $m + 1 = m'$
III. $m + n' = (m + n)'$

Next, the basic properties are proved inductively, with a slight variant of the old order of associativity of sum followed by its commutativity. Dedekind proves first $m' + n = m + n'$ (*Satz* 136) from clauses II and III, silently using even the substitution principle in the successor function from $m = n$ to conclude $m' = n'$, with the intermediate step $m' + n = (m + n)'$ which with clause III gives the result (p. 36).

Then follow proofs of $1 + n = n + 1$ and $m + n = n + m$ that don't use *Satz* 136, and then the standard proof of associativity expressed as $(l + m) + n = l + (m + n)$. The order of things is somewhat redundant: Associativity with the instance $(m + 1) + n = m + (1 + n)$ together with $1 + n = n + 1$ gives at once Dedekind's *Satz* 136. His intermediate step $m' + n = (m + n)'$ in the proof is the more interesting result because it shows that the recursion clause can be applied as well to the first argument of a sum.

As intimated by Hankel, Grassmann's recurrent approach to arithmetic turns computation into a "purely mechanical" process. The same idea was seen already in the work of Leibniz, who set out to prove the arithmetic formula $2 + 2 = 4$, as detailed previously. This equation can be expressed with the successor operation expressed in unary arithmetic as

$$s(s(0)) + s(s(0)) = s(s(s(s(0))))$$

A derivation of Leibniz' example by recursion equations is at the same time a *formal computation* of the value of the sum function for the arguments 2 and 2. One way to express the matter is that the nonnormal form $a + b$ is brought into a normal or canonical form of a zero or successor. In computer science, one would call the sum the program form and the zero or successor the data form. The computation is an example of *term rewriting* by the recursion equations, with such basic questions as *termination* of rewriting, either for at least one reduction sequence or all of them, and the question of *uniqueness* of the result. Term rewriting represents algorithmic computation whenever it always terminates and the result is unique.

2.3. PEANO: THE LOGIC OF GRASSMANN'S FORMAL PROOFS

(A) THE PEANO AXIOMS. In 1889, Giuseppe Peano published a separate little treatise, the 36-page *Arithmetices Principia, Nova Methodo Exposita* (The principles of arithmetic, presented by a new method). It was written in Latin, and the earlier parts received an English translation in the 1967 collection *From Frege to Gödel*, edited by Van Heijenoort. Peano's original is readily available online, and one sees that this booklet consists of a 16-page preface and explanation and a 20-page systematic development that begins with §1: On numbers and on addition. Peano writes in the introduction (Van Heijenoort's translation, p. 85):

> I have denoted by signs all ideas that occur in the principles of arithmetic, so that every proposition is stated only by means of these signs.
>
> $\vdots$
>
> With these notations, every proposition assumes the form and the precision that equations have in algebra; from the propositions thus written other propositions are deduced, and in fact by procedures that are similar to those used in solving equations.

Peano's signs are, first of all, *dots* that are used in place of parentheses, and then P for *proposition, $a \cap b$,* even abbreviated to *ab,* for *the simultaneous affirmation of the propositions a and b, $-a$* for *negation, $a \cup b$* for *or, V* for *truth,* and the same inverted for *falsity* Λ. The letter *C* stands for *consequence,* used inverted as in todays stylized implication sign $a \supset b$. Even if it is read "deducitur" (one deduces), it is clearly a connective, because it is found iterated. For example, Peano's second propositional axiom is

$$a \supset b . b \supset c :\supset .a \supset c$$

There is also the connective of *equivalence, $a = b$,* definable through implication and conjunction as $a \supset b . \cap .b \supset a$.

Peano writes in the preface that in logic he has followed Boole, among others, and for proofs in arithmetic Grassmann (1861), "in arithmeticae demonstrationibus usum sum libro: H. Grassmann." Dedekind's booklet of 1888 is mentioned as a "recent script in which questions that pertain to the foundation of numbers are acutely examined." There is no evidence of the influence of Frege, even if some of the initial statements about the ambiguity of language and the necessity to write propositions only in signs are very close to those in Frege's *Begriffsschrift* of 1879.

Peano says of definitions (Van Heijenoort's translation, p. 93):

> A *definition,* or *Def.* for short, is a proposition of the form $x = a$ or $\alpha \supset .x = a$, where *a* is an aggregate of signs having a known sense, *x* is a sign or aggregate of signs, hitherto without sense, and α is the condition under which the definition is given.

Pure logic is followed by a chapter on *classes*, or sets as one could say. The notation is $a\,\varepsilon\,b$ for *a is a b*, and $a\,\varepsilon\,K$ for *a is a class*.

When Peano proceeds to arithmetic, he first adds to the language the symbols N (*number*), 1 (*unity*), $a + 1$ (*a plus 1*), and $=$ (*is equal to*). The reader is warned that the same symbol is used also for logic. Next, he gives the famous Peano axioms for the class N of natural numbers:

Table 2.1. Peano's axioms for natural numbers

1. $1\,\varepsilon\,N$
2. $a\,\varepsilon\,N. \supset .a = a$
3. $a, b\,\varepsilon\,N. \supset : a = b. = .b = a$
4. $a, b, c\,\varepsilon\,N. \supset \therefore a = b.b = c :\supset .a = c.$
5. $a = b.b\,\varepsilon\,N :\supset .a\,\varepsilon\,N.$
6. $a\,\varepsilon\,N. \supset .a + 1\,\varepsilon\,N.$
7. $a, b\,\varepsilon\,N. \supset : a = b. = .a + 1 = b + 1.$
8. $a\,\varepsilon\,N. \supset .a + 1 -= 1.$
9. $k\,\varepsilon\,K \therefore 1\,\varepsilon\,K \therefore x\,\varepsilon\,N.x\,\varepsilon\,k :\supset_x .x + 1\,\varepsilon\,k ::\supset .N \supset k.$

The reader would have been helped in axioms 2, 7, and especially 8, with its negated equality, had separate signs for equality of numbers and equivalence of propositions been used.

One direction of axiom 7 can be seen as a principle of replacement of equals $a = b$ in the successor function $a + 1$. The remaining direction together with axiom 8 are often referred to as "the axioms of infinity," after Dedekind (1888). With 0 included in N as in Peano (1901), it follows that the positive natural numbers N^+, a proper subset of N, are in a one-to one correspondence with N, which makes N infinite in Dedekind's definition.

The last axiom is the principle of induction. Let k be a class that contains 1, and for any x, let it contain $x + 1$ if it contains x. Then it contains the class N. The implication has the eigenvariable x of the inductive step as a subscript.

The list of axioms is followed by a definition:

10. $2 = 1 + 1; 3 = 2 + 1; 4 = 3 + 1$; and so forth.

What Peano calls a definition contains the same defect as those of Grassmann, Hankel, and Schröder, revealed by the elliptic "etc." or similar, namely that no way is given for inductively producing arbitrary decimal expressions from expressions in pure successor form.

Now follows a list of theorems, the first one with a detailed proof:

11. $2\varepsilon N$.

Proof.

P 1 . ⊃ :	$1\varepsilon N$	(1)
1 [a] (P 6) . ⊃ :	$1\varepsilon N. ⊃ .1 + 1\varepsilon N$	(2)
(1)(2). ⊃ :	$1 + 1\varepsilon N$	(3)
P 10 . ⊃ :	$2 = 1 + 1$	(4)
(4).(3).(2, 1+1) [a,b] (P 5): ⊃ :	$2\varepsilon N$	(Theorem).

It will be very useful to inspect this proof in detail. The justifications for each step are written at the head of each line so that together they imply the conclusion of the line. The derivation begins with P 1 in the antecedent, justification part of an implication, and $1\varepsilon N$ in the consequent as the conclusion. The meaning is that from axiom P 1 follows $1\varepsilon N$. The second line has similarly that from axiom P 6 with 1 substituted for a follows $1\varepsilon N. ⊃ .1 + 1\varepsilon N$. The next line tells that from the previous lines (1) and (2) follows $1 + 1\varepsilon N$. The following line tells that definition 10 gives $2=1+1$. The last line tells that lines (4) and (3) give, by the substitution of 2 for a and 1+1 for b in axiom P 5, the conclusion $2\varepsilon N$. The order in which (4) and (3) are listed is $2=1+1$ and $1 + 1\varepsilon N$. The instance of axiom P 5 is $2 = 1 + 1.1 + 1\varepsilon N :⊃ .2\varepsilon N$. Thus, we have quite formally in the justification part the expression

$$(2 = 1 + 1).(1 + 1\varepsilon N).(2 = 1 + 1.1 + 1\varepsilon N :⊃ .2\varepsilon N).$$

Line (3) is similar. It has two successive conditions in the justification part:

$$(1\varepsilon N).(1\varepsilon N. ⊃ .1 + 1\varepsilon N)$$

There are altogether two instances of logical inference, both written so that the antecedent of an implication as well as the implication itself is in the justification part, and the consequent of the implication as the conclusion of the line. Each line of inference in Peano's work therefore has one of the following two forms, with b a substitution instance of axiom a in the first:

$$a \supset b.$$

$$a . a \supset b :\supset b.$$

After the first detailed example, Peano starts to use an abbreviated notation for derivations that makes it rather hard to read them. The first derivation is written "for the sake of brevity" as

$$P\,1\,.\,1\,[a]\,(P\,6):\supset:\ 1+1\,\varepsilon\,N\,.\,P\,10\,(2,1{+}1)\,[a,b]\,(P\,5):\supset\ Th.$$

An even shorter notation is given as an alternative:

$$P\,1\,.\,(P\,6):\supset:\ 1+1\,\varepsilon\,N\,.\,P\,10\,(P\,5):\supset\ Th.$$

It is left for the reader to figure out the meaning of the notation. The expression stands for a formula, in modern notation, of the logical form

$$A\,\&\,(A \supset B) \supset (B\,\&\,(B \supset C) \supset C).$$

Peano's abbreviation turns a derivation from axioms into a single formula in which the axiom instances together imply the theorem.

After the preceding theorem, there follow other very simple consequences about the equality relation, numbered 12–17. Next comes the definition:

18. $a,b\,\varepsilon\,N. \supset .a + (b + 1) = (a + b) + 1.$

Peano notes (Van Heijenoort 1967, p. 95):

> *Note.* This definition has to be read as follows: if a and b are numbers, and if $(a + b) + 1$ has a sense (that is, if $a + b$ is a number) but $a + (b + 1)$ has not yet been defined, then $a + (b + 1)$ signifies the number that follows $a + b$.

Peano gives as examples of the use of the definition formal computations of the values $a + 2$ and $a + 3$, written as

$$a + 2 = a + (1 + 1) = (a + 1) + 1$$

$$a + 3 = a + (2 + 1) = (a + 2) + 1$$

Combining these two examples, we get

$$a + 3 = a + (2 + 1) = (a + 2) + 1$$
$$= (a + (1 + 1)) + 1 = ((a + 1) + 1) + 1$$

Now begins a section with theorems, clearly ones suggested by those in Grassmann, and with inductive proofs. Number 19 shows that natural numbers are closed with respect to addition, number 22 is a principle of replacement of equals in a sum, by which $a = b \supset a + c = b + c$, and number 23 is the associative law. To arrive at commutativity, Peano proves first as number 24 the lemma $1 + a = a + 1$, then finishes with number 25, which is commutativity of addition, through number 28, which is replacement at both arguments in sums, $a = b . c = d \supset a + c = b + d$. The part on natural numbers finishes with sections on the recursive definition and basic properties of subtraction, multiplication, exponentiation, and division, all of it following Grassmann's order of things in definitions and theorems to be proved (in §§ 2–6). Dedekind's effect on Peano, sometimes suggested, does not concern the development of elementary arithmetic, which must have been well on its way by the time Peano had Dedekind's booklet available.

Peano adds the consideration of classes to Grassmann's work, and therefore he has operations such as M for "the maximum among...," and the same inverted for minimum. Thus, he expresses Euclid's result about the infinity of prime numbers as result 23 in § 3, with Np standing for the class of prime numbers:

23. $M.Np := \Lambda$

Here Λ is the sign of absurdity or the empty class. The whole is read as something like "the maximum among prime numbers is equal to an empty class."

(B) THE STRUCTURE OF DERIVATIONS IN PEANO. From the derivations in Peano's treatise, the following structure emerges:

Peano's formal derivations consist of a succession of formulas that are:

  (i) *Implications in which an axiom implies its instance.*
  (ii) *Implications in which previously derived formulas a and a $\supset$ b imply b.*

As shown in Section 5.1, Russell took over this structure of formal derivations *verbatim* in his 1906 article *The theory of implication*.

Peano likened his propositions to the equations of algebra and his deductions to the solving of the equations. Rather startlingly, Jean van Heijenoort, who edited the book that contains the first English translation of the main part of Peano's 1889 work, instead of figuring out what Peano's notation for derivations means, claims in his introduction that there is "a grave defect. The formulas are simply listed, not derived; and they could not be derived, because no rules of inference are given ... he does not have any rule that would play the role of the rule of detachment" (Van Heijenoort 1967, p. 84). Had he not seen the forms $a \supset b$ and $a . a \supset b : \supset b$ in Peano's derivations, the typographical display of steps of axiom instances and implication eliminations with the conclusion b standing out at the right, and the rigorous rule of combining the antecedent of each two-premiss derivation step from previously concluded formulas?

Van Heijenoort's unfortunate assessment, which becomes much worse if one reads further, has undermined the view of Peano's contribution for a long time, when instead Peano's derivations are constructed purely formally, with a notation as explicit as one could desire, by the application of axiom instances and implication eliminations.

Van Heijenoort's comments on recursive definition in Peano's treatise are also flawed, though not as terribly as those about deduction. He writes about Peano's definition of addition and multiplication (p. 83):

> Peano ... puts them under the heading "Definitions", although they do not satisfy his own statement on that score, namely, that the right side of a definitional equation is "an aggregate of signs having a known meaning".

When introducing his primitive signs for arithmetic, Peano enlisted *unity*, notation 1, and *a plus* 1, notation $a + 1$. Thus, the sum of two numbers was not a basic notion but just the successor, and definition 18 laid down what the addition of a successor $b + 1$ to another number means, in terms of his primitive notions. Peano explained the matter carefully in the note after definition 18, cited earlier. If, as Peano assumes, $a + b$ is a number, i.e., if $a + b \varepsilon N$, then $(a + b) + 1 \varepsilon N$, so the definiens "has a meaning" as Peano writes, and one really wonders what Van Heijenoort may have been thinking here, if anything.

There is even a theorem in Peano's treatise proved by induction, that if a, b are numbers, even $a + b$ is. His misfortune was perhaps to use the same notation for the operation of a successor and for an arbitrary sum.

The notation for natural numbers and their operations was clearly improved in the project "Formulario mathematico," a series of five versions of a book in which mathematics was to be developed within Peano's formalism. The title amounts to something like "Formula collection for mathematics," published in four French versions between 1895 and 1903 with the title *Formulaire de Mathématiques* and *Formulaire Mathématique*, with the fifth and last in Peano's own artificial language *latino sine flessione* as the *Formulario Mathematico* (1908). (N.B. one might wonder what else "Latin without inflexions" could be than Italian slightly contorted.)

In the *Formulario* of 1901, there is a clear stylized implication symbol, identical to one used here, and also the existential quantifier, an inverted uppercase E, but just the axiom by which an instance gives existence (p. 28). The natural numbers begin with 0, and the successor is written as $a+$, "the number that comes after a, the successor of a, a plus," as if the second summand had been left blank (p. 39). The recursion equations for sum are (p. 40):

$\cdot 1 \quad a\varepsilon N_0. \supset .a + 0 = a$

$\cdot 2 \quad a, b\varepsilon N_0. \supset .a + (b+) = (a + b)+$

The definition $1 = 0+$ now gives

$$a + 1 = a + (0+) = (a + 0)+ = a+ \text{ and } a + (b + 1) = (a + b)+1$$

2.4. AXIOMATIC GEOMETRY

The study of axiomatic geometry was revived after the discovery of non-Euclidean geometries. It was found, for example, that Euclid's geometry had left completely open the ordering principles of geometry, for example when a point is between two other points on a line. The pioneering work here was the 1882 *Vorlesungen über neuere Geometrie* (Lectures on recent geometry) by Moritz Pasch. His achievement has been shadowed by Hilbert's development of axiomatic geometry in his *Grundlagen* in 1899, in which Pasch's discoveries are put to use.

Hilbert does acknowledge that his treatment of the ordering principles comes from Pasch, but that was soon forgotten. The latter is instead still remembered for the axiom of Pasch that can be put as: If a line cuts one side of a triangle and does not pass through the opposite vertex, it cuts one of the remaining sides.

Here I shall discuss in detail Hilbert's axiomatization, in particular his first group of axioms that concern the notion of *incidence* and the notion of *parallelism*. Two questions will be addressed in detail, the first about *construction vs. existence*. A curious little finding testifies to Hilbert's changing his mind about the order of the two notions. Secondly, we shall address the question of to what degree Hilbert succeeded in formalizing his geometrical system. Finally, we shall have a brief look at how plane projective geometry is axiomatized by today's standards.

(A) CONSTRUCTION AND EXISTENCE. Hilbert's incidence geometry of the plane has just three axioms:

(I1) For two points A, B there exists always a line a such that both of the points A, B are incident with it.

(I2) For two points A, B there exists not more than one line with which both points A, B are incident.

(I3) On a line there exist at least two points. There exist at least three points such that they are not incident with one line.

Hilbert adds that one always intends expressions such as "two points" in the sense of two distinct points.

Going through the subsequent development, one verifies that Hilbert *never* refers to axiom I1 in the proofs of his theorems. Yet axiom I1 seems quite like the two other axioms in appearance. The explanation of what seems like a strange fact may be that Hilbert was unwilling to revise the texts of his proofs. The preceding formulations are my translations of those of the seventh edition of 1930, and a comparison with the text of the first edition of 1899 reveals an essential difference. The original axioms read:

(I1) Two points A, B distinct from each other determine always a line a; we shall set $AB = a$ or $BA = a$.

(I2) Any two distinct points of a line determine this line; that is, if $AB = a$ and $AC = a$, and $B \neq C$, then also $BC = a$.

The existence of three noncollinear points is guaranteed by axiom 7:

(I7) On every line there are at least two points, on every plane at least three points, not incident with one line, and in space there are at least four points, not incident with one plane.

Thus, Hilbert's original formulation of axiom I1 was in the form of a *construction postulate*, and in 1899 he even had a symbolic notation for the condition required by the construction: the equality symbol crossed over with a thick vertical bar. In the proofs of the theorems, he would simply write AB whenever a line had to be constructed from two distinct points.

From 1903 on, the first two axioms read as follows:

(I1) Two points A, B distinct from each other determine always a line a.
(I2) Any two distinct points of a line determine this line.

These are just like the original formulations, but with most of the notation dropped out. The notation AB is used also for line segments and rays.

The preceding form is how the axioms remained until the seventh edition, when finally "bestimmen" (to determine) gave way to "es gibt" (there are). The construction postulate for connecting lines, as we may call it, was abandoned, and a formulation in terms of implicit existential quantification was given. But Hilbert did not change the practice of justifying steps in proofs by simply writing AB for a line whenever two distinct points were available. It was left to the reader to guess what his notation meant.

Even if the notation of constructions remained until 1930, Hilbert's conceptual change had occurred already around 1900. In the geometry, he still talks of proving geometrical truths from a few simple axioms. The study of the properties of the axioms is based on "the logical analysis of our spatial intuition" (1899, pp. 89–90 and introduction). In his famous "mathematical problems" paper of 1900, the doctrine of existence as consistency is instead very clear (1900a, p. 301):

If we succeed in proving that the properties given to our objects never can lead to a contradiction in a finite number of logical inferences, I will say that the mathematical existence of an object, say a number or function fulfilling certain properties, has been demonstrated.

In another paper of the same year, on the concept of number, Hilbert compares the "genetic" and "axiomatic" methods. In the former, one begins with the number 1, then builds up the sequence of natural numbers and their arithmetic, goes on to rational numbers, and so on. In the axiomatic method, instead, one begins with the assumption of the existence of the things one talks about, like points, lines and planes in geometry. These are related to each other through axioms that have to form a consistent and complete system (1900b, pp. 180–181).

Further particulars testify to Hilbert's changing ideas about mathematical existence. We shall first look at the problem of establishing a construction for the intersection point of two distinct lines and then remark on Hilbert's treatment of geometric construction problems from a more general point of view.

Hilbert mentions the following theorem as an immediate consequence of the incidence axioms (*Satz* 1, p. 6): Two (distinct) lines of a plane have one or no points in common. No proof is given, but certainly Hilbert's idea was that if two distinct lines a and b had at least two distinct points A and B incident, axiom I2 would give $a = AB = b$, which is impossible. (Actually, I found an argument to this effect in Hilbert and Cohn-Vossen 1932, p. 103.) Note that the argument is indirect, it shows only that the lines a and b have *at most* one point in common.

Chapter VII of Hilbert's book is devoted to geometric constructions. There he states that the axioms of group I make it possible to execute the following task 1 (1899, p. 78):

> To connect two points with a line, and to find the intersection point of two lines in case they are not parallel.

The second part is quite problematic, as the concept of parallelism appears only later, in axiom III (p. 10):

> In a plane with a point A outside a line a one and only one line can be drawn that does not intersect the line a; It is called the parallel to a through point A.

Even with this definition, the intersection point construction still remains to be effected, because the concept of parallels refers only to lines obtained by axiom III, so actually by a rule of parallel line construction.

What is needed is a general definition of parallels: that of distinct lines that don't have a point in common. Indeed, if lines a and b are distinct and not parallel, this definition gives that they have a point in common. Its uniqueness is guaranteed by Hilbert's first theorem that we just enunciated.

Hilbert soon changed the axiom of parallels from a construction postulate into an existential axiom (1903 edition, p. 15):

> Let a be an arbitrary line and A a point outside a: There is in the plane determined by a and A only one line b that is incident with A and that does not intersect a.

Only in the 1930 edition do we find the missing definition of parallels (p. 28):

> Explanation: We call two lines parallel if they are in the same plane and don't intersect each other.

Intersection, in turn, must be a concept that applies to two distinct lines that have a point in common. This can be gathered from the explanations following axioms I1 and I2 in the 1930 edition (p. 3):

> A is incident with a line a and also with another line b, we shall also say: the lines a and b intersect, have a point in common, etc.

Putting the preceding observations together, we arrive at the following picture of Hilbert's geometry of incidence and parallelism:

1. The *basic concepts* are equal points, equal lines, and incidence of a point with a line.

2. Two lines *intersect* if they are not equal and there is a point incident with both.

3. Two lines are *parallel* if they are not equal and there is no point incident with both.

The classical disjunction, two distinct lines are parallel or intersecting, is precisely what is needed in order to obtain the two cases concluded in Hilbert's *Satz* 1.

To be precise, in 1903 Hilbert changed the parallel line construction into an axiom that expresses the uniqueness, but not existence, of the parallel to a line through a point. The reason is that his first axiomatization had five groups of axioms: group I for incidence, group II for order, group III for parallels, group IV for congruence, and group V for continuity. In 1903, he changed the order of groups III and IV, claiming he could do the parallel line construction through angle congruence: a line a and a point A outside a are assumed to be given (1903 edition, p. 28):

> Let us draw a line c which goes through A and intersects a, and then a line b through A so that line c intersects lines a and b with the same angles. It follows . . . that the lines a, b don't have a point in common.

The trouble here is that no axiom says there is, for any line and point outside the line, another line through the point that intersects the given line.

Let us see what would be needed to resolve the situation. We can try to use axiom I3 which says that on any line there are at least two points, say B and C for line a. Then, if we succeed in proving that point A is distinct from B, we can construct the line AB. To this effect, assume A and B are equal. Since B is incident with a, and A is outside a, we have a contradiction. Therefore, A and B are distinct. Since lines a and AB have a point in common, it remains to prove that they are distinct lines. Assume they are equal. It follows that A is incident with a, which is impossible. Therefore, a and AB are distinct lines. Now, finally, we have proved the existence of the parallel to a given line through a given point. Therefore, if two lines are distinct and not parallel, we can infer the existence of an intersection point. This inference relies essentially on the purely existential axiom I3.

Striking evidence of Hilbert's change of mind regarding mathematical existence can be gathered from his notes on geometry from the 1890s. Five years before the *Grundlagen der Geometrie*, Hilbert put the axioms as follows (Hilbert 2004, p. 73):

Existence axioms. Better: *Axioms of incidence*

  1.) *Any 2 points A, B determine always one and only one line a.*
    One says that A, B lie on a. a is called the connecting line, goes through A, B.

2.) *Any* 2 *points* A, B *on line a determine the line a.*
   Or in formulas: from $AC = a$ and $BC = a$, $A \neq B$ follows $AB = a$.

Hilbert defines what it means for two lines to have a point in common, but he remarks that "it remains undecided whether 2 lines of a plane ... have a common point at all" (p. 74). Further on, as already mentioned in the previous Chapter, he remarks (p. 75):

> It escapes our experience whether one can always find a point of intersection for two lines. We therefore leave the matter preliminarily undecided and state only: 2 lines of a plane have either one or no points in common.

It is a troublesome feature of Hilbert's *Grundlagen* of 1899 that it retains the axioms of 1894 but claims to solve the task of constructing an intersection point. More generally speaking, Hilbert's treatment of geometric construction problems in his book displays the oddity that the existence of solutions to geometric construction problems is proved purely indirectly. The preceding passages from 1894 are evidence of Hilbert's awareness of the problem of effectiveness of a geometric construction. The question of constructive or nonconstructive existence did not become a sharply defined issue until later, after Brouwer's criticism of the classical law of excluded middle, but by that time Hilbert had decided his way.

(B) DIFFERENT SENSES OF FORMALIZATION. The original idea about formalization is that it consists in the *use of symbols*. Thus, one used to speak of "symbolic logic." One use for symbols is to identify the clearly formal parts. But other means are available for that, as in computer science, where a typeface different from that of the main text is used to indicate expressions in a programming language, as in if A do B else C. Here the words *if, do*, and *else* are not plain English words but rather expressions in a formally defined sentential grammar.

Formalistic philosophy of mathematics introduced the idea of a mathematical system as a collection of finitary rules for the manipulation of concrete symbols. A second, related idea is formalization as *machine executability*. It calls for an explicit syntax for expressions taken as *strings of symbols*. A third idea of formalization is that of *generative grammar*, in which phrase structure is made explicit by formal rules for the generation of expressions.

None of the preceding corresponds precisely to Hilbert's sense of formalization in his geometry. As for the first sense, his geometry in its later editions uses letters as identifiers for points, lines, and planes. It has a symbol for congruence, and a mnemonic sign for angles. Beyond these few symbols, it is written in an informal language familiar to anyone who reads German. As to the second sense, any text can be thought of as a string of symbols, as long as its characters come from some standard set such as ASCII. Then, the idea of formalization as "the use of symbols" is just like saying that we do word processing. Thus, there is more to the formalization of the language of mathematics than writing strings of standard symbols, namely the structure that is left unrevealed when treating text as a string.

Poincaré's review of Hilbert's book from 1902 is a wonderful illustration of the idea of formalization as machine executability. Says Poincaré (1902, pp. 252–253):

> M. Hilbert has tried to put the axioms in such a form that they could be applied by one who doesn't understand their meaning because he never sees a point, line, or plane. It must be possible to follow the reasonings by purely mechanical rules.

Such formalization would be a "puerile exercise," were it not for the question of completeness (ibid.):

> Is the list of axioms complete, or have some that we use unconsciously escaped us? . . . One has to find out whether Geometry is a logical consequence of explicitly stated axioms, or in other words, whether the axioms given to a reasoning machine will make the sequence of all theorems appear.

Poincaré mocked the formalizability of mathematical proofs that he thought reduced the mathematician's task to just waiting and seeing whether proofs of presumed theorems come out. Here we have the first and most typical misunderstanding of Hilbert's aims, even committed by some of those who edit Hilbert's works today (as in Hilbert 2013, p. 41). Hilbert states, in a characteristic passage of his Paris talk, what it takes to find a solution to a mathematical problem:

> Foremost, one should succeed to present the correctness of an answer through a finite number of inferences, on the basis of a finite number of

conditions that lie at the basis of the setting up of the problem and that have to be formulated precisely each time. This requirement of logical deduction by a finite number of inferences is nothing but the requirement of rigour in the carrying through of a proof.

In Poincaré and many others, the formalizability of mathematical proof is mixed with the *decidability* of mathematical theorems. All theorems in a formal system can be numbered from 1 on, just like all days from this day on. To say *possibly some nth day in the future is doomsday* is not unlike saying *possibly some nth theorem in the future is theorem A*. If there is a doomsday to come, someone shall experience it, if not, one would have to wait until the end of time to be sure, and the same with *A*. The intuition of most mathematicians that there is no general decision method despite the possibility of formalization was confirmed independently in 1936 by Alonzo Church and Alan Turing. The negative solution to the *halting problem* for Turing machines has as a consequence that mathematical reasoning with the quantifiers is just like waiting for doomsday: Given a formula *A*, it is possible that it turns out someday to be a theorem, but until that happens, judgment has to be suspended.

Hilbert himself states as the basic principle of his study of geometry: "to express every question so that we could at the same time find out if it is possible to answer it following some prescribed way and using given restricted means." The aim is "to decide which axioms, assumptions or methods are needed in the proof of a geometrical truth" (1899, pp. 89–90). One motive for explicitness was to give a clear sense to the impossibility proofs of geometry. We may describe this sense of formalization as the quest for rigor, or "Strenge in der Beweisführung," as Hilbert liked to say. The need for rigor is felt when one considers an axiom system as an implicit definition of the concepts therein. A purely symbolic development helps keep apart unwarranted steps based on an intended interpretation. We have seen that Hilbert's axiomatizations of geometry until 1930 did not quite reach this standard.

Even in the 1930 edition of Hilbert's *Grundlagen der Geometrie*, many things remain quite vague; for example, the treatment of the ordered plane. It has a concept of a given side of a line, and several results are mentioned, all without proof. To get started with the proofs, one would need to make precise the concepts "same side" and "different side," a principle to the effect that two points on different sides of a line are distinct, and so on. The axiom of Pasch that is Hilbert's only plane ordering axiom will not suffice.

Hilbert's theorem 5 (8 in the 1930 edition), given without proof, states that each line a on a plane α divides the rest of the points of this plane α in two regions, with the property that "each point A of one region determines with each point B of the other region a line AB within which there is a point of the line a." The line construction AB can be effected only if A and B are distinct, and one can wonder how Hilbert thought this should be proved. Let us call the sides "side one" and "side two," and let point A be on side one and B on side two. If A and B were equal, then, since A is on side one, B also would be on side one, and likewise, both would be on side two. The unstated geometric assumption seems to be that no point is on both sides of a line. Moreover, the substitutability of equal points in geometric statements about the two sides of lines is assumed (see von Plato 1997 for details of these criticisms).

The details of Hilbert's geometry are incoherent. The reason lies in part in the imprecision of the formalization that requires implicit appeal to intuition, as with Hilbert's theorem 5. The incoherence is also caused by Hilbert's changing ideas about mathematical existence. These new ideas led to changes that were not consistently applied throughout the book. As said, synthetic geometry never turned into an active field of research after Hilbert. The permanent effect of his *Grundlagen der Geometrie* lies rather in its identification of some of the central foundational problems that a mathematical theory faces.

It is often said that Hilbert formalized the genuinely geometrical principles and left just the logical principles implicit (as in Weyl 1944, pp. 635, 640). It should be clear that such statements are not based on a reading of Hilbert's book, or are much later synthetizations of youthful impressions. Hilbert's first published formulation of the axioms in 1899 shows use of the equality symbol for lines. General principles for equality would have to be made explicit, unless these are considered part of the logical rules, and the same goes for the rules of substitution. A related question concerns the use of diagrams. In a truly formalized geometry, there should be no place for diagrams, except as a practical aid. Diagrams are overdetermined in comparison to the purely geometric assumptions language is able to express. A typical consequence of their use is incomplete case analysis. Hilbert's degree of explicitness in formalization matters can be profitably compared to a forgotten section on projective geometry in an otherwise well-known paper by Skolem from 1920. We shall do that in the next chapter.

(C) AXIOMATIC PROJECTIVE GEOMETRY. For purposes of comparison of standards of formalization, it will be useful to give a rigorous axiomatization of plane projective geometry. The structure of the axiomatization is presented in five parts.

1. **The Domain and Its Basic Relations.** We have two sorts of objects in our domain, *points* denoted $a, b, c, \ldots$ and *lines* denoted $l, m, n, \ldots$. Secondly, we have the basic relations $a = b, l = m$, and $a \in l$ (point incident with a line). The last could be written in any way, say $Inc(a, l)$, as no standard notation for incidence has established itself.

After the domain and basic relations are fixed, we consider the following.

I **General properties of the basic relations**

Reflexivity: $a = a$	$l = l$
Symmetry: $a = b \supset b = a$	$l = m \supset m = l$
Transitivity: $a = b \& b = c \supset a = c$	$l = m \& m = n \supset l = n$

2. **Constructions and Their Properties.** Next, we have a choice of geometric constructions and their properties to consider. We shall introduce the *connecting line* of two points a and b, denoted $ln(a, b)$, and the *intersection point* of two lines l and m, denoted $pt(l, m)$. These are formally defined as functions over pairs. Let the domain consist of points denoted Pt and lines denoted Ln. We then have the functions:

$$ln : Pt \times Pt \to Ln, \quad pt : Ln \times Ln \to Pt.$$

We can now express the incidence properties of constructed objects as the next group of axioms.

II **Properties of constructed objects**

$$a = b \vee a \in ln(a, b) \quad a = b \vee b \in ln(a, b)$$
$$l = m \vee pt(l, m) \in l \quad l = m \vee pt(l, m) \in m$$

The axioms state that the line $ln(a,b)$ is constructed exactly through the points a,b, and similarly for the construction $pt(l,m)$. In the axioms, $a=b$ and $l=m$ express *degenerate cases* of the constructions.

3. **Uniqueness of Constructed Objects.** We want to have the property that any two points on a line $ln(a,b)$ determine it, i.e., that

$$c \in ln(a,b) \& d \in ln(a,b) \supset c=d \vee ln(a,b)=ln(c,d)$$

$$pt(l,m) \in n \& pt(l,m) \in k \supset n=k \vee pt(l,m)=pt(n,k)$$

A simple formulation is

$$a \in l \& b \in l \supset a=b \vee ln(a,b)=l$$

$$a \in l \& a \in m \supset l=m \vee pt(l,m)=a$$

Thoralf Skolem found in 1920 a single axiom from which the uniqueness of both constructions follows:

III **Uniqueness of constructions**

$$a \in l \& a \in m \& b \in l \& b \in m \supset a=b \vee l=m$$

The previous formulations follow as special cases of Skolem's axiom.

4. **Substitution of Equals.** We need to guarantee that equals can be substituted in the basic relations. Transitivity of equality is, from this point, just the axiom by which equals are substituted by equals in the equality relations. For the incidence relation, we have the following.

IV **Substitution axioms for incidence**

$$a \in l \& a = b \supset b \in l$$

$$a \in l \& l = m \supset a \in m$$

Axiom groups I–IV give the *universal theory* of projective geometry.

5. **Existence Axioms.** To the universal axioms there is to be added an axiom of noncollinearity by which there exist at least three noncollinear points:

V **Axiom of noncollinearity**

$$\exists x \exists y \exists z (\neg x = y \& \neg z \in ln(x,y))$$

The greater part of a detailed axiomatization such as this would be left implicit in a standard mathematical use of axiomatic projective geometry, where properties of the equality relations and the substitution of equals are used without mention, and the only explicit axioms state the existence and uniqueness of connecting lines and intersection points, together with the axiom of noncollinearity.

The past couple of decades have seen the rise of computerized systems for the development of truly formal proofs, known as *proof editors*. Proofs are produced interactively in these systems, and what is called a *type checker* controls the formal correctness of each step proposed. A typical experience in this type of work is that some principle in the informal practice of a mathematical theory such as projective geometry is found missing, or that the conceptual order of definitions or the order of proofs of theorems is not correct. A concrete example is given by the computer implementation of a constructive version of the preceding axioms of projective geometry I found in 1995 (The axioms of constructive geometry). It was one of the two earliest theories implemented in the now widely used Coq proof editor and appeared in the work of the late Gilles Kahn (1995).

2.5. REAL NUMBERS

(A) CONTINUITY. The irrationality of $\sqrt{2}$ shows that line segments can be produced by ruler and compass constructions that cannot be measured by rational numbers. The geometric constructions are of infinite precision, and so are the real numbers that came to represent such constructions. The "arithmetization of analysis" started in the middle of the nineteenth century with definitions of convergent series of rationals, the Cauchy sequences, and the ε-δ-definition of continuity in the style of Karl Weierstrass.

Richard Dedekind (1831–1916) was a Berlin mathematician known for two foundational contributions, the 1872 booklet on continuity and irrational numbers and the 1888 booklet on natural numbers and what they are used for. The latter is written in set-theoretic terms that have become standard in mathematics, even if Dedekind himself was thoroughly idealistic in his mathematical philosophy. He writes that sets are "collections of things" and that the latter in turn are "completely determined by everything that can be stated or thought about them" (1888, p. 1). Of the natural numbers, he writes in the preface as an answer posed by his title that "the natural numbers are creations of the human spirit" and that they "serve as means to conceive the distinctness of things [Verschiedenheit der Dinge] more easily and sharply."

The booklet of 1872 introduces the real numbers through *Dedekind cuts*, still a standard part of mathematics instruction today. Dedekind begins with the rationals and poses three conditions on their strict order relation (p. 6):

1. *Transitivity*: Whenever $a > b$ and $b > c$, then also $a > c$.
2. *Density*: Whenever $a > c$, there is a b such that $a > b$ and $b > c$.
3. *Cut*: Each number a divides the rationals into two classes A_1 and A_2 such that numbers in the first class are less than those in the second.

Next, thinking of the continuous geometrical line, Dedekind notes that there are irrational numbers all over so that the line is "infinitely richer in individual points than the domain of rational numbers" (p. 9). His crucial insight about "the essence of continuity" is (p. 10):

> If all points of the line fall into two classes such that each point of the first class is at the left of each point of the second class, there exists one and only one point that brings forth this division of all points into two classes, this cutting of the line into two parts.

The standard remark to Dedekind's definition is that it is *impredicative*, meaning the definition just singles out an element from a totality assumed to exist in advance.

Order on the real line has the same properties 1–3 as the one for rationals, the last about cut divided into two parts. The first is that any real number α divides the reals into two classes, the second that any such division defines a unique real number α. Proofs are given for these properties that reduce them to those of the rationals. Dedekind uses the same case distinctions for reals as for rationals: Any two numbers stand in one of the relations $a < b, a = b$, or $a > b$ to each other. The understanding that such case distinctions are not decidable for reals, unlike for rationals, came several decades later; it has not reached the level of mathematics instruction even today.

Even if Dedekind doesn't see that the equality of real numbers is an infinitistic notion, he has a clear intuition of the need to have the property of continuity for operations on the reals. His last section, section 6 (p. 17), contains a very clear idea that the computation within a prescribed interval of accuracy of the value of a number λ from given numbers $\alpha, \beta, \gamma \ldots$ requires there to be intervals for the given numbers that result in the prescribed accuracy for λ.

The preface of Dedekind's second booklet is somewhat apologetic: He had worked on the foundations of arithmetic in the 1870s but had other duties. Meanwhile, works by others on the natural numbers appeared, of which he mentions Schröder's book of 1873, Kronecker's works, and von Helmholtz' essay of 1887 on counting and measuring, with the unreserved addition that his own approach had been "formed since many years and without any influence from whatever side." Four topics are listed and claimed as his proper main contributions (numbers added):

1. The sharp distinction into the finite and infinite.
2. The concept of the number of things [Anzahl].
3. That complete induction … really proves things.
4. That the definition by induction (or recursion) is determinate and consistent.

Dedekind's book introduces an abstract set-theoretic mode of thinking, with the basic notions of objects, sets, and mappings, conceived independently of Georg Cantor's set theory, as he suggests. I list these notions with a more current terminology, with the emphasis of undefined concepts taken from Dedekind:

1. A set S *consists* of the objects $a, b, c, \ldots$ and is as an object of thought again a *thing*.
2. Two sets S, T are *equal* if they have the same objects.
3. A set S is a *part* of set T if every object of S is an object of T.
4. Each object s of a set S can be thought of as a set and is therefore a part of S.

Next, there follow the basic results about the equality and subset relations, and about the intersection and union of sets.

The second section introduces:

1. *Mappings* (Abbildungen) φ over a set S, with the *image* $\varphi(s)$ and the *image set* $\varphi(S)$.

2. A mapping is *injective* (ähnlich) if the images $\varphi(a)$ and $\varphi(b)$ are distinct whenever a and b are distinct. The *inverse* of such a mapping is written as $\overline{\varphi}$.

3. Two sets S, R are *similar* if there is an injective mapping φ such that $\varphi(S) = R$, meaning if each object in R is the image of some object in S. Then even $\overline{\varphi}(R) = S$.

4. A set is *infinite* if it is similar to a proper subset of itself, otherwise it is *finite*.

Finally, going back to the topic of natural numbers and recursive definitions, we note Dedekind's abstract way of constructing the natural numbers: Any set S with a mapping φ into S and a "ground element," designated by 1, generates a sequence of natural numbers by iteration, $1, \varphi(1), \varphi(\varphi(1)), \ldots$. Natural numbers are the "chain" or closure of the ground element relative to the mapping, designated by N. The properties are that $\varphi(N)$ is a proper subset of N because 1 is not in the image set and that all images are distinct and φ is thus injective. Therefore, N is an infinite set by the two criteria Dedekind had laid down. Schröder's stroke notation is also used for the successor operation. The two properties of the successor mapping are forms of the "infinity axioms" 7 and 8 of Peano's list, with another formulation.

Dedekind gives the recursive definition of functions the following general formulation, with Z_n the set of the first n natural numbers, as in the earlier generation process (§9): Given any set Ω with an element ω and a mapping θ to Ω, there is for every n a mapping ψ_n from Z_n to Ω such that

I $\psi_n(Z_n)$ is a subset of Ω,
II $\psi_n(1) = \omega$,
III $\psi_n(t') = \theta \psi_n(t)$ for $t < n$.

This scheme for ψ_n is formulated as a theorem with an inductive proof by n. Next, a more general formulation is given, the "theorem of definition by induction" numbered as paragraph 126, that for any mapping of a set Ω to itself and distinguished element states the existence of a unique function from a chain N to Ω with the preceding three properties. Addition and multiplication of natural numbers are given as examples of recursive definitions, but Dedekind makes no claim that these are his own, contrary to what many later German authors, in particular Hilbert, suggest.

The scheme of foundational studies from the early twentieth century on, if not earlier, was fairly clear: to build up a solid foundation for the number systems and their functions, in particular the natural and real numbers. The central questions with such foundational systems were, both for arithmetic and analysis, those of rigorous formalization, consistency, completeness, and decidability. The consistency of analysis was placed second in Hilbert's list of important open mathematical problems of 1900:

2. The consistency of the arithmetical axioms.

When the question is one of studying the foundations of a science, one has to put up a system of axioms that contain a precise and complete description of those relations that obtain between the basic concepts of the science in question.

$$\vdots$$

Among the numerous questions that can be posed regarding the axioms, I would like to indicate this one as the most important, namely, *to prove that these are consistent among themselves, i.e., that one can never arrive on their basis, by a finite number of logical inferences, to results that stand in contradiction to each other.*

Few readers seem to have had the patience to go through all of Hilbert's two-page problem; around midway, he writes:

The axioms of arithmetic are essentially nothing else but the known laws of computation together with the axiom of continuity.

$$\vdots$$

I am convinced now that one must succeed with a direct proof of the consistency of the arithmetical axioms, if one goes through in detail and modifies in an appropriate way the known methods of inference in the theory of irrational numbers, with regard to the designated goal.

A dozen presumably reliable sources, not to mention Wikipedia, tell that Hilbert's second problem is the consistency of arithmetic, that is, of the theory of natural numbers. How important it is to read original texts!

By the late 1930s, there were answers to all the questions for the arithmetic of natural numbers, those for completeness and decidability being negative ones. Because of incompleteness, the interest in consistency turned into an interest in the proof-theoretical nature of systems of arithmetic, especially what is called the *proof-theoretical strength* of a proof system. Analysis inherits the latter two negative answers, but the proper formalization and proof theory of analysis are still in their beginnings today; it seems an unending struggle to come to an understanding of the real numbers.

(B) CANTOR'S DIAGONAL ARGUMENT. In 1892, Georg Cantor presented in a short article the *diagonal argument* by which it can be proved that the real numbers cannot be listed one by one, unlike the natural and rational numbers. Here is the simplest possible formulation of the essence of his argument, which except for notation is the same as in Cantor's concise article.

We consider *infinite binary sequences* and assume that they can be numbered one after the other like the natural numbers. Then we have, say:

1. 001011...
2. 010101...
3. 100101...
   ⋮
n. $\underbrace{010...0}_{n}$...

   ⋮

The nth decimal of the nth sequence is the indicated 0. Now form along the *diagonal* the sequence that has for each i the ith decimal changed, from 0 to 1 and the other way around, as the case may be. In this way, one obtains the sequence

  101...1...

Given any of the preceding sequences, say the ith, the new sequence differs from it at the ith decimal, so it cannot be in the list.

One might think: Let's repair the situation by adding the preceding sequence as the zeroth, on top; no use, because one can then diagonalize again. In fact, one can form an infinity of diagonal sequences, say, by shifting the diagonal step by step to the right: change the ith member of the first sequence, the $i + 1$st of the second, etc., to infinity, for each i in turn. There is obviously no limit to how complicated schemes for changing every sequence could be devised.

We ask next how the individual binary sequences come about. There could be a simple arithmetic law that generates a sequence algorithmically, by successive application to 1, 2, 3,.... Such a law is a function f from the positive natural numbers $\mathcal{N}^+$ to the values 0 and 1, in set-theoretical notation

$$f : \mathcal{N}^+ \rightarrow \{0, 1\}$$

The sequence is generated by computing and writing down one after the other the values $f(1)f(2)f(3)$.... Another, completely different procedure would be a random generation of a sequence by tossing a coin, with heads giving 0 and tails 1.

The *set-theoretic notion of a function* is very comprehensive and easily accommodates both of the preceding extremes. The above function $f : \mathcal{N}^+ \rightarrow \{0, 1\}$ is a set of *ordered pairs* (x, y) such that $x : \mathcal{N}^+$ and $y : \{0, 1\}$, with the property of uniqueness: Given any two pairs (x, y) and (x, z), we must have $y = z$. The way functions come into being in the set-theoretic approach is that conditions are written down from which the existence of a set of ordered pairs with the uniqueness property follows. In all but trivial cases, these sets are infinite.

Let us look again at binary sequences as functions $f : \mathcal{N}^+ \rightarrow \{0, 1\}$, and collect all arguments with the function value 0 into a set A and collect those with the value 1 into a set B. These sets are *subsets* of $\mathcal{N}^+$, with the notation $A \subset \mathcal{N}^+$. In set theory, all subsets of a given set such as $\mathcal{N}^+$ can be collected into the *powerset*, in this case denoted $\mathcal{P}(\mathcal{N}^+)$. To each set A is associated its *cardinality*; say with $\mathcal{N}^+$ we have a *denumerably infinite* set with a cardinality Cantor denoted by the Hebrew letter $\aleph_0$ (aleph-nought). Any set that can be put into a one-to-one correspondence with $\mathcal{N}^+$ has this cardinality.

Given a set $\mathcal{A}$, its cardinality can be denoted by $\overline{\overline{\mathcal{A}}}$. Cantor's diagonal argument can be generalized into

$$\overline{\overline{\mathcal{P}(\mathcal{A})}} > \overline{\overline{\mathcal{A}}}$$

In Cantor's set theory, the infinite cardinalities obey a total ordering, and the notion of a first nondenumerable cardinal is considered meaningful. It is denoted by $\aleph_1$. Cantor's *continuum problem* is the question of whether $\overline{\overline{\mathcal{P}(\mathcal{N}^+)}} = \aleph_1$.

Each function $f : \mathcal{N}^+ \rightarrow \{0, 1\}$ determines a set of natural numbers and its complement, written in the notation of set theory as $\{n \,|\, f(n) = 0\}$ and $\{n \,|\, f(n) = 1\}$. On the other hand, each set of natural numbers determines such a function; say, take the elements of the set in order and let $f(n) = 1$ if n is in the set, and $f(n) = 0$ otherwise. Each function $f : \mathcal{N}^+ \rightarrow \{0, 1\}$ gives on the other hand a binary decimal that determines a real number of the unit interval from 0 to 1. Therefore, each subset of the set of natural numbers corresponds to a real number. Cantor's continuum problem can be posed as the following question: Is there a subset of the set of real numbers that is nondenumerable but that does not have a one-to-one correspondence with the set of all real numbers? This was the first of Hilbert's Paris problems in 1900. Twenty-five years later, he made the famous exclamation that "no one shall drive us out of the paradise Cantor created for us."

The *continuum hypothesis* is the assumption that there is no set with a cardinality strictly between those of the natural and real numbers, so that the reals would form the first infinity beyond the natural numbers. In the late 1930s, Gödel showed that the continuum hypothesis can be consistently added to a suitable axiomatization of set theory. He had already surmised by then that the continuum hypothesis would be independent of the known axioms of set theory. That this is so, that the negation of the continuum hypothesis can be equally well consistently added to the axioms of set theory, was proved by Paul Cohen in 1963. The independence of the *axiom of choice* was established by Gödel and Cohen in the same order, first consistency relative to the rest of the axioms in the late 1930s, then independence in 1963.

There is another way to look at the binary sequences, namely as a *tree with binary branchings*. We start from a root and climb either to the left or to the right and then repeat this:

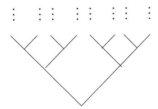

Left can be designated 0 and right 1, so that each branch of the tree corresponds to a binary sequence as here. The diagonal argument shows that there is no way of listing the branches one after the other. Each branch, or equivalently binary sequence, corresponds also to a point in the *unit continuum*. Take a geometric line segment and a binary sequence. If the first digit is 0, take the left half of the line; if it is 1, take the right one. With the second digit, do the same with the half you had, and so on. In the infinite limit, the binary sequence defines a point in the unit continuum.

Given a finite length n, there is a clear sense in which all initial segments of the branches of the binary tree exist. To define an infinite sequence, a function over $\mathcal{N}^+$ is needed. Set theory has its own unlimited answer to the question of what is allowed as a function; other approaches put restrictions on the notion.

(C) **THE CONSTRUCTIVE CONTINUUM.** Emile Borel (1908) gave an interpretation of Cantor's diagonal procedure in which he pointed at an implicit assumption, that the argument assumes that a numbering of binary sequences is given, and Borel specified this as *given by an effective enumeration*. The argument just shows that there cannot be any effective enumeration of binary sequences, and the Cantorian hierarchy of infinite cardinalities collapses. It even happens that an effectively enumerable set can have a subset that is not effectively enumerable. Since 1936, it has been known that the theorems of predicate logic are one such set, because all correctly formed formulas can be enumerated, but there is no effective way of separating out of that set those formulas that are theorems. Still, set theory allows the definition of a function f over the formulas A of predicate logic as follows:

1. Let $f(A) = 0$ if A is a theorem.
2. Let $f(A) = 1$ if A is not a theorem.

It has been proved that such a function cannot be computable. Its existence consists in the inconsistency of the assumption that it does not exist.

Borel's insight was provoked by *Richard's paradox*, which was one of the paradoxes found in the years around 1900, the years of the "foundational crisis" of mathematics. Other paradoxes included Cantor's paradox on the cardinality of the class of cardinal numbers, and Russell's paradox on the class of all classes that do not contain themselves as elements. Richard's paradox can be put as follows. Consider all real numbers of the unit interval definable by a finite number of words. These definitions can be organized into a sequence, according to length and within the same length alphabetically. Now change the first decimal of the first real number, say, if it is less than or equal to 5, increase it by 1, if greater than 5, decrease it by 1. Then do the same for the second real number, and so on, and form a decimal of these new numbers. These last two sentences define a real number of the unit interval in 50 words. On the other hand, the number so defined is different from each and every single real number of the unit interval definable by a finite number of words, which is a contradiction. The similarity to Cantor's diagonal argument for the nondenumerable infinity of the reals is striking.

In the constructive tradition of foundational philosophy of mathematics, mathematical objects are taken to exist as products of thought. This is the tradition Borel shared. In an exposition of his scientific work up to 1912, he wrote that "the mathematical entities we study are real for us only because we think of them" (1912a, p. 128). He admitted that this is philosophically an idealist position. A 1903 paper by Borel on the arithmetic analysis of the continuum starts with the claim: "All of mathematics can be deduced from the sole notion of an integer; here we have a fact universally acknowledged today" (1903, p. 1439). Further, following Kronecker's lead, Borel required that definitions in mathematics be based on "a limited number of integers" (p. 1440).

The principle of finite definability relates directly to Borel's remarkable solution to Richard's paradox (1908b, pp. 1271–1276): Granted that the only mathematical objects we can ever encounter are all definable in a finite number of words, we still may not be able to decide whether a given prescription is a definition of a real number in finite words. For example, we can make the suggested definition depend on the solution of an unsolved mathematical problem. In suggestive terminology, Borel said there is a distinction between denumerable infinity in the classical sense and effective enumerability, decades before there were any systematic theories of effective procedures. An example of a denumerable and also effectively enumerable sequence is obtained from the 40 characters in the (French) alphabet, periods and other punctuation included. Just put all finite strings of these characters in a lexicographical order. Still, the set of real numbers definable by a finite number of words is not effectively enumerable. This is the conclusion one has to take from Richard's paradox, so that an effectively enumerable set may have subsets that are not effectively enumerable.

Borel's views have been echoed in the thinking of constructivist mathematicians at various times. Making counterexamples from the assumed inexhaustibility of unsolved mathematical problems was a favorite move of Brouwer's. The banishment of the higher Cantorian transfinities, and the resolution of the diagonal argument as showing some infinite sets are not effectively enumerable, is the position taken in attempts at basing mathematics on the notion of effective computability or recursiveness. Borel, however, did not create any precise theory of what it means for a procedure to be effective. His notion of calculability was the intuitive one of following a finite prescription. In his "Le calcul des integrals definies" (Borel 1912b), we find some of the basic concepts of computable analysis, and some of Borel's insights into the mathematics that follows from these notions. A real number is calculable if one knows how to obtain arbitrarily good rational approximations to it (1912b, p. 830). Borel says that a representation of a real number as a decimal, for example, is not theoretically important. That particular representation is not "invariant" under arithmetic operations (ibid.). It seems that here Borel hints at the impossibility of obtaining decimal expansions for all calculable real numbers. It was not definitely noticed until later that an argument such as the one in Richard's paradox assumes real numbers have decimal expansions. The insight that this would not always be the case is Brouwer's (1921). Say one can calculate arbitrarily long an expansion $a = 0.000\ldots$ without knowing whether a nonzero number some time appears. For the number $0.5 - a$, one cannot determine even the first decimal before a is expanded into infinity. Borel (1912b) goes on to define a real function as calculable if its value is calculable for a calculable argument. Calculability means that one knows how to obtain arbitrarily good approximations to the function value, given good enough approximations of the argument. Therefore, calculable functions have to be continuous (p. 834). Here we have in fact the basic intuition of Brouwer's famous 1924 *uniform continuity theorem*.

3

THE ALGEBRAIC TRADITION OF LOGIC

3.1. BOOLE'S LOGICAL ALGEBRA

Algebraic logic began in 1847 when George Boole presented his "calculus of deductive reasoning" in a short book titled *The Mathematical Analysis of Logic*. His calculus reduced known ways of logical reasoning into the solution of algebraic equations. The known ways of logical reasoning were not just accounted for but were extended to full classical propositional logic.

Boole's starting point was Aristotle's theory of syllogistic inferences and its later development. Let us recall the propositions in Boole's notation:

Table 3.1. The four basic forms of syllogistic propositions

1. A: Each X is Y.
2. E: No X is Y.
3. I: Some X is Y.
4. O: Some X is not-Y.

Boole considered also more complicated forms, ones that appear in what are known as "hypothetical syllogisms."

Boole's logical calculus assumes as given a universe of objects, denoted 1. Classes of objects in the universe are denoted by $X, Y, Z, \ldots$. Lowercase letters $x, y, z, \ldots$ are "elective symbols" for these classes. The easiest way to understand these symbols is that x is a variable that takes the values 1 and 0 according to whether an object a belongs to the class X. The product xy is used for expressing that an object belongs to both X and Y, the sum $x + y$ for expressing that it belongs to at least one of X and Y, and the difference $1 - x$ for expressing that the object belongs to not-X, that it is a "not-X" type of object.

The reading that is closest to Boole's takes $X, Y, Z, \ldots$ to be subsets of the universe of objects, with not-X given by the complement of a set, product by the intersection, and sum by the union of two sets. We can equivalently use monadic predicate logic, by which the predicates $X, Y, Z, \ldots$ are applied to objects $a, b, c, \ldots$, with elementary propositions such as $X(a)$ as results. Anachronistically speaking, Boole invented the valuation semantics of classical monadic predicate logic and therefore also of classical propositional logic, and this was even before these two systems of logic had a well-defined syntax. The small variables represent valuations over the propositions: We have a valuation function v with $v(X(a)) = x$ and with $x = 1$ whenever $X(a)$ is true, $x = 0$ whenever $X(a)$ is false.

Boole writes properties of valuations in an intuitive way, such as $xy = yx$ or $x(y + z) = xy + xz$. The syllogistic forms of table 3.1 are represented in terms of equations as:

Table 3.2. Propositions represented as algebraic equations

1. A: With X a subclass of Y, we have $xy = x$, so $x(1 - y) = 0$.
2. E: If no X is Y, then $xy = 0$.
3. I: With V the class of those X that are also Y, $v = xy$ gives I.
4. O: Similarly to 3, we have $v = x(1 - y)$.

We are now ready to *reason by calculation*. Consider the first syllogistic inference, called Barbara. The premisses and their algebraic representations are:

$$\text{Each } X \text{ is } Y, \qquad x(1 - y) = 0,$$
$$\text{Each } Y \text{ is } Z, \qquad y(1 - z) = 0.$$

We therefore have $x - xy = 0$ from the first premiss. Multiply $1 - z$ by the left side to get $x(1 - z) - xy(1 - z) = 0$. Since by the second premiss $y(1 - z) = 0$, also $xy(1 - z) = 0$, so that $x(1 - z) = 0$. This just states that every X is Z, or the conclusion of Barbara.

Boole moves to hypothetical propositions such as: If A is B, then C is D. It has the form: If X is true, then Y is true. The four possibilities, X true and Y true, X true and Y false, X false and Y true, X false and

Y false, are represented by $xy, x(1-y), (1-x)y$, and $(1-x)(1-y)$, respectively. It is now clear that any correct inferences of classical propositional logic are validated by Boole's algebraic semantics, of which the "truth tables" popularized by Wittgenstein are a notational variant. Thus, Boole has no difficulty to continue the list of "principal forms of hypothetical Syllogism which logicians have recognized," as in his final example:

If X is true, then either Y is true, or Z is true. But Y is not true. Therefore, if X is true, Z is true.

The disjunction in the first premiss is exclusive and can be given as the term $y + z - yz$. Then the first premiss is $x(1 - y - z + yz) = 0$. The second premiss is $y = 0$, and the conclusion $x(1 - z) = 0$ follows at once by easy calculation.

Boole did not put down any definitive list of algebraic laws that would define what we today call a Boolean algebra. The usual way to introduce such algebras is to start from classical propositional logic and then collect all formulas equivalent to a given formula A into an equivalence class denoted $[A]$. Product, sum, and complement relative to the universe 1 in these classes correspond to the logical operations through the formation of the equivalence classes $[A \& B], [A \vee B]$, and $[\neg A]$.

Boole reduced Aristotelian syllogistic reasoning to calculation, which was a wonderful achievement. Encouraged by the success, he wrote a book with the bold title *An Investigation of the Laws of Thought* (1854). His logic was not able to treat relations but just one-place predicates, however. Today we know that there is no algebra of logic for full predicate logic, in which logical reasoning could be reduced to algebraic computation in the way Boole did for monadic predicate logic.

3.2. SCHRÖDER'S ALGEBRAIC LOGIC

The next important person in the development of the algebraic approach to logic and logic in general was Ernst Schröder. His work is found in the three-volume *Vorlesungen über die Algebra der Logik*, published between 1890 and 1905. He goes beyond Boole in that there is as a basic structure a partial order relation over objects, called "groups," or "domains" (Gebiet), with areas of the blackboard delimited by circles and ovals as a paradigmatic example. The order relation is used to represent logical consequence where Boole reasoned in terms of equalities. There are operations such as product and sum and relative complementation, with an obvious interpretation on the blackboard. The "Gruppenkalkül" amounts to the study of logic in terms of lattice theory, though the latter terminology and its German equivalent "Verbandstheorie" are of later origin. The partial order relation $a \leqslant b$ in a lattice has various readings, one of which is set inclusion and another is logical consequence. Schröder's own symbol for the order is produced by superposing something like a subset symbol and an equality, the latter indicating that the "subsumption" need not be strict, as in $\in$. The axioms are (Schröder 1890, pp. 168 and 170):

   I. $a \in a$.
   II. When $a \in b$ and $b \in c$, then also $a \in c$.

Algebraic laws determine unique lattice operations $a \cdot b$, also written ab, and $a + b$ (product and sum) that correspond to conjunction and disjunction in the logical reading. The principles that govern these operations are written as the "definitions" (p. 197):

   $(3_\times)'$ If $c \in a, c \in b$, then $c \in ab$.
   $(3_\times)''$ If $c \in ab$, then $c \in a$ and $c \in b$.
   $(3_+)'$ If $a \in c, b \in c$, then $a + b \in c$.
   $(3_+)''$ If $a + b \in c$, then $a \in c$ and $b \in c$.

There are in addition two special domains 0 and 1 with $0 \in c$ and $c \in 1$. By setting c equal to ab in $(3_\times)''$, one obtains the standard lattice axioms $ab \in a$ and $ab \in b$. Axiom $(3_+)''$ gives similarly $a \in a + b$ and $b \in a + b$.

One famous problem in Schröder's logic concerns the *law of distributivity*. It is expressed in Schröder's language as $a(b + c) = ab + ac$, and in a dual formulation as $(a + b)(a + c) = a + bc$ (p. 282). Is there a derivation of distributivity in Schröder's Gruppenkalkül? The founder of pragmatism, Charles Peirce, was one of the algebraic logicians, and he believed he had proved the law. However, Schröder found a counterexample, with explicit reference to analogous counterexamples in Euclidean geometry. It consists of three circular areas a, b, c that intersect in a canonical way. With sum as union and product as intersection of the areas, it is readily seen that the dual formulation of distributivity fails in this case (p. 286). The subsumption relation goes only in one way, not two, as would be required by the definition of equality. Meanwhile also Peirce had come to recognize that his purported proof was fallacious (p. 290).

An abstract formulation of Schröder's counterexample to distributivity is given by a lattice that consists of just five distinct elements a, b, c, d, e with the orderings $d \leqslant b, d \leqslant a, d \leqslant c, b \leqslant e, a \leqslant e$, and $c \leqslant e$. A figure will be useful for computing the terms in the distributive inequality:

From the figure, we have the two equalities $a \wedge (b \vee c) = a \wedge e = a$ and $(a \wedge b) \vee (a \wedge c) = d \wedge d = d$. For distributivity, we should have $a \leqslant d$, but $d \leqslant a$ was assumed. By the condition that all the elements be distinct, distributivity fails. This is the standard counterexample to distributivity in today's lattice theory.

The calculus of groups leads to a calculus of classes through the addition of an operation of negation, denoted $a_{\shortmid}$ and interpreted as a complement for areas. Schröder uses the algebraic calculus in the same way as Boole had. The letters indicate properties, and the task is to show what can be inferred from given letters used as assumptions. Here is one example (p. 530):

> Let it be stipulated that every b that is not d is either both a and c or neither a nor c. Further, no c and no d can be a and b simultaneously. To prove that no a is b.

The assumptions are expressed as

$$bd_{\shortmid} \subseteqq ac + a_{\shortmid}c_{\shortmid} \quad \text{and} \quad c \subseteqq (ab)_{\shortmid} \quad \text{and} \quad d \subseteqq (ab)_{\shortmid}$$

One of Schröder's basic observations is that the subsumption $a \Subset b$ is equivalent to the equalities $a = ub$ and $b = a + v$, for any u and v (p. 398). In particular, $a \Subset b$ whenever $ab_{,} = 0$. By this method, the three subsumptions that are assumed lead into what is called a "combined equation":

$$(ac_{,} + a_{,} c)bd_{,} + abc + abd = 0$$

Since $uu_{,} = 0$ and $u + u_{,} = 1$ for any u, c and d can be eliminated from this equation, with the result $ab = 0$ as required. By the preceding, it follows that $a \Subset b_{,}$; that is, that no a is b.

The example shows how logical consequence relations, as expressed by the subsumption relation, are turned into equalities on which algebraic manipulations in the style of Boole can be performed to obtain a result that can finally be read again in terms of consequence.

3.3. SKOLEM'S COMBINATORICS OF DEDUCTION

Thoralf Skolem's famous paper of 1920, titled "Logisch-kombina-torische Untersuchungen über die Erfüllbarkeit oder Beweisbarkeit mathematischer Sätze nebst einem Theoreme über dichte Mengen" (Logico-combinatorial investigations on the satisfiability or provability of mathematical propositions, together with a theorem on dense sets), contains the crowning achievement of algebraic logic. The paper, though, is known not for this but for its first section that contains the famous Löwenheim-Skolem theorem of quantificational logic. The other parts were completely forgotten, together with the algebraic logic of Schröder that got transformed into lattice theory. Part of the reason was the notation. Skolem wrote (xy) for Schröder's $x \Subset y$, the partial order relation $x \leqslant y$, and $\widehat{xyz}$ for the relation Schröder wrote as $xy = z$ and that can be written as $M(x, y, z)$ ("the meet of x and y is z"), and similarly $\underset{\smile}{xyz}$ for Schröder's $x + y = z$, the join relation $J(x, y, z)$.

Skolem wrote in 1913 a master's thesis in Norwegian, titled *Under-sökelser innenfor logikkens algebra* (Investigations on the algebra of logic). Part of the results were published in 1913 in the article *Om konstitutionen av den identiske kalkuls grupper* (On the structure of groups in the identity calculus). The paper begins with an admirably concise account of the algebraic approach to logic. The lattice ordering $a \leqslant b$ has various readings, one of which is set inclusion, another logical consequence. Algebraic laws determine unique lattice operations ab and $a + b$ (product and sum) that correspond to conjunction and disjunction in the logical reading. Negation $\overline{a}$ is defined by introducing a null class 0 and a universal class 1 satisfying an algebraic law that Skolem writes as

$$(a\overline{a} \leqslant 0)(1 \leqslant a + \overline{a})$$

Now the lattice structure becomes Schröder's "identity calculus," or a Boolean algebra in modern terms. Implication, from a to b, is later defined as the supremum of x such that $ax \leqslant b$.

Toward the end of Skolem's 1913 paper, what is today called Dummett's law, $(A \supset B) \vee (B \supset A)$ in logical notation, is expressed as

$$(a \leqslant b) + (b \leqslant a)$$

It is shown to be equivalent to what is called the *disjunction property under hypotheses*:

$$(A \supset B \vee C) \supset (A \supset B) \vee (A \supset C)$$

The same in Skolem's notation is

$$(a \leqslant b + c) \leqslant (a \leqslant b) + (a \leqslant c)$$

A dual is also noted:

$$(ab \leqslant c) \leqslant (a \leqslant c) + (b \leqslant c)$$

The former equivalence is given in Dummett (1959), but not the latter. Dummett's law is a consequence of the linearity of the order relation $a \leqslant b$.

Between the master's thesis and later work to be detailed shortly, Skolem spent the winter of 1915–16 in Göttingen. He was strongly influenced by Leopold Löwenheim's 1915 paper, and it is in Göttingen that he discovered, as a consequence of what is now the Löwenheim-Skolem theorem, the "Skolem paradox": If set theory is formalized, it has an interpretation in a denumerably infinite domain. Before publishing that result, he finished a long paper with an equally long title, *Untersuchungen über die Axiome des Klassenkalküls und über Produktations- und Summationsprobleme, welche gewisse Klassen von Aussagen betreffen.* (Investigations on the axioms of the calculus of groups and on problems of production and summation that concern classes of propositions.) It was Skolem's second paper on logic, dated 26 May 1917 and published in Norway in 1919. The work contains important early results on lattice theory, including the following:

1. A proof of the mutual independence of all the lattice axioms through graphical models, some of them infinite.

2. The same for distributive lattices. In particular, the five-element lattice that is a counterexample to distributivity is given (p. 74).

3. A method for deciding if an atomic formula is derivable from at most two atomic formulas by the axioms for a distributive lattice.

In the paper, Skolem further studies the notion of implication, or the case of a lattice in which the inequality $ax \leqslant b$ has a maximal solution, denoted by $\frac{b}{a}$ in Skolem (p. 78). The resulting structure is in fact what is today called a *Heyting algebra*. The modern notation introduces an arrow operation $a \rightarrow b$, and Skolem's inequality written with the lattice meet and arrow operations reads as $a \wedge (a \rightarrow b) \leqslant b$. Many basic properties of Heyting algebras are proved, for which see von Plato (2007).

It is quite astonishing that Skolem had found an algebraic axiomatization of intuitionistic propositional logic well before its basic principles were definitively clarified by Heyting's axiomatization (1930). Of his motivations for introducing the algebraic axiomatization, what he called "class rings" (Klassenringe), Skolem writes that they are "a natural continuation and generalization of the groups of the identity calculus" (1919, p. 78).

Let us now turn to Skolem's 1920 paper. It gave the axioms of Schröder's lattice theory as *production rules*, in a formalization with the relations $\overparen{xyz}$ and $\underparen{xyz}$ mentioned earlier instead of the operations

$x \wedge y$ and $x \vee y$:

I. For each x, (xx).

II. From (xy) in combination with (yz) follows (xz).

III$_\times$. From $\overset{\frown}{xyz}$ follow (zx) and (zy).

III$_+$. From $\underset{\smile}{xyz}$ follow (xz) and (yz).

IV$_\times$. From $\overset{\frown}{xyz}$ in combination with (ux) and (uy) follows (uz).

IV$_+$. From $\underset{\smile}{xyz}$ in combination with (xu) and (yu) follows (zu).

V$_\times$. From $\overset{\frown}{xyz}$ in combination with $(xx'), (x'x), (yy'), (y'y), (zz')$, and $(z'z)$ follows $\overset{\frown}{x'y'z'}$.

V$_+$. From $\underset{\smile}{xyz}$ in combination with $(xx'), (x'x), (yy'), (y'y), (zz')$, and $(z'z)$ follows $\underset{\smile}{x'y'z'}$.

VI$_\times$. There is for arbitrary x and y a z such that $\overset{\frown}{xyz}$.

VI$_+$. There is for arbitrary x and y a z such that $\underset{\smile}{xyz}$.

Principles $V_\times$ and V_+ contain assumptions such as $(xx'), (x'x)$. It could have helped the reader if Skolem had written these out as $x = x'$, etc., so that one could see that these two principles just state that equals can be substituted in the meet and join expressions. No such concessions are made to the reader.

The production rules are written in ordinary mathematical language, without symbolic notation, but it is still clear that Skolem takes a purely formal and combinatorial view of them that one usually associated with Hilbert rather than with the algebraic logicians (p. 116):

> The validity of a sentence in algebra, based on this axiomatic foundation, consists simply in the possibility of proving the following: Given these and these pairs and triples (xy), $\overset{\frown}{xyz}$, etc, those and those pairs and triples can be derived by possibly repeated and combined applications of the axioms. . . .

> In fact, *the axioms presented are production principles* by which new pairs and triples are derived from certain initial symbols. . . .

> *Here we have a purely combinatorial conception of deduction on which I would like to put emphasis, because it proves to be especially useful in logical investigations.*

Skolem's paper has no formal notation for derivations either. In proving his results, he writes things like "consider one application of the principle of substitution of equals as a last step." No trace is left of how he did his proofs that involve transformations of the order of application of the rules; in his head, on discarded paper? In my paper *In the shadows of the Löwenheim-Skolem theorem: Early combinatorial analyses of mathematical proofs* (2007), I show in great detail that Skolem's results for lattices are based on the permutation of the order of application of the production rules.

Let us look here at Skolem's example of a formal proof. It is the converse to the distributive law, expressed in the language of lattice theory with operations as $(a{\wedge}b){\vee}(a{\wedge}c) \leqslant a{\wedge}(b{\vee}c)$. In Skolem's notation, the assumptions and conclusion are: Let $\overset{\frown}{abd}$ and $\overset{\frown}{ace}$ and $\overset{\smile}{deg}$. Let $\overset{\smile}{bcf}$ and $\overset{\frown}{afh}$. Then (gh).

Skolem's proof text in the language and notation of his relational lattice theory is (p. 117):

From $\overset{\frown}{abd}$ in combination with $\overset{\frown}{ace}, \overset{\smile}{bcf}, \overset{\smile}{deg}$, and $\overset{\frown}{afh}$ follows the pair (gh).

Proof: From $\overset{\smile}{bcf}$ follows by III$_+$ the pair (bf). From $\overset{\frown}{abd}$ follow (da) as well as (db) by III$_\times$. From (db) and (bf) follows by II (df). From (da) and (df) in combination with $\overset{\frown}{afh}$ follows by IV$_\times$ [original has VI$_\times$] (dh). From $\overset{\smile}{bcf}$ follows on the force of III$_+$ the pair (cf). From $\overset{\frown}{ace}$ follow by III$_\times$ (ea) and (ec). From (ec) and (cf) [original has (ef)] follows by II (ef). From (ea) in combination with (ef) and $\overset{\frown}{afh}$ follows by IV$_\times$ [original has VI$_\times$] (eh). From (dh) in combination with (eh) and $\overset{\smile}{deg}$ follows finally by IV$_+$ (gh).

The proof is easier to read if we eliminate *linguistic variation*, often used to break the monotonicity of sentence structure in a proof text. The sentences in Skolem's proof are then:

1. From $\overset{\smile}{bcf}$ follows by III$_+$ (bf).

2. From $\overset{\frown}{abd}$ follows by III$_\times$ (da) and (db).
3. From (db) and (bf) follows by II (df).
4. From (da) and (df) and $\overset{\frown}{afh}$ follows by IV$_\times$ (dh).
5. From $\overset{\smile}{bcf}$ follows by III$_+$ (cf).
6. From $\overset{\frown}{ace}$ follows by III$_\times$ (ea) and (ec).

7. From (ec) and (cf) follows by II (ef).
8. From (ea) and (ef) and $\overset{\frown}{afh}$ follows by IV$_\times$ (eh).
9. From (dh) and (eh) and $\underset{\smile}{deg}$ follows by IV$_+$ (gh).

Now change the conjunctive conclusion (da) *and* (db) from line 2 into the two conclusions separately (lines 7 and 8 in what follows), and the same for line 6 (lines 12 and 13). Then write out the "given pairs and triples" on lines that begin a proof, beginning with $\overset{\frown}{abd}$ declared as given (lines 1–5 below). The result is, with some renumbering of lines and aligning:

1. Given	$\overset{\frown}{abd}$
2. Given	$\overset{\frown}{ace}$
3. Given	$\underset{\smile}{bcf}$
4. Given	$\underset{\smile}{deg}$
5. Given	$\overset{\frown}{afh}$
6. From $\underset{\smile}{bcf}$ follows by III$_+$	(bf)
7. From $\overset{\frown}{abd}$ follows by III$_\times$	(da)
8. From $\overset{\frown}{abd}$ follows by III$_\times$	(db)
9. From (db) and (bf) follows by II	(df)
10. From (da) and (df) and $\overset{\frown}{afh}$ follows by IV$_\times$	(dh)
11. From $\underset{\smile}{bcf}$ follows by III$_+$	(cf)
12. From $\overset{\frown}{ace}$ follows by III$_\times$	(ea)
13. From $\overset{\frown}{ace}$ follows by III$_\times$	(ec)
14. From (ec) and (cf) follows by II	(ef)
15. From (ea) and (ef) and $\overset{\frown}{afh}$ follows by IV$_\times$	(eh)
16. From (dh) and (eh) and $\underset{\smile}{deg}$ follows by IV$_+$	(gh)

Next, apply the following translation of the linear derivation into a tree form:

1. Write the last formula.
2. Draw a line above it.
3. Write next to the line the rule that was used.
4. Write above the line in order the premises of the rule.
5. Repeat until you arrive at the given formulas.

Then you get

$$
\cfrac{
\cfrac{\widehat{abd}}{(da)}\, III_\times \quad
\cfrac{\cfrac{\overbrace{abd}}{(db)}\, III_\times \quad \cfrac{\overset{\frown}{bcf}}{(bf)}\, III_+}{(df)}\, II
}{(dh)}\quad \widehat{afh}\, IV_\times
$$

$$
\cfrac{\widehat{ace}}{(ea)}\, III_\times \quad
\cfrac{\cfrac{\overbrace{ace}}{(ec)}\, III_\times \quad \cfrac{\overset{\frown}{bcf}}{(cf)}\, III_+}{(ef)}\, II
}{(eh)}\quad \widehat{afh}\, IV_\times \quad deg\, IV_+
$$

$$
\frac{}{(gh)}
$$

Now we see the *deductive dependences* in Skolem's proof. The numbers refer to the formulas at the right end of the corresponding line of the preceding linear derivation:

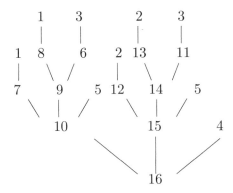

The traditional linear form of a written proof text hides this proper deductive structure of a proof.

Translation of the production rules into tree form gives

$$
\frac{}{(xx)}\, I \qquad
\frac{(xy)\ (yz)}{(xz)}\, II \qquad
\frac{\overset{\frown}{xyz}}{(zx)}\, III_\times \qquad
\frac{\overset{\frown}{xyz}}{(zy)}\, III_\times \qquad
\frac{(ux)\ (uy)\ \overset{\frown}{xyz}}{(uz)}\, IV_\times
$$

With $x = u$ standing for (xu) and (ux), the rules of substitution can be written as

$$
\frac{x = u \quad y = v \quad z = w \quad \overset{\frown}{xyz}}{\underset{\smile}{uvw}}\, V_\times
\qquad
\frac{x = u \quad y = v \quad z = w \quad \overset{\frown}{xyz}}{\underset{\smile}{uvw}}\, V_+
$$

The following can be proved:

The rules of substitution permute down relative to the other rules, and two consecutive instances of a rule of substitution contract into one.

Skolem may have thought such results too obvious to mention. The details can be found in the book *Proof Analysis*, section 5.3(b).

Skolem's treatment of lattice theory as a system of production rules ends with some examples of purely syntactic proofs of independence through *failed proof search*, including the distributive law, or what is written as $a \wedge (b \vee c) \leqslant (a \wedge b) \vee (a \wedge c)$ today (p. 122):

Example 2. The task is to prove that the subsumption $a(b + c) \nleq ab + ac$ (in Schröder's notation), generally valid in the calculus of classes, is not generally valid in the calculus of groups. Translated into our language, this task means the following: Let the triples $\widehat{abd}$, $\widehat{ace}$, $\underset{\smile}{bcf}$, $\underset{\smile}{deg}$, and $\widehat{afh}$ be given. Does the pair (hg) follow from these by axioms I–VI? To investigate this, it suffices for us to apply axioms I–V as long as there appears a set S closed with respect to these axioms. Axiom I gives us the pairs (aa) (bb) (cc) (dd) (ee) (ff) (gg) (hh). Axioms III gives us (da) (db) (ea) (ec) (bf) (cf) (dg) (eg) (ha) (hf), and we get further by II (df) and (ef) and by IV$_+$ also (gf). Now, however, no more pairs or triples can be formed with the help of I–V from the 5 given triples. The pair (hg) does not appear among the pairs obtained. By this, the underivability of (hg) from the 5 given triples by axioms I–VI is proved or, in other words, the unprovability in the calculus of groups of what is known as the distributive law $a(b + c) \nleq ab + ac$.

The lattice-theoretic part is followed by another one on plane projective geometry, and a similar syntactic proof of the independence of what is known as Desargues' conjecture. Skolem's anticipation of proof search methods in algebra and geometry remained unnoticed for some eighty years—a lost opportunity in foundational research, as we shall see when we return to Skolem in Section 6.1.

4

FREGE'S DISCOVERY OF FORMAL REASONING

4.1. A FORMULA LANGUAGE OF PURE THINKING

(A) THE POINT OF FORMALIZATION. Frege is the founder of contemporary logic, through his little book *Begriffsschrift* that came out in 1879. The name stands for something like "writing for concepts," in the sense of a notation, and there is a long subtitle that specifies the notation as "a formula language of pure thinking, modeled upon that of arithmetic." The actual notation in Frege's book made a rather bizarre impression on many, and no one else ever used it. Luckily he had Bertrand Russell among his few readers; some twenty-five years after the *Begriffsschrift*, Russell rewrote Frege's formula language in a style, adopted from Giuseppe Peano, that later evolved into the standard logical and mathematical notation we have today.

In his review of Hermann Cohen's book *Das Prinzip der Infinitesimal-methode und seine Geschichte* (The principle of infinitesimal-method and its history), Frege declares erroneous Cohen's view that only a historical insight can open up the logical conditions of a science (1885, p. 100):

> On the contrary: those logical foundations are always discovered only later, after a considerable coverage of knowledge has been gained. The historical point of departure appears from the logical point of view mostly as something casual.

Even Frege's achievement can be described in a logical light and order. For example, the first volume of his systematic *Grundgesetze der Arithmetik* (1893) is some 250-odd pages long but does not get beyond the mere beginnings of formal arithmetic, and not even to them proper, because of the obsessive drive to define a notion of *Anzahl*, or numerical quantity in the sense of "the number of pages in Frege's book is 254."

Frege's booklet *Grundlagen der Arithmetik* (1884) begins with "One usually answers to the question what the number one is, or what the sign 1 means: well, it is a thing," clearly an evasive answer for Frege. Further down, on the second page of the introduction, he asks: "Is it not shameful for science to remain in the dark about its closest object [the number one] that seems to be so simple? One will be even less in a position to say what a number is." These are the layman's and philosopher's questions never encountered in works on arithmetic.

Inspired by the way equations are written one below the other in arithmetic proofs, Frege wanted to display even the logical connections that obtain between the equations. He believed that the steps in proofs can be made explicit and complete to the degree that "nothing is left to be guessed."

To prevent intuitive notions from creeping into proofs, Frege wanted to carry them through "without words," using only letters and other symbols. He thought that ambiguity and uncertainty about steps in proofs is caused by the varying meanings of words. So proofs should be written in a formula language in which all expressions have an unequivocal meaning. The language is not restricted to arithmetic but just uses arithmetic as a model.

How successful is Frege's method of formalization, and what areas does it cover? What motivated his various choices in its design? For example, why does it have two rules of inference?

From the reactions, only a few in fact, Frege realized that no one had understood either the purpose or the design of his conceptual writing. Perhaps the worst reaction was that of Peano, who wrote that as his logic has just three basic signs compared to Frege's five, his system "corresponds to a more profound analysis" (Frege's *Briefwechsel*, p. 181). The *Begriffsschrift* was indeed written in a spare, terse style that made little compromise toward the reader. Therein lies its beauty for today's readers, who know what astonishing novelties it introduced. In the long preface to the *Grundgesetze* some fifteen years later, Frege was more forthcoming, as is shown by the following four passages.

1. On proofs without words (pp. v–vi):

> The proofs themselves contain no words but are carried through solely in my signs. They present themselves to the eye as a sequence of formulas separated by plain or intermittent lines or other signs. Each of these formulas is a complete sentence together with all the conditions that are necessary for it to hold. This completeness, one that will not tolerate assumptions that could be added tacitly, seems to me to be indispensable for the rigorous carrying through of proofs.

Frege's German for the rigorousness is: *die Strenge der Beweisführung.*

2. On rules of inference fixed in advance (p. vi):

> It cannot be required that all things be proved because that is impossible; but one can require that all sentences that are used without proof are expressly stated to be such, so that one can see clearly on what the whole edifice depends. One must, also, try to diminish the number of these ground laws, by proving all that is provable. Moreover, and here is where I go beyond Euclid, I require that all forms of inference and consequence that come to be used are listed in advance.

3. On gapless chains of inference (p. vii):

> The gapless nature of the chains of inference has as a consequence that each axiom, condition, hypothesis on which a proof depends, or however one wants to name them, is brought to light: and in this way one obtains a foundation for judging the epistemological nature of the theorem in question.

4. On how a proof is conducted (p. viii):

> One is usually satisfied if in a proof each step appears in a correct light, and sure enough so when one just wants to become convinced of the truth of the theorem to be proved. When, however, the question is to have an insight into this appearance in a correct light, the procedure is not sufficient, but one must instead write down each intermediate stage, so that the full light of consciousness may fall upon these.

Up to this page in the introduction, the point of Frege's logicism seems to be just that proofs be decomposed into simple logical steps. That is an incontestable part of Frege's doctrine. The other part is that mathematics, and arithmetic first, can be reduced to logic. The basic concepts of mathematics, especially that of a natural number, have to be given a logical definition, and the principles of proof, that of arithmetic induction especially, even be shown to consist of logical steps.

Frege's logicism springs from his peculiar philosophy of logic, in which *logical truth* is the foundation. In his 1885 article *Über formale Theorien der Arithmetik*, he writes (Frege 1885a, p. 103):

> All arithmetic theorems must be derivable from definitions purely logi-cally... hereby arithmetic is put into opposition with geometry that requires certain axioms proper to it, something no mathematician doubts, the opposites of which would be–purely logically considered–equally possible, i.e., free of contradiction.

This passage is reminiscent of Grassmann, quoted earlier in Section 2.2 (A). Frege lists three consequences from the logical nature of arithmetic, the first of which is that there is no sharp borderline between logic and arithmetic but rather a difference in level of generality. The second is (p. 104):

> There are no peculiarly arithmetic ways of inference that cannot be reduced back to the general ones of logic. Were such a reduction not possible with a form of inference, there would arise the question of the epistemological ground for its correctness.

Arithmetic is not based on any spatial intuition or physical observation; there remains only a logical foundation for it, and the task is especially to bring into light the logical nature of arithmetic inferences, in particular the step from n to $n + 1$ in proofs by induction (p. 104).

The *Begriffsschrift* of 1879 is in three parts:

I. The basic concepts and notation are explained (pp. 1–25).
II. Formal derivations are given in propositional logic (pp. 25–50), in the theory of equality (pp. 50–51), and quantificational logic (pp. 51–54).
III. Predicate logic is applied in the study of sequences, as determined by a binary order relation (pp. 55–86).

Arithmetic proper is not treated, but the last chapter is a preliminary step toward that purpose. It was also the topic that led Frege to discover predicate logic in the first place. As he explains in the preface, the basis of knowledge is a proof that can proceed either "purely logically" or else be based on empirical experience (p. ix). To find out which is the case in arithmetic, Frege "tried how far one can get in arithmetic with just inferences" (p. x):

> It went so that I tried first to reduce the concept of an order in a sequence into a *logical* succession, to proceed then further to the concept of number. So that nothing intuitive should creep in unnoticed, all of it had to depend on having a chain of inferences without gaps.

Ordinary language was a hindrance to precision, and "from this craving, the thought of the concept writing to be presented arose."

(B) FREGE'S FORMULA LANGUAGE. Frege's starting point is the writing of equations one below the other in arithmetic, say the following:

1. $(a + b) + 0 = a + b$
2. $b + 0 = b$
3. $b = b + 0$
4. $a + b = a + (b + 0)$
5. $(a + b) + 0 = a + (b + 0)$

A little training in arithmetic proofs suggests that we have here a proof of the associative law of addition, $(a + b) + c = a + (b + c)$, for the special case of $c = 0$. Equation 1 is an instance of the definition of addition of 0, and so is equation 2. Equation 3 follows from 2 by the symmetry of equality. Equation 4 follows from a principle by which, if in a sum $a + b$, the term b is *substituted* by an equal one, the sum remains the same in value. Finally, equation 5 follows from 1 and 4 by the *transitivity* of equality.

The equations are what Frege calls *contents* or even *conceptual contents* (begriffliche Inhalte). Each content expresses some thought. Just to *consider* this, Frege uses a *content stroke* (Inhaltsstrich) as in ⎯⎯ $7 + 5 = 12$. There is no *judgment* or *assertion* at this point that what is considered is true; that is expressed by adding a vertical *judgment stroke* as in ⊢⎯⎯ $7 + 5 = 12$.

There is a need to consider such content the truth of which is not known, or may even be false, as in an indirect proof, as Frege remarks in the *Begriffsschrift* (p. 4). His example is: "Let two line segments

AB and CD not be equal." In indirect proof, this is shown to be false. Negation is expressed by a graphic mark in the content stroke, to be shown soon.

Next in the *Begriffsschrift* is the central notion of *conditionality* (Bedingtheit) by which logical relations between contents are expressed. With two contents A and B, Frege notes that there are four possibilities of judgment (p. 5):

1. A is admitted and B is admitted
2. A is admitted and B is denied
3. A is denied and B is admitted
4. A is denied and B is denied

Frege now introduces the classical notion of implication, for which his notation is

$$\vdash\!\begin{array}{c} A \\ B \end{array}$$

This looks just like a vertical notation for the implication $B \supset A$ ("B implies A"), but iterated implications show how it really is. Let us take as an example the formula that is written in standard linear notation as

$$(C \supset (B \supset A)) \supset ((C \supset B) \supset (C \supset A))$$

Frege writes this as

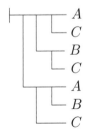

No parentheses are needed. The uppermost formula A is called the *upper* member of the implication, and the other formulas that hit the uppermost horizontal line *lower* members, here the implications $B \supset C$ and $C \supset (B \supset A)$.

Frege's classical implication is based strictly on his truth-functional analysis. He discusses in turn the possible cases. First, the consequence A is admitted. In this case, B plays no role, and "no connection of substance need obtain." The second case is that B is denied, and the result is analogous, namely that $B \supset A$ must be admitted. The most important observation is that one can admit $B \supset A$ "without knowing if A and B are to be admitted or denied."

It follows from Frege's truth-functional analysis that the rule of inference, from

$$\vdash\!\!\!\begin{array}{c} \rule{1em}{0.4pt}\ A \\ \rule{1em}{0.4pt}\ B \end{array}$$

and $\vdash\!\!\rule{2em}{0.4pt} B$ to conclude

$\vdash\!\!\rule{2em}{0.4pt} A$, is justified.

Next to implication, there is a negation sign that is just a little stroke, as in

$$\top\!\!\rule{1em}{0.4pt}\ A$$

Now conjunction and disjunction can be expressed. Frege goes through the latter in detail, with the result that

$$\vdash\!\!\!\begin{array}{c} \rule{1em}{0.4pt}\ A \\ \top\!\rule{0.5em}{0.4pt}\ B \end{array}$$

is admitted exactly when at least one of B and A is admitted and so expresses the inclusive form of "A or B." Conjunction in turn is given by

$$\vdash\!\!\!\begin{array}{c} \top\!\rule{0.5em}{0.4pt}\ A \\ \rule{1em}{0.4pt}\ B \end{array}$$

Frege has a perfect understanding of his formula language. The antecedents in implications are conditions, and several conditions that could be taken conjunctively, as in $C \& B \supset A$, can be expressed as so many iterated implications, as in $(C \supset (B \supset A))$ that is written in Frege's notation as

A conjunctive consequent in an implication, as in $C \supset B \& A$, would perhaps best be divided into two conditionals, $C \supset B$ and $C \supset A$. Similarly, that something is a consequence of a disjunctive condition, as in $C \vee B$, can be expressed as $(C \supset A) \supset ((B \supset A) \supset A)$, or in Frege's notation as

There is no direct way to express a disjunctive consequent, as in $C \supset B \vee A$, but Frege would have to modify it into $C \supset (\neg B \supset A)$, in his notation:

Frege's notation is quite handy for exchanging the order of iterated conditions and for taking contrapositions with any such conditions, as in

The universal quantifier is written so that the variable that is quantified is written in a little notch in the horizontal line at the head of a formula, as in

In Frege's classical logic, existence can now be defined by

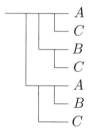

$$\vdash^{a}\!\!\sim\!\!\top A(a)$$

These matters will be detailed in the next section.

Frege's careful distinction between the content stroke, as in ——A, to indicate "the circumstance that A" or "the sentence that A," and the judgment stroke, as in $\vdash$——A, is in more modern terms the distinction between *propositions* and *assertions*; from the former, the latter is obtained by the addition of the assertion symbol, the vertical judgment stroke (Urteilsstrich). This notation has led to the turnstile that is used for the derivability relation.

The two-dimensional nature of Frege's formulas is best seen if we manipulate them a bit. Consider the formula

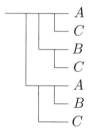

First, turn it 90 degrees clockwise, then straighten the edges, and you get

Next, annotate the nodes:

$$(C \supset (B \supset A)) \supset ((C \supset B) \supset (C \supset A))$$

$$C \supset (B \supset A) \qquad (C \supset B) \supset (C \supset A)$$

$$B \supset A \quad C \supset B \qquad C \supset A$$

$$C \qquad\qquad B \quad A \quad C \quad B \quad C \quad A$$

The result is a two-dimensional tree. Therefore,

Frege's formulas are two-dimensional syntax trees with missing annotations.

Our observation gives a way to translate the standard linear notation for logical formulas into Frege's notation—just a reverse of the preceding in three stages:

1. Construct the syntax tree of the given linear formula.
2. Next, leave out the subformulas and instead connect the line segments that indicate immediate subformula relations.
3. Finally, turn the tree 90 degrees counterclockwise and adjust the Fregean hooks a bit.

In categorial grammars to which syntax trees belong, sentences are analyzed into a form of function and argument. For example, the intransitive verb *think* is taken as a function that takes a noun phrase (an expression that names an individual) as an argument, say *Gottlob*, and then gives as value *think(Gottlob)*. Some "surface transformations" convert the inherently two-dimensional functional form into a standard linear one, as in *Gottlob thinks*. The two dimensions come from having two independent things, the function and the argument, that are combined in the application of the function, as in the scheme

$$\frac{f : N \rightarrow S \quad a : N}{f(a) : S}$$

The idea of analyzing sentences of a language into function and argument, instead of the subject and predicate of traditional school grammar, seems to go back to Frege. It was his conviction that the language of everyday life leads mathematical arguments astray, and it is even more generally mainly responsible for "misunderstandings by others as well as errors in one's own thinking" (Frege 1882, p. 48). The subject-predicate form is not essential to mathematics; in fact, Frege notes that it is quite casual, because any sentence can be made a subject by the addition of a predicate such as: "The fact that...," where some sentence A stands in place of the dots. This is precisely what Frege's judgment stroke $\vdash\!\!\!\!-\!\!\!-\!\!\!- A$ does, with A the subject and $\vdash\!\!\!\!-\!\!\!-\!\!\!-$ the predicate.

(C) **FREGEAN INFERENCES.** Next, we look at the rules of proof of Frege's axiomatic logic, in a modern notation and with an anachronistic definition. Proofs begin with axioms; what they are can be determined from the list of deductive dependences in formal derivations as given at the end of his *Begriffsschrift*.

Definition: (i) *The axioms of logic are logical truths.*

(ii) *If $A \supset B$ and A are logical truths, B is also.*

(iii) *If $A(x)$ is a logical truth for an arbitrary x, $\forall x\, A(x)$ is also.*

Frege made explicit the principles that govern the notion of an *eigenvariable*, the "arbitrary" x that is used in mathematics for making universal generalizations. He noticed that a clear-cut syntactic criterion about free variables is sufficient to warrant generalization, which is one of the great insights in the development of logic. *Deduction* from the axioms is typically organized in a "chain of inferences." In these, any previously derived truth can be used. There is no problem about, even no need to, the combination of derivations, because the premisses of rules are always logical truths.

Instead of hypotheses, all conditions are made into antecedent parts of implications. It is almost as if Frege had understood a much later principle of logical reasoning, sometimes called "the deduction theorem," by which inferences from assumptions or hypotheses H into a claim C can be turned into assumption-free inferences of $H \supset C$. As to the second point, let us see what derivations with just the rule of implication elimination look like: We can best write the derivations after a two-dimensional pattern, as in

$$\frac{A \supset B \quad A}{B}$$

With n steps, this becomes

$$\frac{A_1 \supset (A_2 \supset ...(A_{n-1} \supset (A_n \supset B))...) \quad \vdots \quad A_1}{A_2 \supset ...(A_{n-1} \supset (A_n \supset B))...}$$

$$\vdots$$

$$\frac{A_{n-1} \supset (A_n \supset B) \qquad\qquad A_{n-1} \quad \vdots}{\dfrac{A_n \supset B \qquad\qquad\qquad A_n}{B}}$$

Each of the derivations of the *minor premisses* A_i would exhibit the same pattern, until even the minor premisses are axiom instances. There is no way of altering anything in this static structure; derivation consists in finding the right axiom instances, as topmost formulas in the derivation tree, such that all things match with the pattern of chopping off from the *main premisses* the outermost condition that stands as the head formula at the left.

Finally, let us see how Frege organizes the display of his derivations. The propositional rule of inference is given as

From the two judgments $\vdash\!\!\!\begin{array}{c} A \\ B \end{array}$ *A and* $\vdash B$ *the new judgment* $\vdash A$ *follows.*

This principle is turned into a formal rule of inference at once, on page 8 of the *Begriffsschrift:*

$$\vdash\!\!\!\begin{array}{c} A \\ B \end{array}$$
$$\vdash\!\!\!\!-\ B$$
$$\rule{3cm}{0.4pt}$$
$$\vdash\!\!\!\!-\ A\ .$$

The period is a strange addition, as if a derivation were a complete sentence instead of a formal object with a strict syntax of its own.

Why does Frege write the consequents of implications uppermost? It is certainly contrary to the natural order in which one writes arithmetic equations one below the other in proofs. Let us see how it goes in a combined derivation:

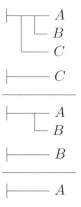

Each time, the minor premiss is right below the condition of the major premiss, with all atomic formulas in the derivation perfectly aligned vertically. These formulas are usually equations, one obtained from another through a suitable substitution, say, and they can become long and complicated. The display makes it possible to check that there is an exact match. Frege's choice leaves open the other option: Change the antecedent and consequent of implications and write the minor premiss first in the rule, an order that would better reflect the actual deductive top-down order of things in arithmetic proofs.

In the construction of chains of deduction, the writing is linear as in the preceding example, and one premiss is usually just referred to by a number, represented schematically by Roman numerals. This is even a necessity because it would be impossible to arrange the minor and major premisses in linear derivations so that they all fit in without having been previously proved.

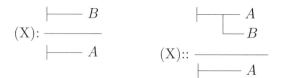

In these figures, (X): means that the major premiss ⊢⌐ A is re-
                                                      └ B
ferred to, and (X):: means that it is the minor premiss ⊢── B.

The *combination* of steps of deduction can be shown schematically,
as in

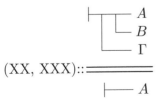

(XX, XXX)::══════

⊢── A

There are two minor premisses left unwritten, ⊢── Γ and ⊢── B.
Frege doesn't tell how to write a derivation if one premiss is a major
one (:) and the other a minor (::) one; perhaps :,:: would do?

In the whole of the *Begriffsschrift* there is only one place in which
a combined deduction is used (viz. in the derivation of formula 102).
Nevertheless, even one place is proof enough that the thing exists.

(D) FREGE'S AXIOM SYSTEM. Chapter II of the *Begriffsschrift* contains
a "presentation and derivation of some judgments of pure thinking,"
68 in number, divided into propositional and predicate logic. Frege
uses a separate word, *Ableitung*, for formal derivation, not the word
Beweis. His axioms are those of the 68 judgments that are given without
proof. At the end of the treatise, there is a complete list of deductive
dependences of the 68 judgments, with which each can be traced back
to ones that are the axioms used in the derivation:

Table 4.1. Frege's propositional axioms in a linear notation

1. $A \supset (B \supset A)$
2. $(C \supset (B \supset A)) \supset ((C \supset B) \supset (C \supset A))$
8. $(D \supset (B \supset A)) \supset (B \supset (D \supset A))$
28. $(B \supset A) \supset (\neg A \supset \neg B)$
31. $\neg\neg A \supset A$
32. $A \supset \neg\neg A$

The choice of these axioms tells something about Frege's understanding
of implication. The first one is a formal expression of his discourse
about implication in case the consequent A is admitted at the outset.

The second axiom is an instance of a principle that is crucial whenever a logical language is extended by new operations, say that of necessity: If it is necessary that $B \supset A$, if it is necessary that B, it should even be necessary that A. One could say this principle expresses that necessity is maintained under the rule of inference of axiomatic propositional logic. In Frege's second axiom, it is required that the addition of conditions, C in the axiom, is maintained under the logical rule of inference, clearly a crucial component in the deduction theorem that surfaced some fifty years after Frege's discovery of formal logic. Frege himself writes that "in case there are connections of substance," the second axiom can be read in terms of "necessary consequence." Finally, axiom 8 is an axiomatic expression of the idea that the order of conditions makes no difference in logic, an idea Frege even formulated as a rule of inference.

In Frege's logic, logical consequence relations are internalized in implications, but Frege is completely at home with the system and thinks about the logical laws in a contentful way, as forms of logical consequence and inference. To give an example of such axiomatic thinking, say axiom 2 has the instance

$$(A \supset (A \supset B)) \supset ((A \supset A) \supset (A \supset B))$$

By axiom 8, the order of the conditions can be changed to get

$$(A \supset A) \supset ((A \supset (A \supset B)) \supset (A \supset B))$$

The condition is easily proved, say from $A \supset \neg\neg A$ and $\neg\neg A \supset A$ by the transitivity of implication, Frege's theorem 5, and so we get the contractive law of logic:

$$(A \supset (A \supset B)) \supset (A \supset B)$$

Here one copy of a repeated condition A is dropped. Experience tells which axioms should be picked out for instantiation to get a derivation of a specific logical law.

In the actual construction of formal derivations, Frege gives explicitly at the left of the derivation the axiom instances used. The whole of his chapter II uses in the propositional part lowercase italic Latin letters for formulas. These, though, are still schematic, as were the uppercase capital Greek letters used in chapter I, because single letters become substituted by compound formulas in the construction of derivations.

In the end, it is cumbersome—Frege would say "sweaty"—to find derivations in axiomatic logic, and additional rules of inference will make the task of actually constructing derivations easier. In the first volume of Frege's *Grundgesetze der Arithmetik* (1893), there is a summary of such possible additional rules, shown here in a linear notation for formulas (§ 48, p. 61):

1. *Fusing together* of two successive content strokes.

2. *Exchange* of lower parts: In an iterated implication, the order can be changed. With a linear notation, we have the upper part D and the exchanged lower parts B and C in the rule:

$$\frac{A \supset \ldots (B \supset (C \supset \ldots \supset D)\ldots)}{A \supset \ldots (C \supset (B \supset \ldots \supset D)\ldots)}$$

3. *Turning over*: A lower member can be exchanged with the upper member if the truth values are reversed. This step is indicated by a "Zwischenzeichen" (intermediate sign), a big cross marked between one formula and another, as in

$$A \supset \ldots (B \supset \ldots \supset C)\ldots)$$

$$\times$$

$$A \supset \ldots (\neg C \ldots \supset \neg B)\ldots)$$

4. *Fusion of equal lower members*: This inference can be written in linear notation schematically as

$$\frac{A \supset \ldots (B \supset \ldots (B \supset \ldots \supset D)\ldots)}{A \supset \ldots (B \supset \ldots \supset D)\ldots)}$$

Exchanges can bring the two occurrences of B next to each other, so this rule is a generalization of the contractive law $(A \supset (A \supset B)) \supset (A \supset B)$.

5. *Change of a Latin letter into a German one*: This is the rule of generalization by which a universal formula is derived. The intermediate sign is a big smile as in $\smile$.

There now follow three rules that correspond to various writings of the implication elimination rule, as in the *Begriffsschrift*, and rules of substitution for Latin and German letters and the elimination of definitions.

In the paragraphs from § 49 on, many examples of derivations by the rules of inference are given.

Frege wrote around 1881 a long article about Boole's logic in relation to his own, with the title *Booles rechnende Logik und die Begriffsschrift* (Boole's computational logic and conceptual writing), in which he made it clear in what way his logic goes beyond that of Boole. Unfortunately, the paper was turned down, first by the *Mathematische Annalen* for being philosophical and then by a philosophical journal. It got published about a hundred years later in Frege's *Nachgelassene Schriften*. The inference to generality is for the first time given as a clear separate rule, to be discussed in the next section.

4.2. INFERENCE TO GENERALITY

The quantifiers, *all* for generality and *some* for existence, are as old as logic itself, but it was Frege, in the *Begriffsschrift*, who got a decisive hold on the principles of reasoning with them. Others, though, were slow in following him, partly because Frege explained generality first by stating that each its instance is a "fact," and only later added the "illuminating" observation by which generality can be inferred from an *arbitrary* instance. It took over fifty years to arrive at a perfect understanding of the quantifiers, in the form of autonomous, purely formulated rules of inference for the universal and existential quantifiers in the work of Gentzen. Meanwhile, a wide spectrum of attitudes toward the quantifiers and quantificational logic emerged, from ambivalent refusal, as in Skolem, to wholehearted platonistic acceptance by set theorists.

Frege's introduction of quantificational logic in the *Begriffsschrift* contains a section 11 titled *Generality* (Die Allgemeinheit, p. 19) that explains the logic of the universal quantifier, in modern notation $\forall x F(x)$. Its assertion means that whatever a is set in place of x, a "fact" $F(a)$ always results. One page later (p. 20), a mere $\forall x F(x)$ without the assertion component is declared to mean that $F(a)$ "is always a fact, whatever one should put in place of a." The quoted passage is, of course, the way a universal assumption is put to use and as such is correct. How, then, to understand universal claims, against assumptions?

What I consider to be the crucial passage on Frege's universal quantifier is found on p. 21, again with the abuse of a modern notation:

It is even illuminating that one can derive $A \supset \forall x\, F(x)$ from $A \supset F(a)$ if A is an expression in which a does not occur and if a stands in $F(a)$ only in the argument places.[1]

In the former passage, Frege is giving what evolved into the standard semantical account of universality: $\forall x\, F(x)$ holds or is true in a domain whenever each instance $F(a)$ is true. Now we have, in all but trivial cases with a finite domain, an infinity of truths, and this went against the grain of finitists such as Skolem and Wittgenstein, as I shall explain. In the latter passage, the one Frege thought illuminating and even put in italics, he is not stating anything about truth but just gives the essential principle for *reasoning about generality*, what became the rule of inference for the universal quantifier in Hilbert and Bernays' axiomatic logic, and the universal introduction rule of Gentzen's natural deduction.

After the quoted passage, Frege justifies the step of derivation in it: If $\forall x\, F(x)$ is denied, "one must be able to give a meaning to a such that $F(a)$ becomes denied." However, because $A \supset F(a)$ was admitted, "whatever a could be, the case in which A is admitted and $F(a)$ denied is excluded" (p. 22). Frege's justification gives his real interpretation of the universal quantifier, namely that there is *no counterexample*. The matter is seen clearly when Frege goes on to show by examples how his logic functions. In § 22, he gives the logical law $\forall x\, F(x) \supset F(a)$ and comments that if $\forall x\, F(x)$ is affirmed, $F(a)$ "cannot be denied. Our theorem expresses this." Instead of simply stating that any instance must hold, he is following the tradition of Aristotle, who explained generality as follows: "A thing is predicated of all of another when there is nothing to be taken of which the other could not be said" (*Prior Analytics* 24b28).

Frege finished volume 1 of his *Grundgesetze der Arithmetik* (Basic laws of arithmetic) fourteen years after the *Begriffsschrift*, in 1893. The propositional machinery is developed somewhat in relation to the earlier account, to make the construction and display of formal derivations

---

[1] A reader of the Van Heijenoort edition will miss Frege's emphasis, for the translation of Frege's *Auch ist einleuchtend* is a bland *It is clear also.*

easier. As for the quantifiers, the universal quantifier is taken into use with the motivation that one needs to make a distinction between the generality of a negation and the negation of a generality (Frege 1893, p. 12). A generality is a truth if each of its instances is a truth, and existence is defined in the usual way (ibid.). The inference to generality is presented inconspicuously twenty pages later, as a discourse about Latin and German (fraktur) letters, the former used as free and the latter as bound variables (p. 32). Frege's notation for the universal quantifier has a horizontal line in front of the formula, with a "notch" that has a fraktur letter in it. The scope of a quantifier is the formula to the right of the horizontal line. The condition is that the letter, the quantified variable, must not remain free anywhere outside the scope, thus making the free variable an eigenvariable (pp. 32–33). On pp. 61–64, as detailed in the previous section, Frege summarizes his axioms and rules and finally presents universal generalization as a rule of inference, by the "smile" inference line, and not just a modification (Verwandlung) of a proposition that is already at hand as he had written earlier.

In sum, Frege presented the two principles, universal generalization and instantiation, in what I take to be a reverse of the conceptual order of things, with a confusion as a result that took some five decades to clear, the final words set by the Göttingen logicians Bernays and Gentzen. In later chapters, I shall consider the reception of Frege's quantification theory through Russell to Skolem, Wittgenstein's second coming as a philosopher, in particular his work on the foundations and philosophy of arithmetic from 1929 to about mid-1930s, the intuitionists Brouwer and Heyting, and end up with the Göttingers Bernays and Gentzen.

4.3. EQUALITY AND EXTENSIONALITY

(A) THE PROPER TREATMENT OF EQUALITY. In between propositional logic and the quantifiers, Frege presents in the *Begriffsschrift* the notion of equality. It is a primitive notion, written as a judgment:

$$\vdash\!\!\!-\!\!\!-\!\!\!-\ (A \equiv B)$$

Frege explains nowhere why the parentheses are needed, instead, he writes (p. 15):

> The signs A and B have the same conceptual content, so that one can substitute everywhere B in the place of A and the other way around.

The formal expression of the substitution principle is Frege's *replacement axiom* 52:

$$
\vdash
\begin{array}{l}
f(d)\\
f(c)\\
(c \equiv d)
\end{array}
$$

Here f is a *function*, explained as follows (p. 16):

> *If in an expression, the content of which need not be judgeable, there occurs a simple or compound sign in one or more places, and we think of it as substituted in all or some of these places by something else, but throughout the same, we shall call the part of expression that remains invariant a function, the substitutable its argument.*

Next to the replacement axiom for equals, Frege has just one more axiom of equality, that of *reflexivity*, numbered 54 (p. 50):

$$
\vdash (c \equiv c)
$$

The first thing proved is the symmetry of equality, a somewhat tricky matter in which the expression in the replacement axiom has to be chosen suitably. As a preparatory step, Frege had already exchanged the two conditions in axiom 52 into

$$
\vdash
\begin{array}{l}
f(d)\\
(c \equiv d)\\
f(c)
\end{array}
$$

This law is numbered 53. To derive symmetry, Frege writes, as he always does, the substitutions at the left of the derivation, with the expression $A \equiv c$ chosen for the function $f(A)$ in the replacement axiom. The whole derivation is (p. 50)

$$
\begin{array}{ll}
54 & \vdash (c \equiv c)\\
(53)\ : & \rule{3cm}{0.4pt}\\
f(A)\vert\ \ A \equiv c & \vdash
\begin{array}{l}
(d \equiv c)\\
(c \equiv d)
\end{array}
\end{array}
$$

The notation at left gives as uppermost the number of the minor premiss $\vdash (c \equiv c)$, axiom 54. Next follows the derived law 53 as a major premiss, left unwritten and just indicated by the colon : and with the substitution of $A \equiv c$ for the $f(A)$ of law 53. The reflexivity is detached by the rule of inference to get a pure formulation of symmetry as an implication.

It baffles me why Frege nowhere proves the transitivity of equality, used as much in arithmetic proofs as symmetry. Had he chosen to actually develop formal arithmetic instead of trying to define the elusive concept of *Anzahl*, he would have needed to prove it.

Frege has reflexivity as an axiom, and a schematic replacement axiom, exactly as it should go. Whether he saw the added finesse that replacement is restricted to atomic formulas and can be shown to be admissible for arbitrary formulas is not easy to decide, but he was decades ahead of anyone else.

(B) EXTENSIONALITY IN A FORMAL LANGUAGE. The *Begriffsschrift* contains, together with its introduction of equality, a discussion of its meaning, that became a cornerstone of philosophical analysis, namely the distinction into *sense* and *denotation* (Sinn und Bedeutung). Frege writes that "equality of content" as in $A \equiv B$ applies to two names, not to their content (p. 13). Two names can have a different sense but the same content; that sameness of content of different names is what equality is meant to express. Frege's example is from geometry (p. 14): Take a circle and mark a point A on its circumference. A ray begins from A and initially intersects the diametrically opposed point B. Now the ray turns and, as it turns, the intersection point B moves on the circumference. After the ray has turned 90 degrees, the point B coincides with the point A. Coincidence in geometry means that the objects occupy the same location, though they need not be the same constructions.

> There corresponds to both of these ways of determination a name. The necessity of a sign of equality of content depends on the following: the same content can be determined in different ways; however, that *two ways of determination* really give *the same* in a specific case is the content of a *judgment*.

Frege sometimes uses *identity* as a synonym for equality, but he makes no attempt at defining or even speaking about an ontological notion of identity. His equality is a purely syntactic notion.

Whenever there is a function, be it propositional or otherwise, it is assumed to be *extensional* in the sense that equals can be substituted in the argument places. The intensional aspect of logic and mathematics is not thereby lost, however, because Frege assumes there is always a clear-cut syntax by which expressions for objects are built, be it arithmetic, geometry, or some other part of mathematics. There are approaches to the foundations of mathematics in which the intensional aspect finds no place. These approaches, especially set theory, are unable to express some very basic notions of mathematics. First of all, there is the notion of a function in the sense of an expression with a free variable, which is clearly explained by Frege, then the application of such a function through the substitution of the variable with a value for the argument, even that a clear notion in Frege, and third the computation of the value of the function. These three notions are made formal in the lambda-calculus of Alonzo Church (1932): *abstraction, application*, and *conversion*, all direct descendants of Frege's ideas and ignored in set theory. The development of theories of formal languages, especially programming languages, and of algorithmic computation, has revitalized the field. Frege's distinction into sense and denotation is taken to correspond to *algorithm* and *value*, as in Moschovakis (1993).

Formally defined syntax and the formal definition of derivations are usually associated with the name David Hilbert. These notions are very clearly present in the *Begriffsschrift* and *Grundgesetze*. The latter was studied in Göttingen in the first years of the twentieth century. What is more, a part of the correspondence between Hilbert and Frege has been preserved. A letter of 1 October 1895 from Frege to Hilbert appears in Frege's *Wissenschaftlicher Briefwechsel* and contains the following passage (Frege 1976, pp. 58–59):

> You told me in Lubeck, if I remember correctly, that you would be inclined to diminish rather than add to the formula-aspect of mathematics. As our discourse was interrupted, I would like to present to you my opinion in this way. The question is, I believe, not at ground about the opposition of spoken words and written symbols, but whether it is better to use theorems and methods of a greater or lesser extent. This opposition seems to me to coincide with the first one only when there is not yet any sufficient symbolism for the methods of greater extent that you rightly preferred. However, where one can express a train of thought completely in symbols, it will appear in this form as shorter and more surveyable than in words.

⋮

One must not equate the use of symbols with a thoughtless mechanical procedure, even if the danger of a bare mechanism of formulas lies closer here as when one uses words. One can think even in symbols. A purely mechanical use of formulas is dangerous 1. for the truth of the results. 2. for the fruitfulness of science. The first danger should be avoidable through a logical completion of the notation. As concerns the second, science would be brought to a halt if a formula mechanism took the upper hand so as to strangle thinking. Even so, I would not see such a mechanism as completely useless or harmful. I believe on the contrary that it is necessary. The natural procedure seems to be the following. What is originally permeated by thinking, reinforces itself with time into a mechanism that takes in part away thinking from the researcher. Just as in music, a host of initially conscious procedures have to become unconscious and mechanical, so that the artist can put his love into the play freed of these things.

⋮

The natural way to arrive at a symbolism seems to me to be to consider the breadth and unsurveyability and imprecision of the language of words as an obstacle in an investigation conducted in terms of words. A language of signs is created to help here, in which the investigation can be carried through in a surveyable and precise way. Therefore: first the drive, then its realization. It would instead be less productive to first create a symbolism and then search for applications. Perhaps the Boole-Schröder-Peano symbolism has gone this way.

Hilbert answered within three days of Frege's writing of the letter (*Briefwechsel*, p. 59):

Your valued letter has interested me extraordinarily; the more I regret that I spoke to you just in passing in Lubeck. Hopefully there will be another time with more opportunity.

At the beginning of the semester, I have the plan to bring your letter into discussion at our mathematical society. I believe that your opinion about the essence and purpose of symbolism in mathematics hits precisely the point.

4.4. FREGE'S SUCCESSES AND FAILURES

(A) FREGE'S SUCCESSES. Many of Frege's discoveries went exactly as they should, with unfailing insight. A list and discussion of some of the central ones follows.

1. *Classical propositional logic and classical implication.* Frege identified the classical propositional calculus as defined through the primitive notions of the conditional and negation, and the semantical criterion of correctness of formulas and inferences through two truth values. Frege's formulation of the truth values was in terms of *admission* and *denial*, rather than simply truth and falsity, and was therefore a step from an abstract notion of logical truth to semantical principles of reasoning. His analysis of the conditional is especially striking, with his discussion clearly covering what later came to be called "the paradoxes of implication" that $A \supset B$ must be admitted if B is admitted, and similarly in the case that A is denied. Frege carefully separates out the case in which there is a substantial connection (*ursachlicher Zusammenhang*) between A and B in an implication (*Begriffsschrift* p. 6):

> The substantial connection that lies within the word "if" is not expressed by our sign even if a judgment of this kind can be made only on such a basis. For this connection is something general, but that is not yet expressed here (see § 12).

In § 12, we read (p. 23):

$$\vdash \!\!\!\overset{a}{\underset{}{\vphantom{|}}}\!\!\!\begin{array}{l} P(a) \\ X(a) \end{array}$$

> means: "whatever one should put in place of a, the case that $P(a)$ is denied and $X(a)$ admitted does not occur."
>
> $\vdots$
>
> One can therefore translate: "if something has the property X, then it has also the property P," or "every X is a P," or "all X's are P's."
> *This is the way in which substantial connections are expressed.*

Russell directly repeats this discourse about the meaning of implication, as we shall see.

Frege does not give any explicit axiom system, but the list of deductive dependences identifies one, as discussed earlier. The question of its completeness seems to be an aspect Frege did not question, as indicated by his explanation of "the derivation of the more composite judgments from simpler ones" (p. 25):

> In this way, one arrives at a small number of laws that contain, if one adds those contained in the rules, the content of all the others, even if in an undeveloped way.

Consistency of at least the propositional axiom system is immediate, because Frege shows how to prove that his axioms must be "admitted" by semantical criteria and that such admission is maintained by the only rule of inference, implication elimination.

2. *Inference to generality.* This was simply Frege's greatest contribution. Russell was the first one to recognize it, first in a general way in 1903 in his book *The Principles of Mathematics.* In it, he notes the appearance of the universal quantifier in Frege (p. 519):

> He has a special symbol for assertion, and he is able to assert for all values of x a propositional function not stating an implication, which Peano's symbolism will not do. He also distinguishes, by the use of Latin and German letters, respectively, between *any* proposition of a certain propositional function and *all* such propositions.

Frege's Latin and German letters stand for free and bound variables, respectively.

Five years later, Russell ends his discussion of *all and any* in section II of his type theory paper of 1908, backed up this time by a formal development, by praising Frege (p. 158):

> The distinction between *all* and *any* is, therefore, necessary to deductive reasoning and occurs throughout in mathematics, though, so far as I know, its importance remained unnoticed until Frege pointed it out.

In Whitehead and Russell's *Principia Mathematica* (1910), Frege's rule of universal generalization received its first precise formulation in terms readable without Frege-exegesis. Thus, it took more than thirty years for this discovery to become understood.

The quantifiers are as old as logic itself. They appear in Aristotle's syllogistic, the theory of the four quantifiers *every, no, some,* and *not some,* what they mean when prefixed to the indefinite form of predication *A is a B,* and what the correct forms of inference are. Even if Frege was proud to present in the *Begriffsschrift* a formalization of the syllogistic inferences in terms of predicate logic, as the final example of his new notation, no formal quantifiers in the modern sense, ones that would bind variables, are needed for their theory. The four kinds of quantified propositions, *Every A is a B, Some A is a B, No A is a B,* and *Not some A is a B,* can be treated simply as atomic formulas, as we saw in Section 1.2. In light of this fact, it should be no surprise that Aristotle's quantifiers and syllogistic played no role in ancient Greek mathematics, contrary to Frege's quantifiers and logic in today's mathematics.

(B) DEAD ENDS IN FREGE. Frege's best-known failure is of course the inconsistency of his foundational system of mathematics, as revealed by Russell's paradox. It put a halt to Frege's logicist system. A less evident failure is Frege's attempt to base arithmetic purely on logic and definitions.

1. *Russell's paradox.* Russell announced the famous paradox to Frege in a letter of 16 June 1902 (*Briefwechsel,* p. 211):

> Let w be the predicate, to be a predicate that cannot be predicated of itself. Can one predicate w of itself? From each answer follows the opposite. Therefore one must conclude that w is not a predicate. Similarly, there is no class (as a whole) of those classes that as a whole do not contain themselves. I conclude from this that a definable set does not form under certain circumstances any whole.

Russell adds a note below his letter in which the paradox is given in Peano's notation, the substance being the contradictory propositional equivalence $w \in w \supset\subset \neg w \in w$.

Russell's paradox can be seen as a consequence of an unlimited *comprehension principle* by which any property $P(x)$ leads to a set of those objects that have the property, $\{x \mid P(x)\}$ in set-theoretical notation. With the property $\neg x \in x$, we get $w = \{x \mid \neg x \in x\}$. If $w \in w$, then w has the defining property $\neg w \in w$, if $\neg w \in w$, then again $w \in w$.

The argument uses the law of excluded middle, here the two classical cases in $w \in w \lor \neg w \in w$, but the development of intuitionistic logic showed that step not to be essential: The logical form of Russell's paradox is that the equivalence $w \in w \supset\subset \neg w \in w$ leads to a contradiction, namely, that the negation $\neg(w \in w \supset\subset \neg w \in w)$ is provable; this, however, is already constructively provable.

Russell's own attempted answer was to develop a *theory of types* in which self-predication is made impossible through a system of layers, the bottom being that of some ground objects, the first level that of their properties, the second that of properties of these properties, etc. Another approach is that of set theory, as in the work of Zermelo, Fraenkel, and Skolem, who put limitations on the comprehension principle in their attempts to find a consistent set theory.

2. *Arithmetic without arithmetic axioms* As explained earlier, Frege did not recognize any specifically arithmetic source of knowledge, in contrast to, say, geometry, which is based on spatial intuition. One way to express the position is that Frege saw no synthetic component in arithmetic, in the Kantian sense, a view he discusses at length in many places in the *Grundlagen* and *Grundgesetze*. Some thirty or forty years after Frege's attempts, the founder of intuitionistic mathematics, L. Brouwer, tried to distill the basics of arithmetic from one synthetic a priori: the succession of "now" that creates the natural numbers. The Kantian doctrine, then, would be that numbers have precisely those properties they inherit from the synthetic a priori intuition of the succession of numbers. In more mundane terms, we have a *zero* and a *successor* with some properties spelled out in axioms, as we saw in Section 2.1 in connection with Grassmann, Schröder, and Dedekind.

The *Begriffsschrift* resulted as a side effect of Frege's attempt at a logical definition of an ordered sequence. He begins with a definition of an expression that contains a property $F(b)$ and a two-place relation $f(b, a)$, the latter described as a "procedure" that gives an element a from a given b. Thus, it can be, as the lowercase letter suggests, a proposition defined by a function, as in $f(b) = a$. Frege's notion has the intuitive reading that "property F is inherited in the f-sequence" (*Begriffsschrift*, p. 58). The notion Frege defines is, in modern linear writing, an abbreviation for the following:

$$\forall x (F(x) \supset \forall y (f(x,y) \supset F(y)))$$

Various consequences of this formula are given, all of them hard to read, but Frege gives verbal formulations, such as (p. 64):

> If x has property F that is inherited in the f-sequence, if y follows x in the f-sequence, then y has property F.

An added footnote gives that "the Bernoullian induction rests on this." The last result of the *Begriffsschrift* is (pp. 86–87):

> If the procedure f is unique, and if m and y follow x in the f-sequence, then y belongs to the f-sequence that begins with m or appears in the f-sequence before m.

It is seen that Frege tries to express in logical terms properties of an order relation such as the linear ordering of y and m in the last theorem. In the *Grundlagen* five years later, it is seen what the point of all this is, for there he defines the concept of *number series* (§ 79, p. 89). The book, though, is written in such a verbose and polemical way that it is difficult to wade through it. For example, § 52 tries to derive foundational results from "German usage," quite contrary to what Frege says elsewhere about the harmful effect of ordinary language. In § 70, he writes that "definitions must prove themselves through their fruitfulness." Two or three pages later, we come to the definitions, in which the basic concepts are "a is subsumed under concept F" and "a stands in relation φ to b." The idea is that there can be a definite number of objects subsumed by a predicate, its numerical quantity, which Frege calls *Anzahl*, something like "numerosity" in a one-word analogy to the German. Further, there can be a one-to-one relation between two such predicates F and G, a condition that defines *equinumerosity*. In general, a numerical quantity is the scope of the concept "to be equinumerous with F" (p. 85). No formulas appear anywhere, but the number 0 is defined through the predicate "to be unequal to oneself" (p. 88, and let's write $a \neq a$ for that). Then, in a useful notation found in Hilbert and Ackermann (1928), we can write $0(F)$ for the numerical quantity 0 of a predicate F, a second-order predicate, and similarly $n(F)$ for numerical quantity n. Note that no definition of numbers in any usual sense has been given so far.

Next comes a tricky passage in which Frege tries to define the relation of immediate succession between numbers (*Grundlagen*, pp. 89–90):

> There is a concept F and an object x under its scope such that the numerical quantity that belongs to the concept F is n and that the numerical quantity that belongs to the concept "under the scope of F but not equal to x" is m.

Frege adds at once that the definition means the same as "n follows immediately m in the natural number sequence." Adopting Frege's polemical style in the *Grundlagen*, this should be marked as a grave error. Frege's definition does not give the passage from m to its successor n but rather the other way around: *given* n, we get its *predecessor* m by excluding one of the objects from the scope of F by the "not equal to x" clause. To wit, Frege defines the numerical quantity 1 as that which "belongs to the concept *equal to* 0." Let's write this as $x = 0$, and there is just one object that falls under the predicate, namely 0. The predecessor of 1 is obtained by adding to clause $x = 0$ the negation $x \neq 0$, and indeed the number of such x is 0. Section 79 gives various paraphrases and results about 0 and 1, among them (p. 91):

1. If a follows immediately 0 in the natural number sequence, then $a = 1$.
3. If 1 is the numerical quantity that belongs to a concept F, if object x falls under the concept F, and if y falls under the concept F, then $x = y$; i.e., x is the same as y.
6. Each numerical quantity except for 0 follows immediately a numerical quantity in the natural number sequence.

Despite the wording, the last one is about the predecessor, not the successor, of a number: It says that *for every* x *except* 0, *there is a* y *such that* x *follows immediately* y, and there is no way to reverse this order of dependence in which x comes first, then y.

No way is given for writing the numbers 2, 3, and so on, though it is clear how these should go: $2(F)$ means that there exist x and y that belong to F, they are distinct, and any z that belongs to F is equal to x or y, etc. Still, whenever in a work on the foundations of arithmetic the expression "etc." or its equivalent appears, we should be on guard: The writing of a few initial cases amounts to an inductive definition, a point made by Wittgenstein that applies to the example Schultz gave (Section 2.1(A)) above. As it stands, *if* we are given a property that defines a number n, Frege's procedure will give the number that *precedes* n, and we can repeat this until we come to 0, but it does not go the other way around.

Frege's discussion of the foundations of arithmetic in the *Grundlagen* begins properly with the folowing question (p. 5): Are numerical formulas provable? After a presentation of Kant's view in which they are unprovable, he illustrates the opposite view through Leibniz' proof of $2 + 2 = 4$: With the definitions $2 \equiv 1 + 1, 3 \equiv 2 + 1$, and $4 \equiv 3 + 1$, we have $2 + 2 = 2 + 1 + 1 = 3 + 1 = 4$. Frege points out a gap in the proof that remains "hidden by the leaving out of parentheses" (p. 7). The proof should be written as

$$2 + 2 = 2 + (1 + 1), 2 + (1 + 1) = (2 + 1) + 1 = 3 + 1 = 4$$

Thus, the associative law of addition $a + (b + c) = (a + b) + c$, instantiated by $c = 1$, is missing (pp. 7–8):

> If one assumes this law, one sees easily that each formula of one-plus-one can be thus proved. Each number can be then defined from the preceding one.... The infinite set of numbers is reduced back through such definitions to one and the increase by one and each of the infinitely many number formulas can be proved from a few general propositions.

Frege refers to the 1840 edition of Leibniz' *Nouveaux essais* by Erdmann, in which the "horizontal curly brackets," the ones shown in Section 2.1, are reproduced. Later editions from the ninteenth century, before the "Academy edition" of 1961, don't reproduce them. Frege gives the proof as follows:

Proof: $2 + 2 \underset{\text{Def. 1.}}{=} 2 + 1 + 1 \underset{\text{Def. 2.}}{=} 3 + 1 \underset{\text{Def. 3.}}{=} 4.$

Therefore: by the axiom: $2 + 2 = 4$.

Even if Frege put this passage in quotation marks, it is different from the text of Leibniz in the Erdmann edition, faithfully reproduced in Section 2.1. Frege had not thought sufficiently deeply about the meaning of Leibniz' brackets.

Pursuing his discussion, Frege mentions Grassmann's aim of arriving at the law $a + (b + 1) = (a + b) + 1$ "through a definition" (p. 8), but he states at once that this definition is circular, a definition of sum through sum. Even putting that objection aside, writes Frege, it could still be said that $a + b$ is "an empty sign" if there is no element in the sequence of natural numbers such that associativity holds. Frege contends that Grassmann merely assumes this is not the case, by which "the rigour is only apparent."

When reading Frege's *Grundgesetze* of 1893, one cannot but think: How is it possible not to see, some thirty years after Grassmann and in full knowledge of his work and the further explanations of Hankel, that the natural numbers have a beginning, 0 or 1, and a step ahead, the successor, and that arithmetic operations are defined recursively and their properties proved inductively? One who stated the matter very clearly was Schröder (1873, pp. 64–65):

> The concept of addition has been determined for two numbers the second of which is equal to 1, and then one can make clear this concept for two arbitrary numbers, by setting by way of definition for arbitrary values of a and b the equation:
>
> (6) $$a + (b + 1) = (a + b) + 1.$$
>
> $\vdots$
>
> This equation contains in fact a *recurrent* definition of the sum of any two numbers. It does not, though, state at once what such a sum means; it traces back instead the understanding of a sum of a and $b + 1$ to that of a and b. For when $a + b$ is a known number of our series of numbers, we have already seen that even $(a + b) + 1$ must be one; then the left side of equation (6) is really explained by its right side.

By rejecting the recursive definition of sum and other arithmetic operations, on page 8 of volume 1 of his nearly 500-page treatise in two volumes, Frege had condemned his attempt at a foundation for arithmetic to fail. There is no answer to questions such as the basic laws for arithmetic operations.

In the *Begriffsschrift*, not a single arithmetic result is proved formally, and the principle of proof by induction especially is left untouched. The same is true of the *Grundgesetze*. It does not bring proofs in formal arithmetic an inch further, even if that was the very purpose of Frege's "concept writing." As impeccable as Frege's insight was in getting the principles of quantificational logic exactly right, the more disastrous was his failure with arithmetic. If he had just followed his advice to Schröder, namely to put the formalism into actual use instead of trying to decide philosophically in advance what a definition is, how numerical quantity is defined, and so on, he would have steered a course past the rocks that sank his attempt. Now, instead, he put decisive weight to objections such as that Schröder's approach does not define what numbers are but just represents them (the German word is *abbilden*), as in the *Grundlagen*, § 43: "It follows that he explains just the *number sign*, not number... what is expressed by the sign, what I had called the number, he assumes to be known."

(C) FREGEANA. Frege's enormous achievement in logic was fully appreciated by Russell. His and Whitehead's *Principia Mathematica* is an apotheosis of Frege's *Begriffsschrift*. Appreciation, though, is not always transitive, and those who read Russell perhaps did not realize sufficiently how his work depended on Frege's invention of quantificational logic.

Russell was led astray with arithmetic in the same way as Frege. Formal arithmetic was developed further in different quarters after the *Principia*, in Skolem's work as an outright reaction to Russell's approach. This is all the more odd if one considers that Russell had taken his notation mostly from Peano, whose arithmetic was in turn based on Grassmann's.

The scholarly appreciation of Frege is mainly the achievement of the Münster logician Heinrich Scholz. He began to collect Frege's manuscripts and correspondence in the 1930s, with the idea of compiling a three-volume publication. A list of the materials was published by Scholz and Bachmann in 1936.

One of Scholz' pieces on Frege has the simple title *Gottlob Frege*, and was published in a daily journal in 1941. His praise of the achievement of Frege, whom he introduces as someone very few know about, includes the following passage (from the collection *Mathesis Universalis*, Scholz 1961, p. 268):

> This Frege was one of the greatest German thinkers. Indeed, one of the greatest Western thinkers in general. A thinker of the range and depth of a Leibniz and a creative force of the rarest potency.

Scholz was an anti-Nazi to whose credit can be counted the saving of more than one logician from concentration camps. He struggled to keep research in logic alive during the Nazi period, making compromises such as publishing some of his own papers, and letting one paper of Gentzen's have a second publication, in the strictly "Aryan" mathematics journal *Deutsche Mathematik*, in exchange for the favor of getting paper for the publication of his own series, the *Forschungen zur Logik und zur Grundlegung der exakten Wissenschaften*.

One year after the essay on Frege, Scholz wrote another similar piece for the eightieth birthday of Hilbert, stating how he had "served the honour of the German spirit," at a time when instead one would think that the only such spirit could be that of bottomless shame (Scholz 1961, p. 290). His list of those who contributed to "the deductive sciences" mentions as contemporaries in theoretical physics only Max Planck, the rest being the unmentionable Jews Einstein, Born, and so on, or dissenters such as Schrödinger. In mathematics, the list jumps over two obvious names, Kronecker and Cantor, both Jewish or of Jewish background, and others who left, such as Weyl. The list of those who contributed to the development of logic includes Gödel, Ackermann, and Gentzen, but none of the expelled Jewish logicians; the name of Paul Bernays, the real driving force of logic in Göttigen, appears only in a reference to "the standard work of Hilbert-Bernays" (p. 289). In these writings, Scholz shows his acceptance of the terms dictated by the Nazis, namely to pass silently over any "un-German" names, be they Jews or others who were threatened or persecuted and left.

When Münster, rather close to England, was bombarded from late 1943 on, Scholz took the Frege papers to what he thought to be the safety of the university. Alas, the library building was hit by a bomb in March 1945, and the primary materials were destroyed. Second copies of most of the papers that Scholz had kept were saved, but the publication project got delayed by decades, until Scholz' death in 1956. All of the extant primary source material from Frege is contained in the *Kleine Schriften* of 1967, the *Nachgelassene Schriften* of 1969, and the *Wissenschaftlicher Briefwechsel* of 1976. The first of these is a very nearly complete collection of Frege's publications, except for a few shorter pieces that relate to the *Begriffsschrift*. These can be found in the 1964 reprinting of the *Begriffsschrift*. The second collection contains papers Frege failed to get published, especially the long article on Boole's logic, and other manuscripts found in Frege's *Nachlass*. The third contains extremely worthwhile correspondence with, among others, Husserl, Hilbert, and Russell. Many letters were lost, though; for example, of the extensive correspondence with Wittgenstein, just a frustrating list of 24 letters with dates has been preserved.

Rudolf Carnap followed some of Frege's lectures in 1910–14. His shorthand notes *Frege's Lectures on Logic*, edited by Erich Reck and Steve Awodey, were published in 2004. Another item that was spared consists of some of Frege's diary pages, and these were published in 1994. Their anti-Semitism and adherence to early national-socialistic ideas caused a considerable stir.

* * *

There are an enormous number of studies on Frege, many of them of lesser value for a logically minded reader. My method for reading accounts of Frege's logic while working on this chapter was that if the rule of universal generalization is not adequately presented as Frege's central discovery, I have little reason to continue reading. Surprisingly, accounts by such otherwise competent logicians as Boolos (1985) failed under the criterion.

Finally, there are even today those who find that Frege's attempt at a purely logical definition of the concept of number can be carried through. Heck's (2011) book is a recent exposition.

5

RUSSELL: ADDING QUANTIFIERS
TO PEANO'S LOGIC

5.1. AXIOMATIC LOGIC

The pursuit of truth in logic had begun with Gottlob Frege and was continued by Russell and the rest, so that axiomatic logic had become the norm by the 1920s. The axioms were supposed to express the most basic logical truths, and just two rules of proof expressed the passage from instances of axioms to the theorems of logic. The latter were, supposedly, the less basic truths, but sometimes theorems were simpler than axioms. The rules of passage were detachment, from $A \supset B$ and A to conclude B, and the rule of universal generalization.

There was another aspect to axiomatic logic, namely a fundamental relativity in the choice of the basic notions. Russell's axiomatization of propositional logic, from the *Principia Mathematica*, uses disjunction and negation as the primitive connectives and became a standard. We recall it here in a slightly modernized notation.

Table 5.1. *Principia Mathematica* style logical axioms

1. $\neg(A \vee A) \vee A$
2. $\neg A \vee (A \vee B)$
3. $\neg(A \vee B) \vee (B \vee A)$
4. $\neg(A \vee (B \vee C)) \vee ((A \vee B) \vee C)$
5. $\neg(\neg A \vee B) \vee (\neg(C \vee A) \vee (C \vee B))$

Implication is defined by $A \supset B \equiv \neg A \vee B$, and its use would make the preceding axioms look a little less bad. Say, the first axiom is

$A \vee A \supset A$ and the last one $(A \supset B) \supset (C \vee A \supset C \vee B)$. The choice of axioms in Russell is motivated by the algebraic tradition of logic of Ernst Schröder, with such algebraic properties of an operation as idempotence (axiom 1), commutativity (3), and associativity (4).

A somewhat neglected paper of 1906, *The Theory of Implication*, is Russell's first contribution to the deductive machinery of logic. Published four years before the *Principia*, it shows clearly the origins of Russell's formal system of proof. In this work, he uses negation and implication as primitives, with a much better-looking axiomatization as a result:

Table 5.2. Russell's 1906 theory of implication

1. $A \supset A$
2. $A \supset (B \supset A)$
3. $(A \supset B) \supset ((B \supset C) \supset (A \supset C))$
4. $(A \supset (B \supset C)) \supset (B \supset (A \supset C))$
5. $\neg\neg A \supset A$
6. $(A \supset \neg A) \supset \neg A$
7. $(A \supset \neg B) \supset (B \supset \neg A)$

The axioms are nicely motivated by intuitive considerations, partly in reference to Frege and partly to Peano, in whose work they appear. Next to the rule of detachment, there is an explicit rule for taking instances of the axioms in formal derivations.

Even if the axiomatization is intuitively understandable, it does not make the provability of theorems of propositional logic any more apparent to Russell than the one based on negation and disjunction. He comments wryly (1906, p. 159):

> In the present article, certain propositions concerning implication will be stated as premisses, and it will be shown that they are sufficient for all common forms of inference. It will not be shown that they are all *necessary*, and it is probable that the number of them might be diminished.

Are the axioms sufficient for $(A \supset (A \supset B)) \supset (A \supset B)$? This is hard to tell, but what is worse, there is no logical content in the statement that the axioms are "sufficient for all common forms of inference." After the axioms, there follow more than 20 pages of formal derivations to bring home the point about sufficiency. The first one, with the original lowercase notation for atomic formulas, and a fractional notation for substitutions in two axioms called Id and Comm, axioms 1 and 4 (p. 169), is

$$\vdash : . \, p \, . \, \supset : p \supset q \, . \, \supset . \, q$$

Dem.

$$\vdash . \, \text{Id} \, \frac{p \supset q}{p} \, . \, \supset\vdash : p \supset q \, . \, \supset . \, p \supset q \tag{1}$$

$$\vdash . \, \text{Comm} \, \frac{p \supset q, p, q}{p, q, r} \, . \, \supset\vdash :: p \supset q \, . \, \supset . \, p \supset q :\supset : . \, p \, \supset :$$
$$p \supset q \, . \, \supset q \tag{2}$$

$$\vdash . \, (1) . \, \supset :\vdash . (2) . \, \supset\vdash . \, \text{Prop.}$$

The structure of derivations in Russell is *identical* to Peano's, as can be readily seen from a comparison to Section 2.2: The first two lines have an axiom that implies its substitution instance, of the form $a \supset b$, and the third line has the form $a \supset ((a \supset b) \supset b)$.

The net effect of the study of axiomatic logic was a scandalously unmanageable logical machinery. Thus, Russell and Whitehead put years into the production of hundreds and hundreds of pages of formal logical proofs within the axiomatic system but had little idea of its properties as a logical calculus. This aspect becomes obvious through the work by Paul Bernays from 1918. He was writing a habilitation thesis on the propositional calculus of the *Principia Mathematica* and wanted to prove the mutual independence of the axioms, exactly as one does in axiomatic geometry *à la Hilbert*. To do this, Bernays invented interpretations of the axioms of logic with more than two truth values. Then, certainly unexpectedly, the fourth axiom resisted attempts at proving its independence by such a semantical method. Bernays began to suspect that it could be a theorem and managed to find a derivation for it from the rest of the axioms.

In the hands of Russell, axiomatic logic had moved far away from its origin with Frege, namely from a formula language for arithmetic in which, as Frege wrote, "everything necessary for a correct inference is expressed in full, but what is not necessary is generally not indicated."

Frege, in contrast to Russell, had a clear idea of the practical meaning of his axioms.

5.2. THE REDISCOVERY OF FREGE'S GENERALITY

(A) THE UNIVERSAL QUANTIFIER IN RUSSELL. Russell's 1903 book *The Principles of Mathematics* contains his first published acknowledgment of Frege's achievement. The book is an old-style synthetic presentation of the foundations of mathematics as a whole, including even a lot of classical mechanics. No explicit logical notation is used, but the treatment is based on Peano's work. Russell thinks he can get along with a single primitive notion in logic, what he called the "formal implication" rendered as "ϕx implies ψx for all values of x" (p. 11). Peano had used the notation $\phi x \supset_x \psi x$ for such an implication with a free variable, typically an eigenvariable in an inductive step from x to its successor x'. That was the nature of Peano's arithmetic: the quantifiers were absent in his formalism.

In the preface to his book, Russell tells that he had seen Frege's *Grundgesetze der Arithmetik* but adds that he "failed to grasp its importance or to understand its contents," the reason being "the great difficulty of his symbolism" (p. xvi). Upon further study, he wrote a lengthy appendix with the title *The logical and arithmetical doctrines of Frege* (pp. 501–522), though with just a disappointing half page dedicated to the formalism of logic.

The universal quantifier is written as (x) in Russell (1906), then detailed in the famous 1908 paper on the theory of types. Its section II is titled *All and any*. Mathematical reasoning proceeds through *any*: "In any chain of mathematical reasoning, the objects whose properties are being investigated are the arguments to *any* value of a propositional function" (p. 156). Still, reasoning with just free variables would not do, for bound variables are needed in definitions (Russell's terminology for free and bound variables is, respectively, "real" and "apparent"). Remarkably, his example is from mathematics proper (ibid.):

We call $f(x)$ continuous for $x = a$ if, for every positive number $\sigma \dots$ there exists a positive number $\varepsilon \dots$ such that, for all values of δ which are numerically less than ε, the difference $f(a + \delta) - f(a)$ is numerically less than σ.

He goes on to explain that f appears in the definition in the *any*-mode, as an arbitrary function, and that σ, ε, and δ instead are just apparent variables without which the definition could not be made. Next, Russell takes into use the 1906 notation for the universal quantifier, $(x)\phi x$, presumably the first such notation in place of Frege's notch in the assertion sign, if we disregard the Π_x notation in Schröder's algebraic logic. Russell's explanation of the universal quantifier, though, is a disappointment, for it is stated that $(x)\phi x$ denotes the proposition "ϕx is always true" (ibid.), a hopeless mixing of a proposition with an assertion that would never have occurred in Frege. Later, in the more formal section VI of the paper, this is corrected when the Fregean assertion sign $\vdash$ is put to use.

Russell's first example of a quantificational inference is: from $(x)\phi x$ and $(x)(\phi x \supset \psi x)$ to infer $(x)\psi x$ (pp. 157–158):

> In order to make our inference, we must go from 'ϕx is always true' to ϕx, and from 'ϕx always implies ψx' to 'ϕx implies ψx,' where the x, while remaining any possible argument, is to be the same in both.

As can be seen, the rule is applied by which instances can be taken from a universal, after which the propositional rule of implication elimination can be applied. Then, since x is "any possible argument," ψx is always true, by which $(x)\psi x$ has been inferred (ibid.). Here we have a clear case of the introduction of a universal quantifier. A further remarkable feature of Russell's example is its purely hypothetical character. He does read the universal propositions in the "is always true" mode, but the argument begins with "Suppose that we know $(x)\phi x$." Thus, we have here a universal assumption that is put into use by the rule of universal elimination. One could add that bound variables are needed not only in definitions but also in theorems, say, in almost any standard result about continuous functions.

Russell's section VI of the type theory paper gives the formal machinery of his type theory. It includes an axiomatization of propositional logic with negation and disjunction as primitives, and implication elimination as the rule, formulated in the logicist manner as: "A proposition implied by a true premiss is true" (p. 170). The existential quantifier is defined through the universal one in the standard way by the equivalence $(\exists x).\phi x . = . \sim \{(x). \sim \phi x\}$, a notation taken from

Peano's *Formulario*, the first appearance of the existential quantifier, not counting Schröder's Σ_x. The axioms and rules for the universal quantifier are

(7) $\vdash: (x).\phi x. \supset .\phi y$

(8) If ϕy is true, where ϕy is any value of $\phi \hat{x}$, then $(x).\phi x$ is true.

(9) $\vdash: (x).\phi x. \supset .\phi a$, where a is any definite constant.

Axiom (7) gives the license to infer to a free-variable expression, used informally in examples we have quoted. In (8), the rule of universal generalization, the expression $\phi \hat{x}$ denotes the propositional function, as opposed to its particular value for an argument, as Russell explains in a footnote. Thus, it is a notation for functional abstraction. As to the last axiom, Russell sees no "infinity of facts" brought in by universal instantiation, contrary to many others who read Frege, but just writes (pp. 170–171): "It is the principle, that a general rule may be applied to particular cases."

Principles (7) and (9) give as contrapositions even the corresponding axioms for the existential quantifier, if it should be chosen as a primitive: a free-variable instance, resp. an instance with a constant, implies existence. As to the rule of inference, existence elimination would be the classical dual of universal introduction, but its formulation is tricky. Starting from Frege's generalization rule, its premiss and conclusion can be turned into the respective contrapositions by which, given $F(a) \supset A$ with a satisfying the same conditions as in Frege's rule, $(\exists x)F(x) \supset A$ can be derived. Existential elimination can be found in this form in Hilbert and Ackermann (1928). That seems like a late date for such a fundamental discovery, and suggests that our mathematical grandparents had no precise idea of what existential assumptions amount to.

(B) QUANTIFIERS IN THE PRINCIPIA. Russell's final word on logic is contained in the first volume of the *Principia Mathematica* that appeared in 1910 and was coauthored with A. Whitehead. The parts about quantifiers up to chapter 9 are elaborations of Russell's earlier work, and I take him to be the principal author of those parts.

Russell's years as an active logician were in fact not that many, less than ten anyway, and his publications of note were few in number. In 1927, in a preface to the second edition of the *Principia*, he makes a remark that is likely to startle a modern reader, namely that Sheffer's stroke, the single connective by which one can axiomatize classical propositional logic, is "the most definitive improvement resulting from work in mathematical logic during the past fourteen years" (p. xiv).

The presentation of logic in the *Principia* is somewhat different from that of Frege and from Russell's 1908 formulation that followed Frege. Part I, titled "Mathematical Logic," begins with section A on "the theory of deduction" (pp. 90–126), followed by a "theory of apparent variables" (pp. 127–160). There are, moreover, things pertinent to propositional and predicate logic in the introductory part, such as on page 3, where we find a typical logicist slogan: "An inference is the dropping of a true premiss."

In first-order logic, as Russell calls it, the *Principia* first has both quantifiers as primitive, alongside negation and disjunction. The reason lies in Russell's reservation about applying the propositional connectives to quantified propositions. Motivated by ideas from type theory, Russell writes that, when applied to propositional and quantificational expressions, respectively, "negation and disjunction and their derivatives must have a different meaning" (p. 127). Russell's way out is not to apply propositional connectives to quantified expressions at all but instead introduce negation and disjunction for quantified propositions as *defined* notions, here with Russell's numbering but with the dot notation replaced by parentheses (p. 130):

$$*9 \cdot 01 \quad \sim (x)\phi x \equiv (\exists x) \sim \phi x$$
$$*9 \cdot 02 \quad \sim (\exists x)\phi x \equiv (x) \sim \phi x$$
$$*9 \cdot 03 \quad (x)\phi x \vee p \equiv (x)(\phi x \vee p)$$
$$*9 \cdot 04 \quad p \vee (x)\phi x \equiv (x)(p \vee \phi x)$$
$$*9 \cdot 05 \quad (\exists x)\phi x \vee p \equiv (\exists x)(\phi x \vee p)$$
$$*9 \cdot 06 \quad p \vee (\exists x)\phi x \equiv (\exists x)(p \vee \phi x)$$
$$*9 \cdot 07 \quad (x)\phi x \vee (\exists y)\psi y \equiv (x)(\exists y)(\phi x \vee \psi y)$$
$$*9 \cdot 08 \quad (\exists y)\psi y \vee (x)\phi x \equiv (x)(\exists y)(\psi y \vee \phi x)$$

The two straightforward cases with the same quantifier in both disjuncts are not listed. The effect of the rules is that all propositions with quantifiers become reduced to prenex normal form, a string of quantifiers followed by a propositional formula. Here we have also an explanation of the use of both quantifiers as primitives.

The axioms for the quantifiers are two:

$*9 \cdot 1 \quad \vdash \phi x \supset (\exists z)\phi z$

$*9 \cdot 11 \quad \vdash \phi x \vee \phi y \supset (\exists z)\phi z$

The latter axiom is needed only for proving the contractive implication:

$$(\exists x)\phi x \vee (\exists x)\phi x \supset (\exists x)\phi x$$

The rule of inference is universal generalization (p. 132): "When ϕy may be asserted, where y may be any possible argument, then $(x)\phi x$ may be asserted." The arbitrariness of y is further explained by: "if we can assert a wholly ambiguous value ϕy, that must be because all values are true." We see in the latter again, as in Frege, that the explanation goes from the truth of the universal proposition to any of its instances, not the other way around.

Formal derivations with the quantifiers in the system of the *Principia* easily become hopelessly tricky. The reason is the curious synthetic nature of reasoning in axiomatic logic, as compared to natural deduction, combined with the ban on propositional inferences with quantified formulas. To have a feasible system of derivation, Russell shows as a preparatory step that propositional inferences with the latter reduce to inferences within the propositional system.

The first example of quantificational inference is the derivation of

$*9 \cdot 1 \quad \vdash (x)\phi x \supset \phi y$

The proof is a formal representation of the following steps of inference that synthesize the conclusion from an instance of excluded middle. First take the derivable propositional formula $\sim \phi y \vee \phi y$, and then use existential introduction to conclude $(\exists x)(\sim \phi x \vee \phi y)$. Now the existential quantifier is moved in the left disjunct, with the result $(\exists x) \sim \phi x \vee \phi y$, and finally, definition $*9 \cdot 01$ is used in the reverse direction to get $(x)\phi x \supset \phi y$.

One would expect that in addition to $*9 \cdot 1$, the rule of existence elimination would be shown to be a derivable rule, but that rule appears nowhere in the *Principia*. With hindsight, we know that the premiss of the rule has to be given by a formula such as $\phi y \supset p$, namely $\sim \phi y \vee p$, assumed to be derived. Now one can generalize to $(x)(\sim \phi x \vee p)$ and then apply definitions $*9 \cdot 03$ and $*9 \cdot 01$ to get $\sim (\exists x)\phi x \vee p$; that is, $(\exists x)\phi x \supset p$.

In paragraph 9 of section B, Russell gives an alternative account of quantification, in which negation and disjunction apply to all propositions. Then existence can be defined in the usual way and its axioms replaced by ones for the universal quantifier, as in his 1908 paper.

In Greek mathematical texts, one finds generality and existence treated informally, but according to our best experts on the topic, explicit quantifiers are practically absent (see, e.g., Acerbi 2010, p. 33). A typical pattern would be to begin a theorem with an assumption like *Given two points A and B such that...* with a claim like *to construct a triangle such that....* These situations are clear, free-variable inferences. Much confusion has been caused by a similar situation in which the given is taken as an assumption about existence. In that case, the free variables act as the eigenvariables of existential elimination, a tricky move, as witnessed by its formal representation as late as 1928. Moreover, such an existential assumption, rendered as $\exists x A(x)$ in modern notation, can have a consequence B in which the eigenvariable does not occur. Then the result can be $\forall x(A(x) \supset B)$ when instead one would expect it to be $\exists x A(x) \supset B$. Thus, sometimes the givens act as free variables that lead to generality of a conditional, whereas at other times they act as eigenvariables in the elimination of an existential assumption. An explanation of what happens here is hidden in Russell's prenex rules for the quantifiers: The application of definitions $*9 \cdot 03$ and $*9 \cdot 01$ to $\exists x A(x) \supset B$ gives at once $\forall x(A(x) \supset B)$ with the, certainly less intuitive, consequence that these two expressions are *logically equivalent.*

Russell gives special emphasis to the notion of formal implication that he took from Peano and defined in the *Principia* as (p. 139)

$$\phi x \supset_x \psi x \equiv (x)(\phi x \supset \psi x)$$

The importance of quantificational logic, as opposed to mere propositional logic, comes out in Russell's opinion as follows (pp. 20–21): In the latter, "material implications" $p \supset q$ between two propositions "can only be *known* when it is already known either that their hypothesis is false or that their conclusion is true." Such implications are useless because "in the first case the conclusion need not be true, and in the second it is known already." Only formal implications serve the purpose of "making us know, by deduction, conclusions of which we were previously ignorant."

Russell's distinction between material and formal implication is found already in Frege's work, though without this terminology, as detailed in Section 4.4(A). Frege remarks in his *Begriffsschrift* that the "substantial connection" in an implication comes out only in the quantificational form, but this does not mean that it could not be there, as he shows through an example (*Begriffsschrift*, p. 6):

One can make the judgment

without knowing if A and B are to be admitted or denied.

The example that follows concerns the Moon's phases. Russell's argument about the "uselessness" of implication in propositional logic is a logical fallacy committed because Russell moves from $p \supset q$ to the classically equivalent $\sim p \vee q$ and then appeals to the disjunction property for the latter, but that property need not hold in classical logic. Contrary to Russell, Frege makes no such mistakes about classical implication in propositional logic.

5.3. RUSSELL'S FAILURES

Russell, like Frege, believed that one could give a foundation to arithmetic by adding suitable definitions to pure logic. The two main problems are that the recursive definitions of arithmetic operations are not counted among the acceptable ones, even using a broad conception of what counts as logic, and that the logical nature of the principle of arithmetic induction remains unexplained. Russell had a monolithic principle by which all concepts have to be defined once and for all, exemplified by his initial resistance to the extension of propositional logic by the quantifiers. It all had to come in one sweep instead of a piecemeal approach, and consequently there are almost no traces of the development of formal arithmetic in the *Principia*.

Russell formulated Peano's logic augmented by Frege's quantifiers in a palatable way. As we shall see, his presentation in *Principia Mathematica* was the one that made Hilbert realize the possibility to actually formalize mathematical proofs. However, Russell tried to effect this formalization in the same wholesale manner as Frege had: through a powerful theory of types that would cover a lot of ground through higher-order quantifiers, explained briefly in Section 4.4(B). As with Frege, the attempt was a failure. What instead turned out not to be a failure was to start with modest theories, such as arithmetic. By the late 1920s, elementary arithmetic and several other theories were presented in an axiomatic form through the formalization of predicate logic in Hilbert and Ackermann's book *Grundzüge der theoretischen Logik* (1928). A conceptual step ahead was required in that logic had to be extended to cover inferences from hypotheses. How this step was made is described in detail in Section 7.2(B).

If Russell had concentrated on the essential mathematical development in Peano's 1889 work, namely its elaboration of Grassmann's proofs into formal ones, he would have achieved a solid logical basis for arithmetic. All the elements were there: Grassmann's recursive definition of arithmetic operations, Peano's axioms for the natural numbers, Peano's notation for propositional logic, and the deductive machinery. Add to this Frege's universal quantifier and the result is what we today call a first-order axiomatization of classical Peano arithmetic. Instead, it happened that arithmetic was formalized by Skolem as an afterthought to the introduction of the quantifiers in *Principia Mathematica*. He left the quantifiers out and thereby discovered primitive recursive arithmetic, the basis of all subsequent work on formal theories of computation, but it takes nothing to put them back, and then again we have a formalization of Peano arithmetic.

After he finished the *Principia*, Russell quit formal work on the foundations of mathematics. His imprisonment as a pacifist toward the end of the First World War gave him time to write the popular book *Introduction to Mathematical Philosophy* that appeared in 1919. It can be seen that by then he was fully aware of the possibility to develop arithmetic along the lines of Peano, with three "primitive ideas" and five "primitive propositions" (p. 5). The former are 0, *number*, and *successor*, and the latter are the standard Peano axioms.

Next the recursive definition of sum is presented (p. 6):

> Taking any number m, we define $m + 0$ as m, and $m + (n + 1)$ as $(m + n) + 1$. In virtue of (5) [the induction axiom] this gives a definition of the sum of m and n, whatever number n may be. Similarly we can define the product of any two numbers. The reader can easily convince himself that any ordinary elementary proposition of arithmetic can be proved by means of our five premises, and if he has any difficulty he can find the proof in Peano.

The observation that the principle of induction justifies the recursive definition of sum and other arithmetic operations was found already in Peano (1889), with a formal proof by induction on n that if m, n are numbers, $m + n$ is also. Russell adds after the passage just quoted "considerations which make it necessary to advance beyond the standpoint of Peano," namely that no unique interpretation of the three primitive ideas is fixed, meaning his conviction and Frege's as well that "our numbers should have a *definite* meaning, not merely that they should have certain formal properties" (p. 10).

In sum, it took thirty years from Grassmann to Peano, another thirty years to Skolem, and all of this could have appeared as a third chapter of Frege's *Begriffsschrift* instead of the inconclusive elaboration of the notion of a sequence. The main reason for Frege's blindness about the possibility of recursive definitions and a formal development of arithmetic on their basis within his quantificational logic seems to be a ridiculous little detail that in the nineteenth century, natural numbers began with 1 instead of 0. With the latter, there are two recursion equations for sum. The first one, $a + 0 = a$, shows that to add 0 reduces to doing nothing, with the symbol + for sum eliminated. In the second equation, to add 1 reduces to the taking of a successor, with the notation $a + 1 = a'$ as in Schröder, and, finally, to add n reduces to the taking of n successors—nothing circular about that and the symbol + completely eliminated after so many recursions that give as a result an expression of the form $0' \cdots '$. But Frege was *folgerichtig*, upright and consequential, and would not just proceed formally and later fix the philosophy behind. Russell shared Frege's rejection of Grassmann's and Peano's arithmetic; moreover, he required that all things be settled once and for all, with no piecemeal extension of theories.

6

THE POINT OF CONSTRUCTIVITY

6.1. SKOLEM'S FINITISM

(A) REJECTION OF THE QUANTIFIERS. The best-known statement of finitism is ascribed to Leopold Kronecker (1823–1891). The source of the statement is not Kronecker himself, at least not directly, but rather his editor, Kurt Hensel, who in the preface of Kronecker's posthumous *Vorlesungen über Zahlentheorie* of 1901 wrote (p. vi):

> He found that one could, and even should, formulate every definition in this field so that it can be decided through a finite number of attempts whether it is applicable to a presented quantity. Similarly, a proof of existence of a quantity would be considered as fully rigorous if it contains at the same time a method by which the quantity can be actually found.

The first part requires all definitions of concepts to be *decidable*, which forces reasoning on a definition to be constructive. Later, when the law of excluded middle became understood, one could allow definitions that are not decidable, say, to be an irrational number, and instead put the restriction in the logical part of reasoning about concepts, say not to reason by the undecidable cases of a rational and an irrational number. The second part is a clear statement of the *existence property* of mathematical proofs, well before the systematic understanding of constructive reasoning that came with the development of intuitionistic logic. These two remarks on Kronecker's principles indicate also in what ways intuitionism can go beyond Kronecker's finitism by relaxing the decidability of definitions and the existence property. An example of the latter will be given in Section 6.3(C).

Hilbert is known to have accused Kronecker of a "Verbotsdiktatur," a dictatorship of denials, but apparently he did not share the immediate continuation of the preceding passage:

Kronecker was far removed from abandoning a definition or a proof completely if it didn't correspond to these highest requirements, but he believed that there was then something still missing and that a completion in this direction would be an important task that would bring our knowledge further at an essential point.

Kronecker's finitistic dictum was the basis for the philosophy of mathematics of Thoralf Skolem (1887–1963), a solitary combinatorial genius who started to work with Ernst Schröder's algebra of logic. His long paper with the matchingly long title, *Untersuchungen über die Axiome des Klassenkalküls und über Produktations- und Summationsprobleme, welche gewisse Klassen von Aussagen betreffen*, was finished in 1917 and published in 1919. In this work, he established many of the basic results of lattice theory, such as the independence of the axioms. He also defined what later came to be called Heyting algebras, lattices with an "arrow" operation that imitates implication, and established many basic properties (see Section 3.3 for some details). These algebras relate to intuitionistic propositional logic in exactly the same way as Boolean algebras relate to classical propositional logic. One cannot but wonder how Skolem was able to determine such a structure, considering that intuitionistic logic did not become formally well-defined until Heyting's (1930) axiomatic presentation.

In 1919, Skolem became acquainted with the *Principia Mathematica*, but his reaction to Russell's universal and existential quantifiers was one of rejection. This is found in his paper on arithmetic as based on recursive definitions, published in 1923 but written in the fall of 1919. Skolem states in the introduction that he wants to show that one can give a logical foundation to elementary arithmetic without the use of bound variables. The real motivation comes out in §4, where the notion of divisibility is considered. Writing in Schröder's notation Σ for the existential quantifier, divisibility $D(a,b)$ can be defined as (p. 160):

$$D(a,b) \equiv \Sigma_x(a = bx)$$

Now comes the essential point:

> Such a definition refers to an infinite—and that is to say unexecutable—work, for the criterion of divisibility consists in *trying out, the whole series of numbers through*, whether one can find a number x such that $a = bx$ obtains.

The situation is saved here because one can limit the search for x: a divisor of a cannot be greater than a. Therefore, universal and existential quantifiers are finitely bounded and can be used, the definition being written with Schröder's notation $+$ for disjunction as (p. 161)

$$D(a,b) \equiv \sum_{1}^{a}{}_x(a = bx)$$
$$\equiv ((a = b) + (a = 2b) + (a = 3b) + \cdots + (a = ba)).$$

The bound variable x has a finite upper bound, and Skolem concludes:

> Therefore this definition gives us a finite criterion of divisibility: one can determine in each case through a finite work—a finite number of operations—whether the proposition $D(a,b)$ holds or not.

In the next section, Skolem defines subtraction $b - c$ in which the problem is to check that $c \leqslant b$; otherwise subtraction is not defined. Thus we have, with $c \leqslant b$ defined as the negation $\overline{(c > b)}$ and a nice overloading of the symbol $+$ as either disjunction or sum (p. 165):

$$\overline{(c > b)} + \sum_x (x + b = c)$$

> Here the propositional summation [existential quantifier] in relation to x should be extended over "all" numbers from 1 to ∞. Though, even here it is not necessary to put into use such an actual infinity.

Toward the end of the paper, Skolem makes a remark that fixes his philosophy (p. 186):

> When a class of some objects is given, one would be tempted to say: These things come in a finite number n means that there *exists* a one-to-one correspondence between these objects and the first n numbers. However, there occurs a bound logical variable in this definition, and no limitation to the finite is given a priori for this variable, unless there is already at the start

a result that states: *the number of possible correspondences is finite.* So, from the point of view of the strict finitism given here, such a result must be proved in advance, for the number of the objects concerned to be definable.

Here we have Skolem's approach that can be put concisely as: *Decidability is the only criterion of existence. All decision procedures have to terminate in a bounded number of steps.*

In elementary arithmetic, the atomic formulas are equations between numerical terms. If the latter don't contain variables, their values can be computed, and thus equality becomes decidable, a situation that can be expressed so that the law of excluded middle, $a = b \vee \neg a = b$, holds for numerical terms. It of course holds in classical logic, but here the point is that it holds constructively. It is an easy exercise to show that the law of excluded middle for arbitrary propositional formulas becomes derivable. Thus, Skolem's recursive arithmetic with its quantifier-free formulas obeys classical logic. In fact, one can interpret classical propositional logic as that special case of intuitionistic propositional logic in which the atomic formulas are decidable (as done in Negri and von Plato 2001, p. 207).

When the quantifiers are added to primitive recursive arithmetic, it becomes full intuitionistic arithmetic, usually called Heyting arithmetic. When the principle of indirect proof is further added, we get classical Peano arithmetic.

Skolem's dissatisfaction with Russell's quantifiers was not just that. In Frege and Russell, there is a totally failed attempt at founding the basic concepts of arithmetic on a heavy logical apparatus. Arithmetic based on recursive definitions, absent from Frege's and Russell's involved theories that don't lead anywhere, is the right way for Skolem. In a paper titled *The development of recursive arithmetic* of 1947 Skolem writes about his primitive recursive arithmetic (p. 500):

> After having studied the Principia Mathematica (Russell-Whitehead 1910–13) in 1919 I discovered that a great part of ordinary number theory could be developed without quantifiers.... A paper containing my considerations was published in 1923, and is so far as I know the first investigation in recursive number theory. The utterance of H. B. Curry (1941) that the recursive arithmetic can be traced back to Dedekind and Peano seems to me rather strange, because the researches of these men had another purpose than to avoid the use of quantifiers.

This passage contains the oddity that neither Dedekind nor Peano had any theory of quantifiers at their disposal, even less one to avoid, as that became accessible only through the very same *Principia* to which Skolem refers. Skolem is clearly upset by Curry's brief remark, (1941, p. 263): "Such a treatment of arithmetic was initiated by Skolem— although the essential ideas were already in Dedekind and Peano."

A later paper of Skolem's, *The logical background of arithmetic* of 1955, contains what Skolem takes to be "the most important views concerning the logical nature of arithmetic which were set forth in the latter half of the 19th century" (p. 541). The only nineteenth-century figure mentioned is Dedekind. However, a reading of Skolem's 1923 paper, in comparison to the treatments of Grassmann, Peano, and Schröder, reveals similarities that are too striking to be casual coincidences. It is good to keep in mind that all of Skolem's early work was based on Schröder's *Vorlesungen über die Algebra der Logik*. The beginnings of Schröder's approach are already seen in the earlier *Lehrbuch der Arithmetik und Algebra*, alongside its clear exposition of Grassmann's recursive arithmetic. Finally, it was noted in Section 2.3 that Peano's 1889 presentation of arithmetic strictly follows Grassmann, so that Peano would have been a possible source for Skolem, despite his belittlement of Peano's presentation of recursive arithmetic. Table 6.1 compares the four treatments.

Table 6.1. Four treatments of arithmetic compared

Grassmann 1861	Peano 1889	Skolem 1923	Schröder 1873
§2 Addition	§1 Numbers and addition	§1 Addition	§6 Number and sum
15 $a+(b+e)=a+b+e$	Def.1 $a+(b+1)=(a+b)+1$	Def.1 $a+(b+1)=(a+b)+1$	$a+(b+1)=(a+b)+1$
20 $e+a=a+e$	Thm.24 $1+a=a+1$	Lemma $a+1=1+a$	§9 $1+a=a+1$
22 $a+(b+c)=a+b+c$	Thm.23 $a+(b+c)=a+b+c$	Thm.1 $a+(b+c)=(a+b)+c$	$a+(b+c)=(a+b)+c$
23 $a+b=b+a$	Thm.25 $a+b=b+a$	Thm.2 $a+b=b+a$	$a+b=b+a$
§3 Subtraction	§2 Subtraction	§5 Subtraction	§31
§4 Multiplication	§4 Multiplication	§3 Multiplication	§21
52 $a.1=a$	$a\times 1=a$	Def.4 $a\cdot 1=a$	$a.1=a$
56 $a.(\beta+1)=a.\beta+a$	$a\times(b+1)=a\times b+a$	$a\cdot(b+1)=a\cdot b+a$	$a.(b+1)=a.b+a$
57 $a.0=0$	–	–	–
Distributivity	$a(b+c)=ab+ac$	Thm.10	(26), (27)
70 $\alpha(\beta\gamma)=(\alpha\beta)\gamma$	Thm.15 $a(bc)=abc$	Thm.11	(28)
70 $\alpha.\beta=\beta.\alpha$	Thm.7 $ab=ba$	Thm.13	(30)
§5 $<,=,>$	§3 Maxima and minima	§2 $<$ and $>$	–
§7 Divisibility	§6 Division	§4 Divisibility	§32

Skolem's order of things is almost identical to Peano's, with just subtraction treated after order and product, as in Schröder's presentation. Schröder's has many more topics because of the inclusion of the "independent treatment." Dedekind, whose little treatise *Was sind und was sollen die Zahlen?* of 1888 was certainly known to Skolem, has on the very first side the following:

> Of the literature known to me I mention the exemplary *Lehrbuch der Arithmetik und Algebra* of Ernst Schröder's (Leipzig 1873) in which one finds even a list of the literature.

The mathematics library of the University of Oslo has had a copy of Schröder's 1873 book in addition to his extensive three-volume work on algebraic logic, the former lost not very long ago. Skolem had put many years into the study and development of Schröder's algebraic logic. It would be very strange if as a student in Oslo he had not at least glanced through a praised book on arithmetic by the same author, perhaps later to forget where the basic ideas of recursive arithmetic came from.

(B) TOWARD A STRICTLY FINITISTIC MATHEMATICS. Skolem wrote another paper at about the time of the one on primitive recursive arithmetic, very famous for its first section that contains the Löwenheim-Skolem theorem. There one finds what to me seems a standard formulation of expressions of predicate logic, with conjunction, disjunction, negation, and universal and existential quantification. No logical rules of inference are given for these, but the reading is clear. Skolem gives a definition of what he calls *Zählaussage*, the clear idea being that these are expressions for things that can be counted, i.e., objects for which the language is that of Russell's "first-order propositions." A counting expression is formed from relations through the five logical operations in which the Schröderian quantifiers Π and Σ range over individuals. The first two examples are (1920, p. 103):

1) $\Pi_x \Sigma_y R_{xy}$. This is in words: There is for every x a y such that the relation R_{xy} obtains between x and y.

2) $\Sigma_x \Pi_y \Sigma_z (R_{xy} + T_{xyz})$. This is in words: There is an x such that a z can be determined for every y so that either the binary relation R obtains between x and y or the ternary T between x, y, and z.

Had Skolem suddenly changed his mind about the Russellian quantifiers? Hardly so. He had found what is now the Löwenheim-Skolem theorem already in 1915-16, while in Göttingen, and he dedicates the first section of the 1920 paper to it. The title is again monstruous, *Logisch-kombinatorische Untersuchungen über die Erfüllbarkeit oder Beweisbarkeit mathematischer Sätze nebst einem Theoreme über dichte Mengen* (Logico-combinatorial investigations on the satisfiability or provability of mathematical propositions, together with a theorem on dense sets), and the contents are a potpourri of three or four separate things. The last one is easiest: Even Skolem himself writes that "the contents of this paragraph are completely independent of those of the preceding ones" (p. 130, footnote). As to the first section, on the Löwenheim-Skolem theorem, that was something on which he wanted to make a statement, namely that nondenumerable infinity is a relative and even fictitious notion. The same is brought forth much more strongly in the 1922 version of the Löwenheim-Skolem theorem, with the explicit criticism of set theory at the end of the paper.

As to sections 2 and 3 of the 1920 paper, they are quite a different matter. Section 2 is titled, in translation, *Solution of the problem to decide if a given proposition of the calculus of groups is provable.* § 3 has the title *A procedure for deciding whether a given proposition of descriptive elementary geometry follows from the axioms of incidence of the plane.* What to make of these? Everyone knows that Skolem formulated and proved in 1920 the result by which predicate logic is unable to distinguish between the denumerable and what is not denumerable: Any consistent collection of first-order formulas admits a denumerable interpretation. Jean van Heijenoort, when delivering the English version of Skolem's argument, noted that the rest of Skolem's paper dealt with "decision procedures for the theory of lattices and elementary geometry, as well as with dense sets," but he left these parts out of the translation. There is no explanation of the contents of these parts, and the same goes for Hao Wang's introduction to Skolem's *Selected Works in Logic.*

In the algebraic tradition of logic, no formal systems of proof were presented. Perhaps that is also the reason why Schröder's Π and Σ are not taken as having established quantification theory. Skolem's sections 2 and 3 are, however, completely different in this respect and present a "purely combinatorial conception of deduction," detailed out in Section 3.3, with explicit formal rules of proof.

It was around 1992 that Stanley Burris found out the following: The second section of Skolem's paper, on Schröder's "calculus of groups," contains a polynomial-time algorithm for the solution of the word problem for lattices. This result was otherwise believed to have stemmed from Cosmadakis (1988) but is now seen as one of the most important early results on lattice theory, with Skolem therefore counted among the founding fathers of the topic. The third section of Skolem's paper contains a solution to a similar derivability problem in plane projective geometry, analyzed in great detail in von Plato (2007) and extended to cover the axiom of noncollinearity in von Plato (2010). Both works are examples of what Skolem preached in his paper on primitive recursive arithmetic: that only the decidable has meaning.

There are some final programmatic remarks in the paper on primitive recursive arithmetic, in which Skolem writes that he is not really satisfied with the logical development of primitive recursive arithmetic, for it is too laborious with its logical notation:

> I shall soon publish another work on the foundations of mathematics that is free from this formal laboriousness. Even this work is through and through finitistic; it is built on the principle of Kronecker by which a mathematical determination is a real determination if and only if it leads to the goal in a *finite* number of attempts.

I believe that the following happened: Skolem was, right before the paper on primitive recursive arithmetic, working toward the new paper on foundations, with chapters on a finitistic treatment of lattice theory and plane projective geometry as examples. At some point, he gave up and just packed together what he had: the Löwenheim-Skolem result, certainly unsatisfying in light of the final remarks in the primitive recursive paper; the two finished chapters on how one should really be a finitist; and the "completely unrelated" result in infinitary combinatorics.

The second section of Skolem's 1920 paper deals with a specific derivability problem in lattice theory. We can see here in action the programmatic statement about the laboriousness of the logical formalism and the unimportance of notation. In the section on lattice theory, Skolem considers what is known today as the *word problem for freely generated lattices.* We have a basic relation of weak partial order, $a \leqslant b$, assumed to be reflexive and transitive (axioms I and II). Next in lattice theory there are usually two operations, a lattice *meet* and *join*. Given two elements, $a \wedge b$ is the meet, the greatest element below a and b. Thus, the axioms are

$$a \wedge b \leqslant a \quad a \wedge b \leqslant b \quad c \leqslant a \ \& \ c \leqslant b \ \supset \ c \leqslant a \wedge b$$

The third axiom states that anything below both a and b is below $a \wedge b$. The axioms for join $a \vee b$ are duals of these. The word problem can now be expressed as: to find an algorithm for deciding if an atomic formula $a \leqslant b$ is derivable from given atomic formulas $a_1 \leqslant b_1, \ldots, a_n \leqslant b_n$ in lattice theory or to show that there is none.

Skolem does not use the lattice operations, well known to him, but gives instead a *relational axiomatization* with two added three-place relations that we can write as $M(a,b,c)$ and $J(a,b,c)$ rendered as *the meet of a and b is c* and *the join of a and b is c*. He does not tell the reader why he does this. The axioms for meet and join just given with lattice operations become the following:

If $M(a,b,c)$, then $c \leqslant a$ and $c \leqslant b$.

If $M(a,b,c)$ and $d \leqslant a$ and $d \leqslant b$, then $d \leqslant c$.

The effect is that the axioms for meet and join use a quantifier pattern as in $\forall x \forall y \exists z M(x,y,z)$, existence axiom VI for meet. In Skolem's paper, all of the axioms are without explicit notation, in the style of: "There is for arbitrary x and y a z such that...."

Skolem's axioms I–V are rules for the production of new atomic formulas from given ones. New "pairs" and "triples," as he calls his atomic formulas for the order relation and the two lattice relations, are produced from the given atoms by these axioms. This means that the "arbitrary" $x, y, \ldots$ in the axioms can be instantiated in any way by any $a, b, \ldots$ known from the given atoms. When instead axiom VI is put to use, with the existential z instantiated, "there appear newly introduced letters." Skolem in fact uses $\alpha, \beta, \gamma \ldots$ as eigenvariables in the elimination of the existential quantifier. It is, on the whole, quite remarkable that Skolem was able to use the elimination rule of the existential quantifier in the proper way. In Göttingen, this insight was won toward the end of the 1920s, it seems. A very clear statement is found in Hilbert (1931b): He gives first the rule of existential introduction and then adds (p. 121): "In the other direction, the expression $(Ex)A(x)$ can be replaced by $A(\eta)$, where η is a letter that has not occurred yet."

It would have been in the line of Skolem's "logic-free" approach to use the meet and join operations, for the axioms could then be written as free-variable formulas, as above. The relational axioms are easily proved from the ones with operations. The other direction is somewhat arduous, but it can be done (see von Plato 2013, p. 175). The advantage of Skolem's relational axiomatization is that there are no function terms but just pure parameters $a, b, c, \ldots$ when the axiomatization is used. He proves in the paper that the existence axioms VI for meet and join are conservative over the rest of the axioms, I–V, for the derivability problem of atomic formulas from atomic assumptions.

It should be clear now that the two sections of the 1920 paper belong to the plan of a new logic-free approach to the foundations of mathematics envisaged in the final remarks of the paper on primitive recursive arithmetic. There is an essential tension, however. Take the meet and join existence axioms, and they are formulated in the typical existential form $\forall x \exists y A(x, y)$. This form is put into perfect, if informal, use in the application of universal and existential instantiation, the former by the use of parameters, as in $\exists y A(a, y)$, the latter through the use of eigenvariables, as in $A(a, \alpha)$.

Perhaps the way out of the dilemma Skolem faced is to think of the axioms not as anything that should be finitely verified by decision procedures but as hypotheses. Very little in this direction can be found in Skolem, or in any of the early constructivistic literature. Following Kronecker's lead, the task was to take the natural numbers as concretely given objects and then to build up the basic structures of mathematics on strictly finitistic grounds. Abstract axiomatics, with no intended interpretation as in lattice theory, does not fit well into this picture. Whatever may have been Skolem's thinking at this time, around 1920, the three papers (1920, 1922, 1923) were his last words on foundations for several years. By his 1928 paper on applications of quantifier elimination, there is no criticism of quantificational logic.

A later echo of Skolem's early critical attitude is found in his *Critical remark on foundational research* of 1955. The Löwenheim-Skolem theorem gives what is often referred to as *Skolem's paradox*, namely that if set theory is taken formally and assumed consistent, it has a denumerable model. For Skolem, this was a decisive critical point that he repeated in his *Remark* (p. 583):

> But the worst phenomenon occurring in the literature are statements about infinite sets of higher cardinality without any comments with regard to the methodological background. As an example I may mention that some authors in connection with the Löwenheim theorem also set forth the inverse theorem that if a logical formula can be satisfied in a denumerable domain, it can also be satisfied in a non-denumerable one, even with an arbitrary cardinal number. What is meant by such a statement? What kind of set theory is used? Is Cantor's set theory still going strong despite the antinomies?

Many authors on model theory take Skolem to be one of the founding fathers of the topic, but his position regarding the model-theoretic definition of truth was disapproval (p. 583):

> As I have often emphasized, the use of quantifiers extended over an infinite range of variation is the somewhat doubtful element of mathematical logic.... If we have a finite range of variation, the interpretation is clear.... This interpretation, however, is of course meaningless when we consider an infinite number of objects, provided that we do not postulate the possibility of carrying out an infinite number of operations.
>
> ⋮
>
> Now there is one interpretation which is quite clear and beyond doubt namely the purely formalist one, usually called the syntactical one. A statement containing quantifiers may be conceived as being merely an expression in a formal system, and its truth may be conceived to mean that it can be derived in the system by given formal rules.

A precise formulation in terms Skolem would accept is: By the undecidability result of Church and Turing of 1936, the definition of what it is to be a theorem of predicate logic is not a strictly finitistic notion.

As these and other late statements about strict finitism show, Skolem never really took in the point of view of the intuitionists by which the boundaries of strict finitism can be superseded by preventing the application of the classical law of excluded middle, with the effect that the bounds of what is meaningfully definable and not just formal stipulation in mathematics can be extended beyond objects with decidable definitions, say to include irrational numbers.

6.2. STRICTER THAN SKOLEM: WITTGENSTEIN AND HIS STUDENTS

(A) WITTGENSTEIN'S LOGIC IN THE *TRACTATUS*. Ludwig Wittgenstein thought he had solved the problems of philosophy in his 1918 little book on logical investigations that got renamed the *Tractatus Logico-Philosophicus* in an English translation, with publication in 1922 under the patronage of Russell.

A careful reader of the *Tractatus* will notice in it the total absence of the notion of inference or deduction. There is instead the semantical method of truth tables by which it can be determined whether a propositional formula is a tautology. How the method is to be extended to the quantifiers is nowhere explained. At 6.1201, the principle of universal instantiation $(x) fx \supset fa$ is simply called a "tautology."

The *Principia* had made it clear that there is no quantificational logic without a *rule* of generalization. Wittgenstein does not see that this rule is crucial, as is shown by the passage 6.1271 in the *Tractatus*, where he states that all of logic follows from one basic law, the "conjunction of Frege's *Grundgesetze*." Rules cannot be part of such conjunctions. Moreover, the *Principia* made it clear that the notion of tautology does not extend to the quantifiers. Therefore, even the rule of detachment is essentially needed. Wittgenstein missed both of these points, and my conclusion is, unfortunately, that the impatient philosopher had never made it to page 130+ of the *Principia*!

Realizing later that there were still things to do beyond the *Tractatus*, Wittgenstein turned to philosophy around 1928, with a great interest in the philosophy of mathematics. Some of his discussions of the time can be found recorded verbatim, through shorthand notes by Friedrich Waismann, in the book manuscript *Philosophische Bemerkungen* dated November 1930 and published in 1964. The notes are on pages 317–346. A comprehensive collection is found in the volume *Ludwig Wittgenstein und der Wiener Kreis* edited by Brian McGuinness.

Wittgenstein went to Cambridge in 1929 and became first a lecturer, then a professor. He prepared long manuscripts on the basis of his lectures that have been published many years after his death in 1951. He also dictated shorter pieces to his students and friends, such as one known as *The Blue and Brown Books*, with several more of these still to be published today.

(B) A PIED PIPER'S PAINS WITH UNIVERSALITY AND EXISTENCE. Wittgenstein's book manuscripts, such as the *Philosophische Grammatik* that was written around 1933, contain lengthy discussions of themes related to logic. Regarding the quantifiers, it emerges from these discussions that Wittgenstein was at great pain to understand them. As in the *Tractatus*, there is no trace of the rule of universal introduction, but quantifiers are instead simply logical expressions of a certain form. Generality is first taken as a "logical product" and existence as a "logical sum," the latter written, with f a predicate, as (p. 269):

$$fa \vee fb \vee fc \vee \dots$$

Generality covers all cases, but its explanation as a "product" of instances becomes infinitistic, and that was not acceptable to Wittgenstein (p. 268). In the absence of a rule of generalization, one gets at most that a universality implies any of its instances. Likewise, existence cannot be a summing up of all the disjunctive possibilities for its introduction, because there is an infinity of them. The dual to universal generalization is existential elimination, and in its absence, one gets only that an instance implies existence.

In the *Philosophische Grammatik*, Wittgenstein discusses at length an example, in translation the phrase *The circle is in the square*, illustrated by a drawing of a rectangle with a circle inside it (p. 260). It is clearly correct to say that there is a circle in the square, but the statement does not fix any specific circle. Wittgenstein sees that there is a generality behind existence and ponders the matter on page after page, all because he does not know that there should be a rule of existential elimination, the one Skolem used in an informal way, Bernays wrote in an axiomatic form, and finally Gentzen stated as a pure rule of natural deduction. Wittgenstein's "generic circle" is correctly presented through the eigenvariable of an existential elimination.

Wittgenstein's first works in his "second period" as a philosopher of logic and mathematics include two specific achievements, both of them somewhat cryptic and clarified only decades later. The first is a constructivization of Euler's proof of the infinity of primes, reconstructed in detail in Mancosu and Marion (2003). The second discovery derives from Wittgenstein's careful reading of Skolem's paper on primitive recursive arithmetic. Both are directly relevant to Wittgenstein's modest understanding of the quantifiers, and to his philosophy of mathematics.

Reuben Louis Goodstein followed Wittgenstein's lectures in Cambridge in 1931–34 and started work on a topic to which I will soon turn. Meanwhile, in 1939, he published an article titled "Mathematical systems" in the well-known philosophical journal *Mind*. It was a statement of what he took to be Wittgenstein's philosophy of mathematics. The article contains many exclamations and positions that should perhaps best be described as silly, but there are even indications that Wittgenstein was not displeased with it, contrary to some writings of other pupils of his such as Ambrose (1935).

In his paper, Goodstein maintains that the inference from $\neg\exists x\neg A(x)$ to $\forall x\, A(x)$ is intuitionistically legitimate. The converse implication is intuitionistically provable, so with the claimed inference, the universal quantifier could be defined by the existential one. Instead, this particular argument against intuitionism and for the "strict finitism" of Wittgenstein and Goodstein is just fallacious. In Goodstein (1951, p. 49), written under Wittgenstein's influence around 1940, it is stated that "some constructivist writers maintain that... a 'reduction' proof of universality is acceptable." In Goodstein (1958, p. 300), we find stated that Brouwer rejects indirect existence proofs, here $\neg(\forall x)\neg P(x) \to (\exists x)P(x)$, "whilst retaining the converse implication $\neg(\exists x)\neg P(x) \to (\forall x)P(x)$." In other words, if $(\exists x)\neg P(x)$ turns out to be impossible, a reduction gives $(\forall x)P(x)$; certainly not anything Brouwer or any other constructivist thinker would have ever proposed.

We shall soon see that Wittgenstein goes even further than Goodstein by denying the meaningfulness of the distinction between direct and indirect existence proofs.

The reason for Goodstein's mistake is somewhat subtle. The intuitionistically invalid implication $\neg\exists x\neg A(x) \supset \forall x\, A(x)$ is perhaps at first sight rather close to $\neg \exists x\, A(x) \supset \forall x \neg A(x)$. The latter is intuitionistically provable and was in fact one of the first examples of intuitionistically correct inference that Gentzen gave when he presented the calculus of natural deduction in his thesis (1934–35). The argument is very easy. To prove the implication, assume $\neg \exists x\, A(x)$. To prove $\forall x \neg A(x)$, try to prove $\neg A(x)$ for an arbitrary x. For this, in turn, assume $A(x)$ and try to derive a contradiction. It comes in one step: by rule $\exists I$, $\exists x\, A(x)$ follows, a contradiction. Therefore $\neg A(x)$ follows from the assumption $\neg \exists x\, A(x)$. The variable x is not free in the assumption, so rule $\forall I$ gives $\forall x \neg A(x)$, and $\neg \exists x\, A(x) \supset \forall x \neg A(x)$ can be concluded.

One could think that if $\neg \exists x\, A(x) \supset \forall x \neg A(x)$ is intuitionistically provable, it makes no difference to have $\neg A(x)$ under the negated existence, and $A(x)$ under the universal, instead of the other way around as in the preceding proof, but this is not in the least so. With $\neg A(x)$ in place of $A(x)$, we do get from what was proved earlier as an instance $\neg \exists x \neg A(x) \supset \forall x \neg \neg A(x)$, but the double negation cannot be deleted.

Alice Ambrose studied with Wittgenstein during the same years as Goodstein. The spell of "the oracle" is revealed in some of her letters, as in the following passages from a letter to a friend early in 1934 (McGuinness 1995, p. 219):

> The class started off too large, and remained so. After about two weeks, he decided he'd meet 5–7 people of his own choosing, twice a week. Fortunately this year Wittgenstein likes me, and we get on just fine… At present I am one of the chosen who go and wait for the oracle to speak…… The rest of the class was green with envy, and it was an awfully awkward mess.

Ambrose's June 1934 letter to Wittgenstein shows the direction in which the pied piper of Cambridge guided his students (McGuinness 1995, p. 231):

> I'm hoping that next year you'll talk some more about proofs in mathematics, induction, general and existential propositions. I have with me now Hilbert and Bernays' new book on foundations. The start seems pretty much of a muddle—that muddle about mathematical and meta-mathematical languages.

In 1935, Ambrose published in *Mind* the two-part article "Finitism in mathematics," in which "finitism" stands for intuitionism. She presents "the main issue in the controversy now current among mathematicians" in the form of four questions, with just the numbering added (1935, p. 186):

1. Is every mathematical problem solvable?
2. Is the law of excluded mean valid universally, that is, for reasoning on finite and infinite classes alike?
3. Is the *reductio ad absurdum* method of conclusion in all cases legitimate?
4. Is there justification for the use, in mathematical analysis and theory of sets, of the phrases 'all' and 'there exists' when they involve infinite classes or processes?

Ambrose is rather well versed in the recent literature, including the development of intuitionistic logic, Glivenko and Heyting (1930) among others, and the ε-axiom from Hilbert (1923) and its later elaboration by Bernays, but she obviously has not read the *Grundlagen der Mathematik* beyond the introductory chapters, therefore missing the detailed exposition of predicate logic, its rules of inference, and Gödel's completeness result—and she would not read them, as will become clear, even if she was clearly the most independent of Wittgenstein's students.

Ambrose's discussion is "guided throughout by certain suggestions made by Dr. L. Wittgenstein in lectures delivered at Cambridge in 1932–35" (p. 188). Thus, we read (p. 197):

> Common to both Weyl and Hilbert is the error of analyzing a general form as a conjunction and an existential form as a disjunction, that is, as logical product and logical sum. Wittgenstein has pointed out that there are cases where $(x).fx$ is a logical product, but these occur only when the subject class can be defined as a list.
>
> ⋮
>
> The error of finitists and formalists alike is in making general and existential forms of infinite range equivalent to infinite logical products and sums. Then they point out that "infinite product" and "infinite sum" are meaningless, which they are; but this analysis was unwarranted in the first place. Some of the extremes of the finitists are traceable to this analysis.

Next, Ambrose turns to the difference between, in her notation, $\overline{(x).\phi x}$ and $(\exists x).\overline{\phi}x$. One has the clear impression that she got the connections right, without Goodstein's erroneous claim by which universality is constructively definable through existence and negation (p. 198):

> I should suggest using the symbol $\overline{(x).\phi x}$ for all cases where a "non-constructive" use of "not all" is involved, and the symbol $(\exists x).\overline{\phi}x$ for all cases where the "constructive" use is involved.

In the second part of the article, Ambrose writes that "the formalists have assumed as legitimate those general and existential forms which by hypothesis cannot be demonstrated contradictory" (p. 317). There-fore, the formalists infer from $(\exists x).\overline{\phi}x$ to $(x).\phi x$, a step that is clearly presented as a purely classical one, not constructively acceptable *pace* Goodstein.

Wittgenstein's lectures during the Easter term of 1935, edited by Ambrose, were published in 1979. They contain comments on Ambrose's article, in which Wittgenstein contrasts the formulas $(\exists x)\phi x$ and $\sim (x) \sim \phi x$ through an example, *there are three 7's in the develop-ment of π* (pp. 194–195):

> The author suggested that there are two methods of proof, (1) showing three 7's between two places in the development, (2) showing that it is self-contradictory that there not be three 7's. $(\exists x)\phi x$ (where three 7's have been exhibited) should be given as the result of one proof, and $\sim (x) \sim \phi x$ (meaning that it is self-contradictory that there not be three 7's) as the result of the other.

This would be just fine, but now comes the surprise: Wittgenstein dismisses this clear distinction between a constructive and an indirect existence proof by turns of phrase so foolish that I cannot reproduce them here.

The second part of Ambrose's article contains clear intimations of Wittgenstein's verificationist philosophy that prevented him from accepting predicate logic, as in "with general forms of infinite range, 'verification by exhibition of instances' is meaningless" (p. 319). That objection against universal quantification has force only if the method of proof by generalization is absent. Another characteristic passage is (p. 333):

> To say that a form has meaning is to say that in the symbolic system an answer to the question whether it is true, or false, is provided for. If a question is asked for which no answer is provided in the system (that is, for which no answer *can* be given), then this question is meaningless.

This view "or a similar one" is ascribed directly to Wittgenstein, and again, the passage shows that Wittgenstein was unaware of the need for rules of proof for the quantifiers. Moreover, the passage suggests that completeness would be a condition for meaningfulness; Gödel's 1930 completeness theorem for classical predicate logic had remained unnoticed by Wittgenstein and his students.

Rather than being satisfied by the discussion of foundations of mathematics sympathetic to his ideas, Wittgenstein was greatly angered by the paper. Ambrose added a footnote to the second part of her paper, at a passage where she stated that "if we do not know what is meant by the statement that p is demonstrated, we do not know what is meant by p" (p. 319):

> This is a view which I understood Dr. L. Wittgenstein to put forward in his lectures and which but for them would never have occurred to me. It is only in this sense that any view which I have put forward can be said to have been "guided by suggestions made by him". In stating on p. 188 of my last article that my views were so guided, this was all I meant to say. That is, I did not intend to claim either that I had understood him correctly or that inferences which I drew from what I understood him to mean would follow from his actual views. Any reader who finds mistakes or absurdities in my views must not suppose that he is responsible for them.

Wittgenstein had proposed through a third person that Ambrose write another article that he could accept, but Ambrose replied to this in a letter of 16 May 1935 that "it is doubtful whether what I write... will be satisfactory to you—unless you dictate the material... there is no point in giving a quotation from you with my name on it" (McGuinness 1995, p. 240).

Ambrose's upright refusal to play the role of an intellectual puppet in front of Wittgenstein must have been a completely new experience to the philosopher. Instead of answering the letter, he sent it back with the words (ibid., p. 241): "Don't destroy this letter, it might interest you one day to reread it." To his friend G. E. Moore, he wrote (p. 242):

> I think you have no idea in what a serious situation she is . . . serious because she is now actually standing at a crossroad. One road leading to a perpetual misjudging of her intellectual powers . . . the other would lead her to a knowledge of her own capacities and that *always* has good consequences.

With these words, Wittgenstein distanced himself from what appears to be his most talented student, one who in contrast to many others showed signs of intellectual independence. Wittgenstein's struggles with the quantifiers make plain the misjudgment of *his own* intellectual powers; there simply wasn't any way by which his introspective method and failed appraisal of what had been achieved by "the formalists" in Göttingen could have led to a command of the principles of quantificational logic. His score consisted of plain misconceptions, played in front of Ambrose, Goodstein, and the rest of his students. One month after the letter to Ambrose, in June 1935, Wittgenstein toyed with the idea of moving to the Soviet Union or of starting to study medicine in England—at the age of forty-eight, and a sign of desperation under any normal circumstances. Might it be that he had regrets about the Ambrose affair, a glimpse of the limits of his own capacities?

Wittgenstein and his students were not alone with their problems. The correspondence between Heyting and Oskar Becker gives ample illustration of how difficult it was to get intuitionistic logic right, even for people who tried hard (see Van Atten 2005).

A tentative conclusion can be drawn from the preceding little story: Wittgenstein's refusal of the quantifiers, even the intuitionistic ones, in favor of a strict finitistic verificationism, was based on a misunderstanding of the nature of the intuitionistic quantifiers and on the inability to learn from what others had accomplished.

We now come to Wittgenstein's second specific discovery.

(C) FROM INDUCTION TO RECURSION. In 1945, there appeared in the *Proceedings of the London Mathematical Society* a long article titled "Function Theory in an Axiom-Free Equation Calculus." The main idea of the work was to recast primitive recursive arithmetic in an even stricter mold than the quantifier-free calculus of Skolem. Even logic had to go, and the venerated principle of arithmetic induction as well. The latter is replaced by a principle by which two recursive functions defined by the same equations are the same (p. 407): "If two functions signs 'a', 'b' satisfy the same introductory equations, then '$a = b$' is a proved equation." A footnote added to this principle tells the following: "This connection of induction and recursion has been previously observed by both Wittgenstein and Bernays." The author of the paper, this time not in the least silly, was Wittgenstein's student Goodstein. The full story of his paper can be recovered through the correspondence he had with Paul Bernays. In the opening letter of 29 July 1940, he writes:

> The manuscript which accompanies this letter gives some account of a new formal calculus for the Foundations of Mathematics on which I have been working for the past six years.

Unfortunately, the original version of the paper is not to be found. The most we know are some comments by Bernays such as the following from his first letter to Goodstein, dated 28 November 1940:

> Generally my meaning is that your attempt could be quite as well, and perhaps even better appreciated, if you could deliver it from the polemics against the usual mathematical logics which seem to me somewhat attackable, in particular as regards your arguments on the avoidability of quantifiers. Of course in your calculus, like in the recursive number theory, quantifiers are not needed. But with respect to the "current works on mathematical philosophy" the thesis that "the apparent need for the sign '(x)' arose from a confusion of the two different uses ... of variable signs" can hardly be maintained. In fact the possibility of taking $f(x) = 0$ instead of $(x)(f(x) = 0)$ consists [$sc.$ exists] only, if the formula in question stands separately and not as a part of a complex logical structure.

> So for instance the negation of $(x)(f(x) = 0)$, that is $\sim (x)(f(x) = 0)$, has of course to be distinguished from the proposition $(x). \sim (f(x) = 0)$; if here the sign "(x)" is left out, then really a confusion is arising. Thus there neither is the possibility of taking simply the proposition $\sim (f(x) = 0)$ instead of $(Ex). \sim (f(x) = 0)$ (or else one would have to add artificial conventions).

Here again we see Wittgenstein's claim that predicate logic is based on a confusion on the part of "the formalists."

Bernays mentions further in the letter that he had presented in 1928 at the Göttingen Mathematical Society "the possibility of taking instead of the complete induction the rule of equalizing recursive terms satisfying the same recursive equations," a discovery he left unpublished. His first letter to Goodstein is ten pages long, typewritten single-spaced, and it displays his full command of Goodstein's calculus. Goodstein was enormously impressed as can be seen from his letters and thankfully revised his paper and cleared it of polemics, adding all the references to a literature that had been unknown to him; quite embarrassingly, even the extensive treatment of recursive arithmetic in the first volume of the *Grundlagen der Mathematik*, § 7, pp. 287–343 belonged there.

Before going on to the replacement of induction by recursion, a brief word about the disposal of logic is in order. Goodstein, as well as Bernays before him (*Grundlagen*, pp. 310–12), noticed that in primitive recursive arithmetic, propositional logic can be reduced to equational reasoning. This idea can be traced back to Gödel (1931) in which it is shown that the operations of propositional logic give back recursive relations when applied to given recursive relations. Equations $a = b$ can be turned into equivalent equations of the form $t = 0$, and now conjunction and negation can be defined: $a = b \& c = d$ turns into $t + s = 0$, and $\neg a = b$ into $1 - t = 0$, with subtraction truncated so that whenever $t > 0$, we have $1 - t = 0$.[1]

The Wittgensteinian background of Goodstein's "logic-free" and "induction-free" arithmetic calculus is not mentioned in the book *Recursive Number Theory* that Goodstein published in the prestigious yellow-covered logic series of the North-Holland Publishing Company in 1957. Instead, when Wittgenstein's book manuscript *Philosophische Grammatik* came out in 1969, one could find his discovery of the way from proof by induction to proof by recursion equations clearly stated, and developed to some extent, mainly through a few

---

[1] Some details: Define the predecessor function p by $p(0) = 0$ and $p(a + 1) = a$, then the "truncated" subtraction $a \dot- b$ by $a \dot- 0 = a$ and $a \dot- (b + 1) = p(a \dot- b)$. Finally, define the absolute distance between a and b by $d(a,b) = (a \dot- b) + (b \dot- a)$. Now $d(a,b) = 0$ whenever $a = b$, and negation can be defined by $\neg a = b \equiv 1 \dot- d(a,b) = 0$, and conjunction further by $a = b \& c = d \equiv d(a,b) + d(c,d) = 0$. A similar development is found in Curry (1941).

examples (*Philosophische Grammatik*, pp. 397–450). The text was written between 1932–34, the years during which Goodstein attended Wittgenstein's lectures. The crucial discovery comes out on the very first page devoted to the topic (ibid., p. 397), where Wittgenstein considers the associative law for sum in elementary arithmetic, denoted by A:

$$(a + b) + c = a + (b + c) \qquad A$$

Skolem's 1923 paper on primitive recursive arithmetic, Wittgenstein's source for the topic of elementary arithmetic, gives the standard inductive proof for A, based on the recursive definition of sum by the recursion equations

$$a + 0 = a$$
$$a + (b + 1) = (a + b) + 1$$

If one counts the natural numbers from 1 on, the second equation gives the base case of the inductive proof. For the step case, one assumes A for c and proves it for $c + 1$, namely $(a + b) + (c + 1) = a + (b + (c + 1))$. By the recursion equation, the left side is equal to $((a + b) + c) + 1$. Then, applying the inductive hypothesis to $(a + b) + c$, one gets $((a + b) + c) + 1 = ((a + (b + c)) + 1$, and finally, by two applications of the recursion equation in the opposite direction, $((a + (b + c)) + 1 = a + ((b + c) + 1) = a + (b + (c + 1))$.

In the *Grammatik*, p. 397, Wittgenstein gives the proof as follows:

What Skolem calls the recursive proof of A can be written as follows:

$$\left. \begin{array}{l} a + (b + 1) = (a + b) + 1 \\ a + (b + (c + 1)) = a + ((b + c) + 1) = (a + (b + c)) + 1 \\ (a + b) + (c + 1) = ((a + b) + c) + 1 \end{array} \right\} B$$

We have to put emphasis on Wittgenstein's words "can be written," for this is not Skolem's proof by induction but rather another proof that Wittgenstein goes on to explain in the following words:

In the proof [B], the proposition proved clearly does not occur at all.—
One should find a general stipulation that licenses the passage to it.

This stipulation could be expressed as follows:

$$\left.\begin{array}{ll} \alpha & \varphi(1) = \psi(1) \\ \beta & \varphi(c+1) = F(\varphi(c)) \\ \gamma & \psi(c+1) = F(\psi(c)) \end{array}\right\} \qquad \begin{array}{l} \Delta \\ \varphi(c) = \psi(c) \end{array}$$

When three equations of the forms α, β, γ have been proved, we shall say: "the equation Δ has been proved for all cardinal numbers."

Here we see the essence of the argument: Two functions φ and ψ that obey the same recursion equations are the same function. Wittgenstein himself writes (*Grammatik*, p. 398):

> I can now state: The question whether A holds for all cardinal numbers shall mean: Do equations α, β, and γ hold for the functions

$$\varphi(\xi) = a + (b + \xi), \quad \psi(\xi) = (a + b) + \xi$$

Wittgenstein's principle can be considered, as in the letter of Bernays quoted earlier, a "rule of equalizing recursive terms." Taken as a rule, it is readily seen to be a derivable rule in *PRA*: Given its premisses for two functions φ and ψ, the conclusion follows by the principle of induction. These premisses are:

$$\begin{array}{ll} \alpha & \varphi(1) = \psi(1) \\ \beta & \varphi(c+1) = F(\varphi(c)) \\ \gamma & \psi(c+1) = F(\psi(c)) \end{array}$$

We want to derive $\varphi(x) = \psi(x)$ for an arbitrary x. Equation α gives the base case of induction. For the inductive case, assume $\varphi(y) = \psi(y)$. By β, we get first $\varphi(y+1) = F(\varphi(y))$, and by the inductive hypothesis $F(\varphi(y)) = F(\psi(y))$. Application of the equation γ gives next $F(\psi(y)) = \psi(y+1)$, so the equation $\varphi(y+1) = \psi(y+1)$ follows. By the principle of induction, $\varphi(x) = \psi(x)$ follows for arbitrary x.

By the preceding, we see that Wittgenstein's rule contains the essential steps that lead from y to the successor $y+1$, namely the inductive step, in a somewhat disguised form.

The path from Wittgenstein's uniqueness principle for recursion equations to induction is similar. Assume as given the premisses of induction, $\varphi(1) = \psi(1)$ and $\varphi(y) = \psi(y) \supset \varphi(y+1) = \psi(y+1)$ for an arbitrary y, and the task is to prove $\varphi(x) = \psi(x)$ for arbitrary x.

The recursive functions φ and ψ are defined by some recursion equations that for the successor case have the forms

$$\beta \qquad \varphi(c+1) = F(\varphi(c))$$
$$\gamma \qquad \psi(c+1) = G(\psi(c))$$

If $\varphi(y) = \psi(y)$, then $\varphi(y+1) = \psi(y+1)$ by the assumption that the inductive step is given. By the recursion equations, this latter equation gives at once $F(\varphi(y)) = G(\psi(y))$. Therefore, when the arguments $\varphi(y)$ and $\psi(y)$ of F and G are equal, the values are equal, so that we get $F(x) = G(x)$ for any x. We have now altogether

$$\alpha \qquad \varphi(1) = \psi(1)$$
$$\beta \qquad \varphi(c+1) = F(\varphi(c))$$
$$\gamma \qquad \psi(c+1) = F(\psi(c))$$

By Wittgenstein's uniqueness rule, $\varphi(x) = \psi(x)$ follows for any x, i.e., we have reached the conclusion of induction.

Wittgenstein's book does not reveal the motive for preferring proofs by recursion equations to proofs by induction, but in 1972, Goodstein published the paper "Wittgenstein's Philosophy of Mathematics," in which the matter is explained. In reference to the *Philosophische Grammatik* that had come out three years earlier, Goodstein recalls Skolem's inductive proof and then adds (p. 280):

> In his lectures Wittgenstein analysed the proof in the following way. He started by criticizing the argument as it stands by asking what it means to *suppose* that (1) [associativity] holds for some value C of c. If we are going to deal in suppositions, why not simply suppose that (1) holds for any c.

Goodstein now gives a very clear, intuitive explanation of why Wittgenstein's method works: With $c = 0$, $(a + b) + 0 = a + b = a + (b + 0)$. Thus, the ground values of Wittgenstein's φ- and ψ-functions are the same, here $\varphi(0) = \psi(0)$ with the natural numbers starting from 0 instead of 1 as in the 1920s. For the rest, when c grows by one, $\varphi(c)$ and $\psi(c)$ obtain their values in the same way, here, both growing by 1, by which $(a + b) + c$ and $a + (b + c)$ are always equal.

As the quoted clear recollection on the part of Goodstein shows, Wittgenstein was led to propose a finitism that was even stricter than that of Skolem, in that *assumptions with free variables* were to be banned. These assumptions are a crucial component in inductive inference, where one assumes a property $A(n)$ for an arbitrary natural number n and then shows that the successor of n has the property, expressed as $A(n + 1)$. However, the assumption $A(n)$ is a far cry from assuming, say in the case of associativity, that the inductive predicate "holds for any c" as Goodstein suggests at the end of the quotation. It is the simplest error in inference with the quantifiers to assume $A(x)$ and then to conclude $\forall x\, A(x)$. The eigenvariable condition in universal generalization is that x must not occur free in any assumption on which its premiss $A(x)$ depends, but here one must keep in mind that if $A(x)$ itself is an assumption, it depends on itself so to say, and thus x is free in an assumption. More generally, to assume $A(x)$ is not the same as to assume $A(x)$ is provable, and only the latter gives $\forall x\, A(x)$. No amount of philosophical reflection in Wittgenstein can replace the command over quantificational inferences that results from Gentzen's formulation of the quantifier rules in terms of natural deduction.

(D) **TURING'S SCRUPLES.** Wittgenstein's lectures on the foundations of mathematics in Cambridge during the first half of 1939 were reconstructed by Cora Diamond in 1975 on the basis of four sets of notes by participants. These lectures were graced by the presence of Alan Turing, who, as a reader of the lectures soon notices, had something to comment on almost every lecture. Turing was to be absent from one lecture, for which reason Wittgenstein announced that "it is no good my getting the rest to agree to something that Turing would not agree to" (pp. 67–68). The lectures show no progress on the part of Wittgenstein as regards the understanding of the principles of quantificational logic. The remarks about generality, existence, and the circle-in-the-square example are in substance the same as they were in 1933 (as on pp. 268–269). Moreover, Wittgenstein's pretense— witnessed by Bernays' comments on Goodstein's lost manuscript—has not changed (p. 270): "If Russell gives an interpretation of arithmetic in terms of logic, this removes some misinterpretations and creates others. There are gross misunderstandings about the uses of 'all', 'any', etc." Sad to say, these misunderstandings were all Wittgenstein's, caused by his inability to learn from others.

Remarkably, Turing kept silent during Wittgenstein's show of *sui ipsius et multorum ignorantia* at the points in which the quantifiers were discussed. His reaction is instead seen in a manuscript he was working on in the early 1940s. It bears the title *The reform of mathematical notation and phraseology* and can be seen in manuscript form on the pages of the Turing archive. Two of the central points were: 1. "Free and bound variables should be understood by all and properly respected." 2. "The deduction theorem should be taken account of." Turing then gives examples of constants and variables and adds:

> The difference between the constants and the free variables is somewhat subtle. The constants appear in the formula as if they were free variables, but we cannot substitute for them. In these cases there has always been some assumption made about the variable (or constant) previously.

The deduction theorem is the main way of handling free variables in axiomatic logic. Turing's example, slightly rephrased, is: Let the radius a and volume v of a sphere be given. Then $v = \frac{4}{3}\pi a^3$.

> The 'deduction theorem' states that in such a case, where we have obtained a result by means of some assumptions, we can state the result in a form in which the assumptions are included in the result, e.g., 'If a is the radius and v is the volume of the sphere then $v = \frac{4}{3}\pi a^3$.' In this statement a and v are no longer constants.

Turing says that "the process converts constants or 'restricted variables' into free variables."

There are passages in the manuscript version of Turing (1948), available at the Turing archives, that suggest that Turing had at least some knowledge of Gentzen's system of natural deduction. It is a pity he did not use it in the explanation of free and bound variables. In the example, there is a typical "Let" phrase about given a and v, an instance of the form $S(x, y)$ that states that some x and y are the radius and volume of a sphere. Eigenvariables a and v are put in place of x and y to get the assumption $S(a, v)$, and then the result $v = \frac{4}{3}\pi a^3$ is derived. The deduction theorem introduces the implication $S(a, v) \supset v = \frac{4}{3}\pi a^3$ with no assumptions about a or v left, so that generalization gives $\forall x \forall y (S(x, y) \supset y = \frac{4}{3}\pi x^3)$. The situation is the same with induction. Once an assumption $A(n)$ has been made and $A(n + 1)$ proved, implication introduction, or "the deduction theorem" in Turing's axiomatic terminology, is used to infer that

$A(n) \supset A(n + 1)$, no longer dependent on the assumption $A(n)$, so that the second premiss of induction $\forall x(A(x) \supset A(x + 1))$ can be inferred. Here we have it, had Turing just cared to explain the correct use of free-variable assumptions to Wittgenstein, but there are no comments by anyone in the last lecture of 1939, in which Wittgenstein briefly discusses primitive recursive arithmetic.

(E) **FINAL REFLECTIONS ON WITTGENSTEIN'S METHOD.** Wittgenstein's method of philosophizing is perhaps best seen in the shorthand notes of Friedrich Waismann, collected in Wittgenstein (1967). With the immediacy of the shorthand, one can see Wittgenstein in direct action there, so to say. The basis there is the conviction that *the problems of philosophy are problems of usage of language.* They are not proper scientific problems that could be stated and solved through genuine discoveries, but rather they dissolve themselves in the best of cases. Competence in language and attention to meaning are the only prerequisites in philosophy.

Wittgenstein set himself a trap with the specific conditions of his philosophical method: His analysis of language was based on truth and falsehood that one establishes by verification. Verifiable atomic sentences without logical structure are at the base, and logically compound sentences inherit their verifiability by the truth table decision method of classical propositional logic. There is no place for inference because each sentence stands on its own feet by its method of verification. This criterion appears again and again in the Waismann notes as the crucial one for meaningfulness.

In Wittgenstein's method, there is no place for hypotheses or assertions either. The former assume something to be the case, but meaning comes only from verification. Similarly, to assert something can at best be the résumé of a verification.

Consider again Wittgenstein's example from Section 6.2(B): *The circle is in the square.* In the *Philosophische Grammatik* and many other places, he is at pains to try to figure out what can be meant by the statement. For a solution to the problem, we must relax some of his principles. First, we have to put the verification principle to rest and accept hypothetical sentences next to categorical sentences, and an account of meaning under hypotheses. Therefore, we need a notion of inference from hypotheses with the property that whenever the hypotheses turn out to hold, what has been inferred also holds. Next, we reflect on the meaning of assumptions and conclusions in inference, and the result is: If the circle is assumed to be in the square, the usage of the sentence goes through the elimination of an existential assumption. This matter needs to be learned either from mathematical practice or from logic books. If it is an assertion, we should have that whenever it has a categorical proof, the proof goes through an instance. There are fine examples in logic books that show how it goes, as with the existence property of intuitionistic arithmetic. There is a difference between an ordinary discourse and a disciplined scientific one in that the latter requires training in the specific meanings of the concepts involved.

6.3. THE POINT OF INTUITIONISTIC GEOMETRY

(A) APARTNESS RELATIONS. Brouwer had in 1924 the idea of replacing the equality relation of two real numbers by *apartness*, $a \neq b$. The properties of apartness are, in logical notation:

1. $\neg\, a \neq a$ *irreflexivity*
2. $a \neq b \supset a \neq c \vee b \neq c$ *apartness axiom, co-transitivity*

The second axiom is notable in a constructive context because it has a disjunction in a part of the formula (positive part) that cannot be rewritten in a constructively equivalent way without disjunction. It follows that whenever we have established $a \neq b$, any third real number c can be taken and the two cases $a \neq c$ and $b \neq c$ can be formed.

By putting a for c in the second axiom, we get the instance $a \neq b \supset a \neq a \vee b \neq a$, with the first disjunct negated in axiom 1. Therefore, symmetry, $a \neq b \supset b \neq a$, follows.

Equality is a defined notion:

$$a = b \equiv \neg\, a \neq b$$

Reflexivity of equality is immediate from the definition, and symmetry and transitivity follow as contrapositions of symmetry of apartness and of axiom 2, the latter in the "Euclidean" form $a = c \,\&\, b = c \supset a = b$, by Euclid's axiom in the *Elements* that says: "Two things equal to a third are equal among themselves."

The idea of an apartness relation is that one replaces an "infinitely precise" notion such as the equality of reals by a "finitely precise" apartness, in the sense that if $a \neq b$ happens to be the case, a finitely precise computation of the values of a and b verifies $a \neq b$, whereas the verification of $a = b$ requires infinite precision, whatever that may be. The "finite precision" idea validates the apartness axiom: Given that a and b are apart, if $a \neq c$ cannot be decided, c must be apart from b, and similarly for $b \neq c$.

(B) APARTNESS IN GEOMETRY. Brouwer's idea was applied in the work of his student Arend Heyting, in his 1925 doctoral thesis *Intuitionistische axiomatiek der projektieve meetkunde* (Intuitionistic axiomatics of projective geometry). A shorter German version of the work was published in 1927 as an article in the *Mathematische Annalen*, with the title *Zur intuitionistischen Axiomatik der projektiven Geometrie.* As in Brouwer, Heyting's notation and terminology are idiosyncratic; the reason is Brouwer's idea that mathematics had to be purified of old, bad connotations through new notation and a new terminology that in part consists of neologisms. Brouwer's notation for apartness is *a#b*. Heyting uses the word *species* for the intuitionistic notion of a set. He gives an axiomatization for space that consists of points, with lines, planes, and other geometric objects construed from them. I describe here the plane part that is structured as follows:

1. The basic objects are *points* that form the geometric space. Other geometric objects are given as *species of points*.
2. There are *apartness* and *equality* relations for points A and B, written as $A \omega B$ and $A \sigma B$, respectively, and definitions of similar relations for species of points.

Lines are species of points. There are altogether 16 geometric axioms, grouped from I to X; I shall describe the first ten. There are five axioms for points, after the first one, by which the space consists of points (p. 493):

Axiom I. The space is a mathematical species.
Axiom II. (*Axiom of separation.*)

IIa. $A \sigma B$ (*A coincides with B*) and $A \omega B$ (*A is apart from B*) are invertible relations between the points A and B.

IIb. $A \sigma A$ for every point A.

IIc. The relations $A \sigma B$ and $A \omega B$ are mutually exclusive.

IId. If $A \omega B$ is inconsistent, $A \sigma B$ holds.

IIe. If there is between the points A and B the relation $A \omega B$, then for every point C either $A \omega C$ or $B \omega C$.

The word for inconsistent in IId is "ungereimt," a somewhat odd word construction. No logical notation is used, but it can be seen that by axioms IIc and IId, coincidence $A \sigma B$ is equivalent to the negation of apartness $A \omega B$ and therefore a superfluous notion. Two theorems follow, the first the usual transitivity of equality, the second the substitution of equals in apartness, or *from $A \omega B$ and $B \sigma C$ follows $A \omega C$*.

There follow now three axioms for lines, after one that produces two distinct points (ibid.):

Axiom III. Two points separate from each other can be determined.

Axiom IV. (*Axiom of lines.*) Lines are point species with the following properties:

IVa. If point P belongs to line l, then every point that coincides with P belongs to l.

IVb. Two points apart from each other determine a line that contains them both (i.e., one can determine a line l that contains both of them, and every line that contains both of them is identical to l), their *connecting line*.

IVc. Every line contains at least three points apart from each other.

These axioms are followed by three lengthy definitions and axiom V (p. 494):

Definition. Point P *is apart* from the point species α (also *lies outside* the point species α), if it is apart from each point of this species; we write in this case $P \omega \alpha$.

The point species α *coincides* with the point species β or $\alpha \sigma \beta$ if every point of α coincides with some point of β and every point of β with some of α.

The point species α *is apart* from the point species β if α contains a point that is apart from β; we write then $\alpha \omega \beta$. This relation is not invertible.

Axiom V. A point can be determined outside every line.

The conceptual order of things in the axioms and definitions contains a crucial if somewhat subtle error that makes the axiomatization constructively too weak, as will be made clear in the next subsection.

Axiom IVb is clearly a construction postulate. An "infinitely precise" object is constructed that has some ideal, "infinitely precise" properties. Axioms III and V are clearly different from construction postulates, as they don't determine any unique new geometric objects from given ones. What should we make of these axioms? Axiom IVc instead is formulated in the language of existence. What difference does that make to axioms III and V? Answers to these questions will be helped by some notation and a little formal work.

(C) HEYTING'S ERROR IN CONCEPTUAL ORDER. A formal notation and some logic will be helpful to set things in order. We take *points* and *lines* and their apartness relations as primitives, with a new basic relation and notation:

$$a \notin l \qquad point\ a\ is\ outside\ line\ l$$

Incidence becomes a notion defined analogously to equality:

$$a \in l \equiv \neg\, a \notin l$$

Heyting's definition of "outside" applied to a and l gives:

$$a \notin l \equiv \forall x (x \in l \supset a \neq x)$$

A properly formulated constructive form of the principle of substitution of equals in the outsideness relation is

$$a \notin l \supset a \neq b \vee b \notin l$$

The correctness of this axiom is seen similarly to the apartness axiom. If $a \notin l$ and $b \in l$, the conclusion $a \neq b$ of Heyting's definitional axiom follows at once, but not the other way around: there is no way to arrive at the constructively correct disjunctive conclusion $a \neq b \vee b \notin l$ from $a \notin l$ by Heyting's axiom. It is in substance $a \notin l\ \&\ b \in l \supset a \neq b$, analogous to $a \notin l\ \&\ a = b \supset b \notin l$. This latter standard substitution principle of equals in the outsideness relation is constructively too weak.

Connecting lines of two distinct points were originally introduced by a postulate in Hilbert (1899) with the notation AB and the *condition of non-degeneracy* $A \neq B$. This postulate was changed into a purely existential axiom in later editions, written in the present notation as

$$\forall x \forall y \exists z (x \neq y \supset x \in z \ \& \ y \in z)$$

A formal notation for the line construction is $ln(a,b)$, the *connecting line* of points a and b. The connecting line has the ideal properties expressed by the axioms $a \neq b \supset a \in ln(a,b)$, $a \neq b \supset b \in ln(a,b)$. A further property required is that the constructed object be *unique*:

$$a \neq b \ \& \ a \in l \ \& \ a \in m \ \& \ b \in l \ \& \ b \in m \supset l = m$$

Specifically, with $m \equiv ln(a,b)$, we get

$$a \neq b \ \& \ a \in l \ \& \ b \in l \supset l = ln(a,b)$$

Heyting's axioms III and V differ from axioms for constructions; they are in some sense *purely existential*, not replaceable by constructions. They can be formulated as

III: $\exists x \exists y . x \neq y$ and $V : \forall x \exists y . y \notin x$

In axiom V, the sorts of objects can be read off from the way the relation is written, with a point as the left member and a line as the right member in the outsideness relation.

The standard intuitionistic explanation of the quantifiers gives

If $\forall x \exists y \, A(x,y)$, there is a function f such that $A(x, f(x))$

This kind of explanation fails for the axiom $\forall x \exists y . y \notin x$. Heyting's axioms III and V are instead rather like the *things required* in the proof of Euclid's first proposition: "On a given line segment to produce an equilateral triangle." The proof starts with: "Let AB be the given line segment." This is clearly an existential assumption, and A and B act as *eigenvariables* in an inference from the assumption. As a result, the construction applies to *any line AB*.

Heyting's axioms III and V express tasks that can be met by "finitary capacities," as in:

Two distinct points a,b can be determined.
To mark three points a,b,c such that they form a triangle.

The points *a,b,c* are "generic," i.e., the only thing we know about them is $a \neq b$ and $c \notin ln(a,b)$. Existential axioms such as

Heyting's $\exists x \exists y . x \neq y$ and $\forall x \exists y . y \notin x$ have precisely this effect when the existential quantifiers are instantiated by eigenvariables. Heyting's geometry gives at the same time an example of an intuitionistic theory in which the existence property fails; a formal proof of this failure can be given for the axiom system of the next subsection.

(D) THE AXIOMS OF INTUITIONISTIC PROJECTIVE GEOMETRY. We finish this section with a summary of an up-to-date axiomatization of intuitionistic projective geometry that is strictly parallel to the classical axiomatization in Section 2.4(C):

Basic relations $a \neq b, l \neq m, a \notin l$

Apartness axioms for point and line apartness:

$$\neg\, a \neq a, \neg\, l \neq l$$
$$a \neq b \supset a \neq c \vee b \neq c, \quad l \neq m \supset l \neq n \vee m \neq n$$

Connecting line and *intersection point* constructions:

$$ln(a,b),\, pt(l,m), \quad \text{conditions } a \neq b, l \neq m$$

Incidence properties:

$$a \in ln(a,b),\; b \in ln(a,b),\; pt(l,m) \in l,\; pt(l,m) \in m$$

Substitution of equals:

$$a \notin l \supset a \neq b \vee b \notin l, \quad a \notin l \supset l \neq m \vee a \notin m$$

Uniqueness:

$$a \neq b \,\&\, l \neq m \supset a \notin l \vee a \notin m \vee b \notin l \vee b \notin m$$

The classical contrapositive is *Skolem's axiom*, as in Section 2.4(C):

$$a \in l \,\&\, a \in m \,\&\, b \in l \,\&\, b \in m \supset a = b \vee l = m$$

The classical versions of substitutions are:

$$a \in l \,\&\, a = b \supset b \in l, \quad a \in l \,\&\, l = m \supset a \in m$$

Noncollinearity:

$$\exists x \exists y \exists z (x \neq y \,\&\, z \notin ln(x,y))$$

The last axiom is purely existential; still, the existence property obviously fails. Assuming it held, there would be a derivation of

$a \neq b$ & $c \notin ln(a,b)$ for some three points a, b, c. Then even $a \neq b$ would be derivable. By the general result in von Plato (2010), such a derivation would have the *subterm property:* all terms in the derivation are visible in the open assumptions, here none, or the conclusion. It is seen at once that there can be no derivation with this property for the formula $a \neq b$.

6.4. INTUITIONISTIC LOGIC IN THE 1920S

(A) HEYTING'S INTUITIONISTIC LOGIC. Discussions about the logic of Brouwer's intuitionistic mathematics were conducted in the pages of the *Bulletin* of the Royal Belgian Academy during the last years of the 1920s. In the midst of these discussions, Heyting figured out in 1928 the proper axiomatization of intuitionistic propositional and predicate logic.

There were three predecessors to Heyting, all unknown to him. In chronological order, Skolem (1919) had found a structure that is known today as Heyting algebra, one that relates to intuitionistic logic in the same way as Boolean algebra to classical logic. The matter was known to Tarski and others when in the latter part of the 1930s they figured out the algebraic semantics of intuitionistic logic (see my 2007 for details).

A second precursor to Heyting was the later famous probability theorist Andrei Kolmogorov in his Russian paper of 1925. At that time, the Communist Workers' Academy of Moscow hosted a seminar on foundations of mathematics; besides Kolmogorov, the participants included another later famous probability theorist, Yakov Khintchine, who in the proceedings of the said academy published a paper titled "Ideas of intuitionism and the struggle for content in contemporary mathematics." One message of the paper, as well as that of Kolmogorov, was that Brouwer can be considered the winner of the foundational debate. These views echoed Weyl's 1921 paper "On the new foundational crisis of mathematics" that provoked clamor because of its enthusiastic support of Brouwer's ideas. Later signs of Kolmogorov's intuitionism can be seen; for example, he wrote a preface for the Russian translation of Heyting's 1934 little book on intuitionism and proof theory. At this time, in 1936, the publishers of the translation took distance to Kolmogorov's intuitionistic convictions.

It was unfortunate that Kolmogorov did not publish his results of 1925 in an accessible way. It is also strange that he did not explain them to, say, Heyting, with whom he was in correspondence, or to Bernays during his stay in Göttingen in 1930–31. Kolmogorov's starting point is the axiomatization of Ackermann (1924), with its groups of axioms for each connective. A classical axiomatization is designated by a Gothic $\mathfrak{H}$, for Hilbert, and an intuitionistic one by $\mathfrak{B}$, for Brouwer. Kolmogorov now adds double negations in front of each subformula of a given formula and shows that the transformed formula is derivable in $\mathfrak{B}$ whenever the original one is derivable in $\mathfrak{H}$. The likely source of the idea is Brouwer's observation by which two out of three negations can be deleted, as in $\neg\neg\neg A \supset \neg A$. Kolmogorov writes that all axiomatizations of mathematics known to him obey the result; that is, under the translation, classical theorems proved from classical axioms turn into intuitionistic theorems proved from intuitionistic axioms. As to the quantifiers, Kolmogorov saw clearly the need for a rule of generalization that "cannot be expressed symbolically"; in other words, a proper rule that cannot be rewritten as an axiom. Further, he saw the genuinely classical case of relations between quantifiers, the implication $\neg\forall x\, A(x) \supset \exists x \neg A(x)$. There is just one blemish: the two independent quantifiers require a second rule, that of existence elimination. — The comments of Hao Wang to the contrary, in the introduction to the English translation of Kolmogorov's paper in Van Heijenoort (1967, p. 414), reveal an embarrassing ignorance of intuitionistic logic.

A third precursor to Heyting's 1930 paper was Bernays, who after a talk by Brouwer in Göttingen found out in 1925 that it is sufficient to leave out the law of double negation from a suitable axiomatization of classical logic, such as the one in table 7.1 below (see Troelstra 1990, pp. 6–7).

Brouwer was enthusiastic about Heyting's work and concluded that he would not have to finish such a work himself (see Van Dalen 2013, p. 607). There has been for some reason a general belief that Brouwer was somehow against the formalization of intuitionistic logic, but this is a clearly erroneous idea.

Heyting's axiomatization has 11 propositional axioms, not quite like *Principia Mathematica* as has often been said, even by Heyting himself, but more like the Hilbert-Bernays axioms in which all connectives and quantifiers are present, to be detailed later. The original version from 1928 is lost, but the published version refers to Glivenko's paper of 1928 that has a handsome set of axioms.

Implication elimination was not the only propositional rule, but there was also an explicit rule of conjunction introduction. It could be dispensed with, but the proofs would then be "even more intricate" (*verwickelt*). A little reconstruction shows that this is the case.

Heyting's first theorem is $\vdash \cdot a \wedge b \supset a$, and the proof he gives is

[2.14] $\vdash \cdot a \supset \cdot b \supset a : \supset :$

[2.12] $\vdash \cdot a \wedge b \supset \cdot b \supset a \cdot \wedge b \cdot$ [2.15] b

The last symbol is a misprint; it should be a. The notation uses abbreviations and the dot notation in place of parentheses, with the axioms referred to by square brackets. When axioms are written, they are indicated by a double turnstile $\vdash\vdash$, and a more detailed rendering of Heyting's derivation, by his own conventions, would be

[2.14] $\vdash\vdash \cdot a \supset \cdot b \supset a$

[2.12] $\vdash\vdash \cdot a \supset \cdot b \supset a : \supset : a \wedge b \supset \cdot b \supset a \cdot \wedge b$

$ \vdash \cdot a \wedge b \supset \cdot b \supset a \cdot \wedge b$

[2.15] $\vdash\vdash \cdot b \wedge \cdot b \supset a \cdot \supset a$

The axioms in use here are

[2.14] $\vdash\vdash \cdot b \supset \cdot a \supset b$

[2.12] $\vdash\vdash \cdot a \supset b \cdot \supset \cdot a \wedge c \supset b \wedge c$

[2.15] $\vdash\vdash \cdot a \wedge \cdot a \supset b \cdot \supset b$

A complete proof without abbreviations has 11 lines, with the formulas written for clarity with parentheses. I have organized it so that each rule instance is preceded by its premises in a determinate order. [1.3] is implication elimination, [1.2] conjunction introduction, and the rest are axiom instances:

1. [2.14] $\vdash a \supset (b \supset a)$
2. [2.12] $\vdash (a \supset (b \supset a)) \supset (a \wedge b \supset (b \supset a) \wedge b)$

3. [1.3] $\vdash a \wedge b \supset (b \supset a) \wedge b$

4. [2.11] $\vdash (b \supset a) \wedge b \supset b \wedge (b \supset a)$

5. [1.2] $\vdash (a \wedge b \supset (b \supset a) \wedge b) \wedge ((b \supset a) \wedge b \supset b \wedge (b \supset a))$

6. [2.13] $\vdash (a \wedge b \supset (b \supset a) \wedge b) \wedge ((b \supset a) \wedge b \supset b \wedge (b \supset a)) \supset$

$$(a \wedge b \supset b \wedge (b \supset a))$$

7. [1.3] $\vdash a \wedge b \supset b \wedge (b \supset a)$

8. [2.15] $\vdash b \wedge (b \supset a) \supset a$

9. [1.2] $\vdash (a \wedge b \supset b \wedge (b \supset a)) \wedge (b \wedge (b \supset a) \supset a)$

10. [2.13] $\vdash ((a \wedge b \supset b \wedge (b \supset a)) \wedge (b \wedge (b \supset a) \supset a)) \supset (a \wedge b \supset a)$

11. [1.3] $\vdash a \wedge b \supset a$

A reader who undertakes to turn Heyting's proof outlines into formal derivations will soon notice two things:

1. There is so much work that one lets it be after a while.

2. The way to proceed instead becomes clear because one starts reading Heyting's axioms and theorems in terms of their intuitive content.

Formally, axioms or previously proved theorems of the forms $a \supset b$ and a lead to another theorem b by a step of implication elimination. In practice, one sees implications $a \supset b$ as rules by which b can be concluded whenever a is at hand. The situation is particularly tempting if the major premiss is an axiom such as $a \supset a \vee b$. The step to deleting the axiom and implication elimination, to conclude $a \vee b$ directly from a, is short, as in:

1. $\vdash\vdash a \supset a \vee b$

2. $\vdash a$

3. [1.3] $\vdash a \vee b$

It would suffice to delete line 1 (as Frege actually did) and to change the reference to rule [1.3] into an explicit disjunction introduction rule. Thus, with the experience of actually using Heyting's logical calculus for proofs, one reads his first theorem $a \& b \supset a$ as: From $a \& b$ follows a. Axioms such as the ones with a disjunction, $a \supset a \vee b$, $b \supset a \vee b$, and $(a \supset c) \supset ((b \supset c) \supset (a \vee b \supset c))$ turn into what are now the familiar natural deduction rules of disjunction.

Heyting gave also a formal presentation of intuitionistic predicate logic, in the tradition of axiomatic logic, in his three-part article of

1930. The axioms and rules for the quantifiers, found in the second part of his article, were taken from the book by Hilbert and Ackermann (1928). At that time, Brouwer had already presented many properties of intuitionistic logic in his papers, first of all of propositional logic, but even of predicate logic, especially in connection with his famous counterexamples to classical logical laws.

(B) BROUWER'S AND HEYTING'S COUNTEREXAMPLES. When Brouwer developed intuitionistic mathematics, mainly in the 1920s, he did not create any separate formal intuitionistic logic but had otherwise a perfect command of what was to count as intuitionistically valid in logic. Thus, Brouwer's 1928 paper *Intuitionistische Betrachtungen über den Formalismus* contains the, in fact quite astonishing, insight that the predicate-logical formula $\neg\neg\forall x(A(x) \vee \neg A(x))$ is not intuitionistically valid. First, Brouwer makes note of the "intuitionistic consistency of the law of excluded middle"; that is, that the assumption of the inconsistency of excluded middle, $\neg(A \vee \neg A)$, leads to an "absurdity" by which $\neg\neg(A \vee \neg A)$ holds intuitionistically (p. 50). Then he shows by induction that the principle can be generalized to any finite "combination of mathematical properties," as expressed by $\neg\neg((A_1 \vee \neg A_1)\&\ldots\&(A_n \vee \neg A_n))$. However, in Brouwer's terminology, "it turns out that the ... multiple law of excluded middle of the second kind [for an arbitrary instead of finite collection of mathematical properties] possesses no consistency." This is just Brouwer's way of expressing the intuitionistic failure of $\neg\neg\forall x(A(x) \vee \neg A(x))$. The counterexample consists of points of the intuitionistic unit continuum for which $\neg\neg\forall x(A(x) \vee \neg A(x))$ holds only if one of $\forall x\, A(x)$ and $\forall x \neg A(x)$ holds, the latter of which need not be the case (p. 52). The counterexample comes from the bar theorem by which it is not possible to divide the unit continuum in any nontrivial way. An accessible brief explanation of the bar theorem and its special case the fan theorem is found in Coquand (2004).

The intuitionistic failure of what is usually called "the double negation shift" $\forall x \neg\neg A(x) \supset \neg\neg\forall x\, A(x)$ follows from Brouwer's counterexample to $\neg\neg\forall x(A(x) \vee \neg A(x))$. The law $\forall x \neg\neg(A(x) \vee \neg A(x))$ is easily proved, for we get by propositional logic that $\neg\neg(A(x) \vee \neg A(x))$ is provable with x arbitrary. Were the double-negation shift an intuitionistic theorem, we could shift the double negation ahead of the universal quantifier in $\forall x \neg\neg(A(x) \vee \neg A(x))$ to conclude the formula $\neg\neg\forall x(A(x) \vee \neg A(x))$, against the unprovability of the double negation of the "universal" excluded middle.

In Heyting's original paper on intuitionistic logic of 1930, *Die formalen Regeln der intuitionistichen Logik*, there are very few formal results. One such result is the independence of the propositional axioms, another the unprovability of $A \vee \neg A$. For the rest, Heyting just states things such as the "correctness" of the following (p. 44; here and later, the standard notation of today is used for uniformity):

$$\neg A \vee B \supset (A \supset B)$$

$$(A \supset B) \supset \neg(A \& \neg B)$$

$$A \vee B \supset \neg(\neg A \& \neg B)$$

The claim of correctness can be shown by formal derivations of these formulas in intuitionistic logic. However, when Heyting states that "the converses are all unprovable," there is not much at all to back up this claim. We could, say, consider for the last mentioned the special case of the converse, with $B \equiv \neg A$:

$$\neg(\neg A \& \neg\neg A) \supset A \vee \neg A$$

The formula $\neg(\neg A \& \neg\neg A)$ is intuitionistically provable, so by the unprovability of excluded middle, even $\neg(\neg A \& \neg B) \supset A \vee B$ must be unprovable. The unprovability of the converse of the first formula is seen similarly, by setting $B \equiv A$ in it, with the resulting formula $(A \supset A) \supset \neg A \vee A$, and the second is seen by setting $A \equiv \neg\neg B$. Heyting doesn't make note of these possibilities.

On p. 50, we find the following formula, with an indication of a formal proof:

$$(A \vee \neg A) \supset (\neg\neg A \supset A)$$

"When the law of excluded middle holds for a definite mathematical proposition A, then even the reciprocity of the complementary species holds for A." This observation is followed by another, "This theorem cannot be inverted," left to Heyting's authority.

The relation between excluded middle and double negation is somewhat involved. Heyting, like Brouwer, sees that if $\neg\neg A \supset A$ is assumed as a general logical law, the law of excluded middle follows, simply by the intuitionistic derivability of $\neg\neg(A \vee \neg A)$, to which the double negation law is then applied, as in Brouwer (1928, somewhat hidden in the "fourth insight," p. 49). Still, $(\neg\neg A \supset A) \supset A \vee \neg A$ is not a theorem of intuitionistic logic.

The preceding discussion makes it clear that Brouwer and Heyting had a deep grasp of the properties of intuitionistic logic, but they had little means of proving their points.

In the second installment of Heyting's 1930 essay, with the slightly modified title *Die formalen Regeln der intuitionistichen Mathematik*, the following are proved in intuitionistic predicate logic (p. 65):

6.77. $\exists x \neg\neg A(x) \supset \neg\neg\exists x\, A(x)$

6.78. $\neg\neg\forall x\, A(x) \supset \forall x \neg\neg A(x)$

The proof sketches for these are followed by a passage that we quote in full. It contains the first use of Brouwer's notion of *choice sequences* in logic:

Remark. The inverses of both of the above formulas are not provable. For 6. 77, this follows easily from the meaning of $\exists x$. For 6. 78, we show it by an example already used by Brouwer (Math. Ann. 93, p. 256), of a set A of all choice sequences that consist of only the signs 1 and 2, with 2 always followed by 2. We associate to each natural number an element of A in the following way: To 1 belongs the sequence of ones throughout; to 2 belongs 2222..., to 3 belongs 1222..., to 4 belongs 1122..., etc.

Let a mean the sentence: "x is associated to a natural number"; let $\forall x\, a$ mean: "a holds for each element of A." Then $\forall x \neg\neg a$ holds, for were an element of A not associated to any number, there would have to occur in it neither ever, nor once, a 2. However, even $\neg\forall x\, a$ holds, for if there were a natural number associated to each element of A, there would be a natural number z such that, for each two elements that are equal in the z first signs, the same number would be associated (Brouwer, Math. Ann. 97 p. 66); so the same number would be associated to all elements that begin with z ones.

However, $\forall x \neg\neg a \,\&\, \neg\forall x\, a$ is by 4.521 [a previous derived formula] incompatible with the inverse of 6.78.[2]

What kinds of principles about choice sequences are hidden in this argument?

(C) HEYTING'S COUNTEREXAMPLE IN DETAIL. We now consider Heyting's direct counterexample to the double-negation shift in detail. The set of choice sequences S consists of infinite sequences of 1's and 2's with the condition that whenever there is a first occurrence of 2, all successive members of the sequence are 2's. The variables $x, y, z, \ldots$ range over S, and $n, m, k, \ldots$ over $\mathcal{N}$. The notation $\bar{x}_k$ stands for the initial segment of x of length k.

If we add 0 as a root element, the choice sequences can be depicted as branches in the following tree:

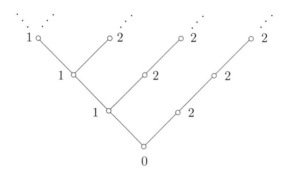

---

[2] The German original reads: *Bemerkung.* Die Umkehrungen der beiden vorstehenden Formeln sind nicht beweisbar. Für 6.77 folgt dies leicht aus der Bedeutung von (Ex). Für 6.78 zeigen wir es an dem schon von Brouwer benuzten (Math. Ann. 93, S. 256) Beispiel der Menge A aller Wahlfolgen, welche nur aus den Ziffern 1 und 2 bestehen, während nach einer 2 immer wieder eine 2 folgt. Wir ordnen jeder natürlichen Zahl ein Element von A zu in folgender Weise: Zu 1 gehört die Folge von lauter Einsen; zu 2 gehört 2222..., zu 3 gehört 1222..., zu 4 gehört 1122..., usw.

a bedeute den Satz: "x ist einer natürlichen Zahl zugeordnet"; $(x)a$ bedeute: "a gilt für jedes element von A." Dann gilt $(x)\neg\neg a$, denn wenn ein Element von A keiner natürlichen Zahl zugeordnet wäre, so müsste in ihm weder niemals, noch einmal eine 2 auftreten. Es gilt aber auch $\neg(x)a$, denn wenn jedem Element von A eine natürliche Zahl zugeordnet wäre, so gäbe es eine solche natürliche Zahl z, dass je zwei Elementen, die in den ersten z Ziffern übereinstimmen, dieselbe Zahl zugeordnet wäre (Brouwer, Math. Ann. 97 S. 66); also wäre allen Elementen, die mit z Einsen anfangen, dieselbe Zahl zugeordnet.

$(x)\neg\neg a \wedge \neg(x)a$ ist aber nach 4.521 unverträglich mit der Umkehrung von 6.78.

Next we define a function f over $\mathcal{N}^+$ as follows:

$$f(1) = \langle 111...\rangle$$
$$f(2) = \langle 222...\rangle$$
$$f(3) = \langle 122...\rangle$$
$$f(4) = \langle 1122...\rangle$$

$$\vdots$$

$$f(n) = \langle \underbrace{1...1}_{n-2}22...\rangle$$

Property $A(x)$ (Heyting's a) is: *To x is associated a natural number.* In terms of f, we can write $A(x) \equiv \exists n. f(n) = x$.

Lemma 1. $\forall x \neg\neg \exists n. f(n) = x$.

Proof. Assume $\exists x \neg \exists n. f(n) = x$. With y the eigenvariable in existence elimination, we have the assumption $\neg \exists n. f(n) = y$.

Assume 2 occurs in y. *Then it can be decided* when a first occurrence of 2 takes place: $y = \langle \underbrace{1,...,1}_{k}, 2, 2,...\rangle$. Now $f(k+2) = y$ against assumption.

Therefore, 2 does not occur in y, and $y = \langle 1, 1,...\rangle$ and $f(1) = y$ against assumption.

Conclusion: $\neg \exists x \neg \exists n. f(n) = x$, or equivalently, $\forall x \neg\neg \exists n. f(n) = x$.

QED.

There are two specific points here rendered in *italics*. First, if the digit 2 occurs in a choice sequence from the collection $\mathcal{S}$, it is assumed decidable when it occurs for the first time. Second, if it doesn't occur, it is concluded that all digits must be 1's. What sort of principles about choice sequences warrant these steps? First, if there is 2 in the sequence y, it can be found. Nothing, however, gives a bound on how far one has to inspect the initial segments of y. The second principle is that if 2 does not occur in y, then $y = \langle 1, 1,...\rangle$. As there are just two alternatives, this seems a justified step.

Heyting's argument somehow seems to go in two directions. First, there is a sequence associated to each natural number. Then he considers sequences and asks what numbers are associated to them.

Easy "Brouwerian counterexamples" can be used to show that not every sequence constructed by the choice conditions stipulated need have an associated natural number. Example: Let the first member of g be 1 and the nth member 1 if $2n$ is the sum of two primes, and 2 otherwise. As long as Goldbach's conjecture remains unsettled, no number is associated to g.

Lemma 2. $\neg \forall x \exists n. f(n) = x$.

Proof. Assume $\forall x \exists n. f(n) = x$. By the bar theorem,

$$\exists k \forall y \forall z (\bar{y}_k = \bar{z}_k \supset f(k) = y \, \& \, f(k) = z)$$

In particular, with $\bar{y}_k = \bar{z}_k = \underbrace{\langle 1, \dots, 1 \rangle}_{k-2}$, $y = z$, but this need not be so.

Therefore, $\neg \forall x \exists n. f(n) = x$. QED.

The fan theorem is actually sufficient for the result, because the tree of choice sequences pictured has a binary branching.

Theorem 3. $\forall x \neg\neg A(x) \supset \neg\neg \forall x \, A(x)$ *is not intuitionistically valid.*

Proof. Let $A(x)$ be $\exists n. f(n) = x$. By lemma 1, $\forall x \neg\neg \exists n. f(n) = x$, and by lemma 2, $\neg \forall x \exists n. f(n) = x$. QED.

Heyting's article established intuitionistic logic as a separate discipline in 1930. Choice sequences were taken into use in studying its properties right away, but this aspect seems not to have been discussed in previous literature, unless I have missed something. It is not easy to judge what Heyting's arguments with choice sequences amount to in detail. At least we have that the space of choice sequences is not discrete, for the equality of two choice sequences cannot be decidable, as the Goldbach example shows.

(D) **FAILED RECOGNITION OF A CRUCIAL FACT.** There is a letter from Gentzen to Heyting, written 23 January 1934, in which Gentzen writes as follows (the letter is included in the collection Gentzen 2017):

My dissertation will appear in the Mathem. Zeitschrift under the title "Untersuchungen über das logische Schliessen." I prove therein a quite general theorem about intuitionistic and classical propositional and predicate logic [the cut elimination theorem]. The decision procedure for intuitionistic propositional logic results as a simple application of this theorem. One can also show with it the intuitionistic unprovability of simple formulas of predicate logic, such as $(x)\neg\neg Ax . \supset \neg\neg(x)Ax$. I have not studied how far, in the end, one could go. I am now working with the proof of the consistency of analysis that has been since 2 years my real aim.

Gentzen must have studied in detail Brouwer's 1928 paper as well as Heyting's 1930 work in which the double-negation shift and the intuitionistic unprovability of $\neg\neg(x)(Ax \vee \neg Ax)$ appear. Gentzen obtained unprovability results syntactically from an analysis of cut-free derivations in sequent calculus, thereby showing that these results belong to pure intuitionistic logic, rather than depending on the intuitionistic theory of real numbers. Heyting was initially enthusiastic about natural deduction, as is shown by his series of Dutch papers from 1935 on (see von Plato 2012 for details). It is difficult to understand why he later completely ignored these developments. Imagine what it would mean if things were other than envisaged by Gentzen: The use of intuitionistic reals would be essential for showing the failure of classically provable formulas of predicate logic. Would not intuitionistic predicate logic be incomplete if that were the case? So, why wasn't Heyting positively alarmed by Gentzen's sensational discovery of a method of syntactic unprovability in intuitionistic predicate logic?

* * *

The discussion of Wittgenstein and his students in this chapter is out of all reasonable proportions in comparison to their altogether rather modest achievements in logic and foundational study. Maybe I can be excused somewhat by personal circumstances, having shared an office with Wittgenstein experts for years at an institution in which his literary heritage was kept, and holding the chair of one of the three literary executors of his will.

As with Frege, the secondary literature on Wittgenstein knows no limits, but remarks similar to those on Frege studies apply: My method for reading accounts of Wittgenstein's logic in the *Tractatus* was to check if the categorical absence of the notion of inference was noted as one essential feature, and, most importantly, if Wittgenstein's failure to understand quantificational logic was adequately presented. Not a single account passed this test, but again, I haven't read all of this literature.

7
THE GÖTTINGERS

7.1. HILBERT'S PROGRAM AND ITS PROGRAMMERS

(A) FROM HILBERT'S PROBLEMS TO HILBERT'S PROGRAM. David Hilbert presented his famous list of open mathematical problems at the international mathematical congress in Paris in 1900. First on the list was Cantor's continuum problem, the question of the cardinality of the set of real numbers. The second problem concerned the consistency of the arithmetic of real numbers; that is, of analysis and so on. These problems are generally recognized and have been at the center of foundational research for more than a hundred years, but few would be able to state how Hilbert's list ended—namely with a twenty-third problem about the calculus of variations—or so it was thought for a hundred years until some notes in Hilbert's hand were found in old archives in Göttingen that begin with:

> As a 24th problem of my Paris talk I wanted to pose the problem: criteria for the simplicity of proofs, or, to show that certain proofs are simpler than any others. In general, to develop a theory of proof methods in mathematics.

The 24th problem thus has two parts: a first part, about the notion of simplicity of proofs, and a second one that calls for a theory of proofs in mathematics. Just like the problems that begin the list, what can be named *Hilbert's last problem* has been at the center of foundational studies for a long time.

When Hilbert later started to develop his *Beweistheorie* (proof theory), its aims were much more specific than the wording of the last problem suggests. He put up a program that aimed to save mathematics from the threat of inconsistency, by which one would also solve the foundational problems for good. These problems had surfaced already by 1899, at the time of Hilbert's *Grundlagen der Geometrie*.

Hilbert's new start with proof theory is found in his 1917 talk in Zurich titled "Axiomatisches Denken." There we find enlisted the following foundational problems, with just the numbering added to Hilbert's formulations (p. 153):

1. The question of the *solvability in principle of any mathematical question.*
2. The problem of a subsequent *controllability* of the result of a mathematical investigation.
3. The question of a *criterion of simplicity* of mathematical proofs.
4. The question of the relation between *content and formalism* in mathematics and logic.
5. The question of the *decidability* of a mathematical question through a finite number of operations.

This list can be profitably compared to the list from 1900 in the beginning of Chapter 2. The third problem is a later echo of Hilbert's last problem of 1900. Consistency was the main problem then; now it is articulated into "a great domain of difficult epistemological problems of a mathematical character," of which the preceding list is representative. With problem 5, Hilbert is very careful to distinguish an indirect proof of finiteness, one that "cannot be reformulated so that we obtain a bound that can be given," from a direct proof that gives "a finite number that is below a limit that can be given before the computation" (p. 154).

With the axiomatic approach, in the lack of an intuitive interpretation of the domain of objects, the principles of proof would have to be made explicit. In Hilbert's early work, there is clearly the idea that the mathematical parts of axiomatizations, as in geometry or real arithmetic, have been laid down, and that the rest is logical reasoning from the axioms. In the Paris lecture of 1900, he emphasizes repeatedly that such reasoning, i.e., mathematical proof, is by its nature something finite: "This requirement of logical deduction by means of a finite number of processes is simply the requirement of rigor in reasoning." Similarly with problems: "Their solution must follow by a finite number of purely logical processes." Hilbert's early work on foundations, as in the 1905 paper *Über die Grundlagen der Logik und Arithmetik*, suffered from a lack of precise logical machinery. He saw the possibility to formalize even the logical processes of proof rather late, by 1917–18, when Russell's *Principia Mathematica* became known in Göttingen. In his paper on axiomatic thinking, Hilbert described the *Principia* in the following terms (p. 153):

> This way, prepared since a long time and not least by the profound investigations of Frege, has been pursued most successfully by the astute mathematician and logician Russell. One could see in the completion of this grandiose Russellian enterprise of *axiomatization of logic* the crowning of the task of axiomatization as a whole.

(B) THE HILBERT SCHOOL. Around 1917–18, Hilbert's work on logic and arithmetic took a new turn through the Russellian influence (as emphasized in Mancosu 2003). As further intimated earlier, a study of the *Principia Mathematica* in Göttingen made Hilbert realize, as the editors of Hilbert's recently published lectures put it, that "the axiomatic method he had developed in the work on geometry in the late 1890s could be extended to the logic of the *Principia* and that the latter could provide the foundation for all of mathematics" (Hilbert 2013, p. 35). The foundation should proceed according to a program set up by Hilbert, with the foundational questions listed above in the center.

Bernays describes the role of the *Principia* as follows (1935, p. 201):

> The thing delivered by this work is the elaboration of a surveyable system of conditions for a deductive development of logic and mathematics done together, as well as a demonstration that this construction actually succeeds.

A recent essay by Reinhard Kahle, "David Hilbert and the *Principia Mathematica*," stresses again how enormously impressed Hilbert was by Russell's rendering of Frege's logic. At last, clarity was brought to inferences with the quantifiers, though the absorption of the novelties took time. For example, Hermann Weyl's 1918 book *Das Kontinuum* is still quite muddled about the quantifiers, as we shall see in a while.

Hilbert's visit to Zurich led to the employment of Paul Bernays in the newly conceived Hilbert program. Bernays soon became the driving force behind the formal developments in logic and arithmetic in Göttingen, as is shown by his masterly *Habilitationsschrift* of 1918 on Russell's logic. Thanks to him, there was by 1928 a finished formal system for predicate logic that little by little put an end to all misunderstandings about the quantifiers and served as a logical foundation of arithmetic. The image one gets from publications of the time is somewhat different. The work of Hilbert, Bernays, and Wilhelm Ackermann of the 1920s on the logic of the *Principia Mathematica* is presented in the following articles: Hilbert (1922, 1923), Ackermann (1924), Bernays (1926), Hilbert (1927), and Hilbert (1931b). The paper by Bernays was drawn from his *Habilitationsschrift* and included the discovery that one of the propositional axioms of the *Principia* is in fact a theorem, as well as proofs of independence of the remaining propositional axioms through valuations that have more than two truth values.

Around 1918, when work based on Russell began in Göttingen, Russell's axioms with disjunction and negation as primitives were used. By the mid-1920s, Hilbert and others in Göttingen gave axiomatizations for all the standard connectives separately. The motivation for the connectives was the same as in axiomatic studies in geometry: to separate the role of the basic notions, especially negation. This move proved its worth when the axioms of intuitionistic logic were figured out, as in Heyting (1930). Bernays had in fact found the right axioms already in 1925, as he wrote in a letter to Heyting (found in Troelstra 1990).

(C) LOGIC IN GÖTTINGEN PRIOR TO THE PRINCIPIA. Before proceeding further with the development of logic in Göttingen, I shall describe through an example the state of the art prior to the Russellian impact, namely through a discussion of Hermann Weyl's treatment of arithmetic in his well-known booklet *Das Kontinuum* dated 1917 and published in 1918. Weyl had been studying and teaching in Göttingen until close to the First World War. He mentions the *Principia*, but only through a reference to its type theory (p. 35).

Weyl's treatise begins in a way reminiscent of Wittgenstein's *Tractatus*:

> A *judgment* asserts a *state of affairs:* If this state of affairs obtains, the judgement is *true*, otherwise it is untrue.

Simple or *primordial* judgment schemes have free variables (the German is "Leerstellen"), as in $E(x)$ and $U(x\,y)$. These lead to *combined* schemes, through the operations of negation, conjunction,

disjunction, substitution in argument places, and existential quantification. The last one is written as in $E(*)$ and read as *there is an object x
such that E(x)*. Universality is defined as the inexistence of a negating
counterexample.

A *proper judgment* is one that has no free variables. It obeys classical
logic, such as the laws of excluded middle, contradiction, and double
negation, written in Weyl's notation as (p. 10)

$$U + \overline{U} \qquad \overline{U \cdot \overline{U}} \qquad \overline{\overline{U}} = U$$

Weyl's example of a quantified judgment reveals the awkwardness of
his variable-free existential quantifier notation. The task is to write a
typical mathematical statement, *there is for every x a y such that U(x y)*.
Weyl posits (p. 7)

$$U(x *) = A(x)$$
$$\overline{A}(x) = B(x)$$
$$\frac{B(*)}{B(*)}$$

The last formula, an abbreviation for something like $\overline{U}(**)$, can be read
with some good will as $\neg \exists x \neg \exists y U(x\, y)$.

Arithmetic is based on a single relation written as $\mathcal{F}(n, n')$, a relation
that holds precisely when n' is the *successor* of n, with the natural
numbers beginning with 1. The number 1 is characterized by the
property that it is not the successor of any number (p. 9):

$$\overline{\mathcal{F}(*, x)} = I(x)$$

The uniqueness of 1 is just given "as a fact." Next, 2 is characterized by

$$I(y) \cdot \mathcal{F}(y, x) = \mathcal{F}_2(y, x) : \quad \mathcal{F}_2(*, x) = II(x)$$

"Analogously 3, 4, etc" writes Weyl, not realizing that the "etc" hides
an inductive step in his definition of numbers.

The logical definition of numbers in the style of the doctrine
of numerical quantity (Anzahllehre) is not put to any real use in
Weyl: later in the book, arithmetic operations are defined recursively,

with a reference to Dedekind (1888), and the basic laws are proved inductively (pp. 39–44). Weyl writes that "the elementary truths about numbers" can be proved logically through complete induction from the following two axioms: "there is to every number a unique successor, and every number except 1 has a unique immediate predecessor" (p. 44).

Logical inference is treated in the same somewhat cavalier way as logical structure, again reminiscent of the *Tractatus* (pp. 9–10):

> There are among the categorical judgments [without free variables] ones that we recognize as true purely on the basis of their logical structure— independently of the category of objects in question, what the simple properties and relations mean, and what objects have been used to 'fill in' by principle 5 [substitution for free variabes]. These judgments are true purely because of their formal (logical) construction (so that they don't possess any 'material content'). We shall call them *(logically) self-evident*.

Judgment V is a *logical consequence* of judgment U if $U \cdot \overline{V}$ is absurd, i.e., if $\overline{U \cdot \overline{V}}$ is self-evident. Then, if $\overline{U \cdot \overline{V}}$ and U are self-evident, even V is. Inference consists in the composition of such elementary steps. Thus, Weyl's only rule of inference is propositional, and his quantification theory is therefore incomplete.

In mathematics, logical inference proceeds from axioms that "are recognized as true by an immediate insight" (p. 10). Weyl contrasts his view with one in which axioms are taken as "posits" and theorems are mere consequences of the axioms in a "hypothetico-deductive game" (p. 12). He makes two observations, one a conceptual error, the other intriguing. First is what we would call completeness, namely "the conviction that all categorical, generic, and true judgments ... can be derived from the axioms by logical inferences is an article of scientific *faith*." Should a proof of completeness succeed, "truth and falsity could be decided by a definite method of decision ('in finitely many steps'): mathematics would be, in principle, *trivialized*." Here Weyl makes the same errors as some commentators on Hilbert: a mere formalization, even a complete one, need not give any decision method, because of the lack of "a limit that can be given before the computation," as Hilbert put it in 1917.

Weyl's second point also relates to completeness: The hypothetico-deductive game view relies on completeness, but this is dubious. The example is Dedekind's theory of real numbers in which $\alpha < \beta$ holds for two reals if there exists a rational r such that $\alpha < r$ and $r < \beta$ (p. 12):

> The judgments $\alpha < \beta$ and $\beta \leqslant \alpha$ do not, however, build complete alternatives, because it can very well happen that neither the existence nor the inexistence of such a rational number r follows from the arithmetical axioms. The view under discussion can be carried through only if one knows: The axioms are *consistent* and *complete* in the sense that one and only one of two 'contradictory' categorical judgments U and $\overline{U}$ is a logical consequence of the axioms. That, however, we don't *know* (even if we perhaps believe in it).

7.2. LOGIC IN GÖTTINGEN

(A) LOGICAL AXIOMS IN THE STYLE OF GEOMETRY. In Bernays (1918), a step back from the relativism of the Russellian axiomatizations had been taken, in that all the standard connectives, conjunction, disjunction, implication, and negation, were present. Bernays hoped to pinpoint the role of negation in logical axiomatics, motivated by the possibility of using negation and any one of conjunction, disjunction, or implication as basic connectives. He asks (p. 19) whether "it is possible to set up a negation-free axiom system from which all negation-free formulas provable in our calculus, and only these, can be derived." A negation-free system is one in which conjunction and implication are "not abbreviations, but symbols for basic operations ... the question has a positive answer" (ibid.). Instead of giving the answer, Bernays sets out to derive the fourth of Russell's axioms and to show that the rest are mutually independent. In the latter part of the *Schrift*, he explores various possibilities for replacing axioms by rules, for example, one that is rather peculiar (p. 47): There is just one axiom, and among the rules there is one that allows replacing a subformula of the form $\alpha \vee \beta$ by $\beta \vee \alpha$.

The formal derivations are written in a linear form, with a stacking of the premisses of the rule of detachment before an inference line.

Disjunction is just concatenation:

$$\begin{cases} \overline{\alpha}\ \beta \\ \alpha \end{cases} \over \beta \qquad\qquad \begin{cases} \alpha \to \beta \\ \alpha \end{cases} \over \beta$$

The notation at the right is given as an alternative.

Formal derivations start appearing on p. 22 of Bernays' work. They always have the major premiss of the rule first, then the minor, and the conclusion below the inference line. Repeated applications of the rule are displayed so that this arrangement is kept. It of course soon happens that not all derivations can be arranged under the pattern, unless previously derived formulas are numbered and can be used as premisses in rule instances. A good example in this respect is the derivation of Russell and Whitehead's axiom 4 (p. 26), as in his table 5.1. The very first derivation of the paper is of the formula $X \to YX$, that is, of $\overline{X}(YX)$, by axioms 2, 3, and 5 (p. 22). The first line in the derivation comes from "basic formula 5 with XY substituted for X, YX for Y, and $\overline{X}$ for Z":

$$\begin{cases} (XY \to YX) \to \{\overline{X}(XY) \to \overline{X}(YX)\} \\ XY \to YX \qquad \text{(basic formula 3)} \end{cases} \over {\begin{cases} \overline{X}(XY) \to \overline{X}(YX) \\ \overline{X}(XY) \qquad \text{(basic formula 2)} \end{cases} \over \overline{X}(YX)}$$

A two-premiss rule of syllogism, the use of which is indicated by a double curly bracket, is shown to be derivable (p. 23):

$$\begin{cases}\begin{cases} \alpha \to \beta \\ \beta \to \gamma \end{cases} \\ \alpha \to \gamma \end{cases}$$

In the proof of derivability of this rule, the premisses appear as formulas assumed to be derivable, so it is in fact an admissible rule, a notion weaker than derivability. In the rest of the derivations, this rule replaces the rule of detachment.

The negationless axioms Bernays alludes to can be gathered from various sources. One place is the first volume of the *Grundlagen der Mathematik* of 1934 (p. 66), but earlier sources exist, such as Hilbert (1928, 1931a) and the earliest I have found that had the axioms for conjunction and disjunction, the doctoral thesis of Hilbert's student Wilhelm Ackermann (1924) (with whom he actually studied) will be discussed in the next Section. Zach (1999, p. 352) reports that the axioms presented here in table 7.1 stem from a lecture course Hilbert and Bernays taught jointly in 1922–23.

Table 7.1. The Hilbert-Bernays axioms

1. $A \rightarrow (B \rightarrow A)$
2. $(A \rightarrow (A \rightarrow B)) \rightarrow (A \rightarrow B)$
3. $(A \rightarrow B) \rightarrow ((B \rightarrow C) \rightarrow (A \rightarrow C))$
4. $A\&B \rightarrow A, \quad A\&B \rightarrow B$
5. $(A \rightarrow B) \rightarrow ((A \rightarrow C) \rightarrow (A \rightarrow B\&C))$
6. $A \rightarrow A \vee B, \quad B \rightarrow A \vee B$
7. $(A \rightarrow C) \rightarrow ((B \rightarrow C) \rightarrow (A \vee B \rightarrow C))$
8. $(A \rightarrow B) \rightarrow (\neg B \rightarrow \neg A)$
9. $A \rightarrow \neg\neg A$
10. $\neg\neg A \rightarrow A$

These axioms appear beautiful to anyone with experience in elementary logic. The first three are, in terms of Gentzen's sequent calculus, axiomatic equivalents to the rules of weakening, contraction, and cut. Next are what look like axiomatic equivalents to the rules of natural deduction for conjunction and disjunction, and then the axioms of negation. Some of the implicational axioms go back to Frege. Sometimes an axiom that corresponds to a rule of exchange in sequent calculus is seen, as in Hilbert (1923).

Bernays wrote in 1918 that there is an axiom system for his negation-free fragment of classical propositional logic. If so, axioms 1–7 of table 7.1 are not sufficient, for they axiomatize the negationless fragment of intuitionistic logic that falls short of the corresponding classical fragment, witness a purely classical negationless theorem such as $(A \supset B) \vee (B \supset C)$.

(B) HILBERT-ACKERMANN. The development of logic in Göttingen, excluding the work on separate axioms for each connective and what is known as Hilbert's ε-substitution method, is summarized in the Hilbert-Ackermann book of 1928, *Grundzüge der theoretischen Logik*, mostly written by Bernays. The axiomatization is a clear step back, counterbalanced by many novelties, which we shall present, after a look at the axiomatic deductive machinery.

Here is a summary of the 120-page first edition of Hilbert-Ackermann: Chapter 1 presents propositional logic with axioms from the *Principia Mathematica*, chapter 2 is a short presentation of monadic predicate logic, chapter 3 presents predicate logic, equality included, and the first half of chapter 4 presents higher-order logic, with examples of its use in attempts at defining the basic concepts of arithmetic in purely logical terms (§ 2), in set theory (§ 3), and in a study of the logical paradoxes (§ 4). The second half of chapter 4 is dedicated to type theory, including the "ramified" version and the axiom of reducibility.

The path to the axioms of propositional logic in Hilbert-Ackermann is a rather tortuous one. First, some propositional equivalences are ascertained by showing that the equivalent formulas in question always have the same truth values (p. 5). These are gathered into rules of logic that bear an analogy to algebra, used for the production of a conjunctive normal form (pp. 9–10). Finally, certain rules of inference are shown to be correct (p. 11), written here in modernized notation:

b1) $A \lor \neg A$ "is always correct"

b2) if A is correct, $A \lor B$ is correct

b3) if A is correct and B is correct, $A \& B$ is correct

As mentioned, Hilbert-Ackermann presents a system of classical logic with negation and disjunction as primitives in the propositional part. The axioms are written with disjunctions and implications, the latter with the standard classical definition $A \to B \equiv \neg A \lor B$.

The axioms, those for quantifiers included, are (Hilbert and Ackermann 1928, pp. 22 and 53):

Table 7.2. The Hilbert-Ackermann axioms

a) $A \lor A \to A$

b) $A \to A \lor B$

c) $A \lor B \to B \lor A$

d) $(A \to B) \to (C \lor A \to C \lor B)$

e) $(x)Px \to Ph$

f) $Ph \to (Ex)Px$

The rule of inference is *modus ponens*, and for the quantifiers there are two rules, to be detailed soon (p. 54).

Among the novelties of Hilbert-Ackermann are 1. a definition of derivability under assumptions in axiomatic logic, 2. a clear notion of derived rules, 3. explicit rules for the two quantifiers, and 4. a clear formulation of the completeness question of quantificational logic.

1. *Derivability under assumptions.* By the mid-1920s, it had become clear that axiomatic logic had to be extended from its original goal in Frege and Russell, which was to grind out more logical truths from the axioms of logic that are the given logical truths. Hilbert, Bernays, and Ackermann wanted to apply logic in mathematics by admitting mathematical axioms as premises in logical derivations. These axioms were not given truths but instead hypothetical elements that could turn out true in one application, and fail in another. In the second part of the 1920s, there thus evolved the idea of derivations from mathematical axioms. Hilbert and Ackermann's book of 1928 demonstrates clearly this change. The task of its § 9 (pp. 18–22) is to extend the pure logic of connectives to axiom systems:

> Let there be given a finite number of axioms $A_1, A_2, \ldots A_n$. The question whether a given propositional formula C is a logical consequence of these axioms can be always answered by the methods presented so far. Such is the case if and only if $A_1 \& A_2 \& \ldots A_n \rightarrow C$ is a logically generally valid formula.

The passage shows a clear awareness of the need for a *deduction theorem* when pure logic is extended to axioms that are treated as assumptions (*Voraussetzungen* in Hilbert and Ackermann, see their p. 18). The 1928 book was based on Hilbert's lectures held almost ten years earlier and elaborated by Bernays in later lecture series.

2. *Derivable rules.* A remarkable aspect of Hilbert-Ackermann is its clear awareness of the role of derivable rules (p. 24):

> In the derivation of formulas, it is recommendable to collect into *derived rules* certain operations that repeat themselves often . . . the proof of the rule consists in giving in a general form the procedure by which the passage is carried through in each single case by the basic rules.

Among such derivable rules, there is one by which $A \vee B$ can be concluded if A has been proved, and another that is the rule of cut in terms of sequent calculus.

3. *Rules for quantifiers*. Hilbert-Ackermann has the two familiar quantifier axioms $(x)F(x) \to F(y)$ and $F(y) \to (Ex)F(x)$ and the following two rules, even these the work of Bernays, as acknowledged by the authors (pp. 53–54):

> Let $B(x)$ be an arbitrary logical expression that depends on x, and A one that doesn't. If now $A \to B(x)$ is a correct formula, also $A \to (x)B(x)$ is. One obtains similarly from a correct formula $B(x) \to A$ the new $(Ex)B(x) \to A$.

Both rules are quite particular, reminiscent of the rules of sequent calculus. In the rule for existence, one has first that A follows under the assumption with an eigenvariable and then from existence. In the case of the first rule, if A is provable, the rule gives the simple rule of generalization, from $B(x)$ for x arbitrary to infer $(x)B(x)$. A similar dual rule for existence would be to instantiate $(Ex)A(x)$ by an eigenvariable, as in Skolem (1920), and even in Hilbert (1931b). In the preceding formulation, the rule does not guide the use of eigenvariables.

In Hilbert-Ackermann, both quantifiers are treated independently. Readers of the *Grundzüge* would not know the reasons that were somewhat particular. First of all, in Hilbert's first "post-Russell" paper of 1922 that aims explicitly at the justification of arithmetic, the logic is restricted to implication and universal quantification, with negation treated by taking in addition to equality $a = b$ also inequality $a \neq b$ as a primitive. The rules for universal quantification seem to be, even if the statement is anything but clear, universal generalization and instantiation, but no use is made of them (p. 167). The paper is on the whole a testimony to Hilbert's modest level of sophistication in purely logical questions.

Hilbert may have sought advice, for his next paper, of 1923, is much clearer. At the suggestion of Bernays, there is just one "transfinite axiom," i.e., one that contains quantifiers, written $A(\tau A) \to A(a)$. The reading is (p. 156): To each $A(a)$ is associated a determinate object

τA such that if $A(\tau A)$, then $A(a)$ for all objects a. Hilbert now writes "definitional axioms" for the quantifiers:

1. $A(\tau A) \to (a)A(a)$
2. $(a)A(a) \to A(\tau A)$
3. $A(\tau \overline{A}) \to (Ea)A(a)$
4. $(Ea)A(a) \to A(\tau \overline{A})$

The first two axioms give some sort of license for inference to and from free-variable propositions. Thus, the former would be better rendered as a rule of inference, rather than as an implication: Whenever $A(a)$ for an arbitrary a, $(a)A(a)$ can be concluded. Hilbert does not elaborate on the third and fourth axioms; however, if in 1 we put $\overline{A}$ in place of A and take contrapositions, we get

$$\overline{(a)}\overline{A}(a) \to \overline{\overline{A}}\ (\tau \overline{A})$$

The antecedent is classically equivalent to existence, and the double negation in the consequent can be deleted, which brings us Hilbert's definitional axiom 4. Axiom 3 is obtained similarly from axiom 2.

This method was used in Ackermann's 1924 paper, with the τ-operator for universality changed to its dual ε-operator for existence. The same is found in Hilbert's 1927 paper (Mancosu 1998, p. 460), with the axiom

$$A(a) \to A(\varepsilon(A))$$

Here, as Hilbert writes, "$\varepsilon(A)$ stands for an object for which $A(a)$ certainly holds if it holds for any object at all." The idea of such a variable binder already appeared in Peano (1889).

With the ε-operator, we now have the equivalences dual to axioms 1–4:

$$A(\varepsilon(A)) \leftrightarrow (Ea)A(a)$$

$$A(\varepsilon(\overline{A})) \leftrightarrow (a)A(a)$$

In Hilbert's last paper, the "Beweis des tertium non datur" of 1931, both quantifiers appear in a standard formulation, as in the Hilbert and Ackermann book.

4. *Completeness and decidability of quantificational logic.* Section 9 of chapter 3 shows that predicate calculus is consistent. Toward the end, two pages are used for showing that $(Ex)F(x) \rightarrow (x)F(x)$ is not derivable in the system (pp. 66–68). At the end of this proof, it is noted that predicate calculus is by the underivability of the said formula not syntactically complete, i.e., lacks what is called *Post completeness.* Finally, completeness in the usual sense is given as an open problem (p. 68):

> Whether the axiom system is complete in the sense that all logical formulas correct in each domain of individuals really are derivable in it, is a question still unresolved.

The last of Hilbert's foundational problems, decidability, is formulated in semantical terms (p. 73): to decide for each formula if it is valid resp. satisfiable. Further down, it is described as "the main problem of mathematical logic" (p. 77). Section 12 discusses special cases of the decision problem. It then gives the Löwenheim-Skolem theorem, turned upside-down, however (p. 80):

> *Every expression that is valid for denumerable domains has this property for every other domain.*

This formulation is known in model-theoretic literature as "the upward Löwenheim-Skolem theorem." Löwenheim's original formulation is described in Hilbert-Ackermann as "the dual formulation" by which each formula is either contradictory or satisfiable in a denumerable domain, all of it clearly the work of Bernays, who put his words in this way in an article submitted for publication in 1927, the Bernays-Schönfinkel work on the decision problem. A finitistic proof of Löwenheim's result would yield a semantical decision method. To what extent Hilbert understood that is hard to say.

After these explanations, completeness in the sense of the derivability of each valid formula is considered plausible ("zu vermuten"), which gives another formulation of the decision problem.

To what extent Hilbert presumed the decision problem to have a positive solution is seen from the following passage (p. 77):

> Whereas the decision problem was easily solved for the propositional calculus, finding a general decision procedure in the functional calculus is a still unsolved difficult problem.

The German words are "ein noch ungelöstes schwieriges Problem." Jacques Herbrand seems to have shared Hilbert's expectation that the problem could be solved, as is seen from the closing sentence of his thesis written in 1928, but he also prepared for the opposite in a later note from 1931 (Herbrand 1930 p. 188, 1931 p. 257):

> The solution of this problem would yield a general method in mathematics and would enable mathematical logic to play with respect to classical mathematics the role that analytic geometry plays with respect to ordinary geometry.

> Perhaps this is impossible; perhaps it can even be proved metamathematically that it is impossible.

Von Neumann probably had the former view. A few years later, when it was realized that standard arithmetic can be formalized with predicate calculus, decidability was as a matter of course taken as impossible. This is seen in a 1931 note by the very young Gerhard Gentzen and in the book by Hilbert-Bernays; we shall come to these in a while.

As said, chapter 4 of Hilbert and Ackermann's book goes over to higher-order logic. It contains two topics that illuminate something we have already discussed, and something to come.

The logicist definition of number. Section 2 of chapter 4 is dedicated to what Bernays called "Anzahllehre," the logicist doctrine of numerical quantity of Frege and Russell that we just met with Weyl. Numbers can be defined as properties of predicates: 0 is the property of predicates F given by the formula $(\overline{Ex})F(x)$, "there is no x that has F," as Hilbert-Ackermann explain, and abbreviated as $0(F)$; 1 is defined as a property of those F for which we have, with $\equiv(x,y)$ the two-place equality relation (p. 86):

$$(Ex)[F(x)\&(y)(F(y) \rightarrow \equiv(x,y))]$$

This formula is abbreviated as $1(F)$. There follows another definition, that of $2(F)$, with a negation on top of the first $\equiv$:

$$(Ex)(Ey)\{\overline{\equiv}(x,y)\& F(x)\& F(y)\&(z)[F(z) \rightarrow \equiv(x,z) \vee \equiv(y,z)]\}$$

Hilbert and Ackermann, or perhaps it is again just Bernays alone, write that "the importance of this representation of numbers depends on the fact that the higher-order predicates that form the numbers can be expressed to the full with the help of the logical symbols" (ibid.).

Next, the *equinumerosity* (Gleichzahligkeit) of two predicates F and G, with the notation $\text{Glz}(F, G)$, is defined as the existence of a one-to-one relation R between those objects for which F resp. G hold, a rather involved formula in second-order logic (p. 87):

$$(ER)\{(x)[F(x) \rightarrow (Ey)(R(x,y)\&G(y))]\&$$

$$(y)[G(y) \rightarrow (Ex)(R(x,y)\&F(x))]\&$$

$$(x)(y)(z)[(R(x,y)\&R(x,z) \rightarrow \equiv(y,z))\&$$

$$(R(x,z)\&R(y,z) \rightarrow \equiv(x,y))]\}$$

Two predicates F and G are defined to be *incompatible* (unverträglich), a second-order relation found already in § 1:

Unv(F, G) is to mean the same as $(x)(\overline{F}(x) \text{ v } \overline{G}(x))$, i.e., $(x)(\overline{F(x)\&G(x)})$.

Numerical equations can be expressed and proved within this purely logical formulation; for example, the equation $1 + 1 = 2$ now amounts to the following provable second-order formula (ibid.):

$$(F)(G)([\text{Unv}(F, G)\&1(F)\&1(G)] \rightarrow 2(F \text{ v } G))$$

The next level of abstraction introduces the general notion of a number, the third-level predicate Φ, for which the following is posed (ibid.):

$$(F)(G)\{(\Phi(F)\&\Phi(G) \rightarrow \text{Glz}(F, G))\&[\Phi(F)\&\text{Glz}(F, G) \rightarrow \Phi(G)]\}$$

With the abbreviation $\mathfrak{Z}(\Phi)$ (fraktur Z for *Zahl*) for this formula, we have (p. 88):

A number is a predicate of predicates Φ that has the property $\mathfrak{Z}(\Phi)$.

Written without abbreviations, this definition would be close to half a page in length. One cannot but bring to mind what Hilbert had stated at the beginning of his Paris problem list, that "what is clear and easily comprehensible attracts us, what is wicked frightens us away" (das Klare und leicht Fassliche zieht uns an, das Verwickelte schreckt uns ab). Moreover, there is no reward for the labor in the form of easy proofs, on the contrary.

The attempted definition of the concept of a number is a failure, however. When applied to a finite domain of n members, all numbers beyond n will be equated, a fact described as an "Übelstand" (a bad state of things). The only remedy suggested is an axiom of infinity, by which the original aim of a purely logical foundation of arithmetic is defeated.

There is a blind spot in this whole affair of the logicist attempt at defining numbers and the rest of the foundations of arithmetic in purely logical terms, which can be exposed as follows.

Say we want to prove $n + m = m + n$ and, to do this, let n and m be any given numbers. What do their definitions look like? There is no definition of $n(F)$ and $m(F)$ that would let us start to work with the expression $n + m$. We saw earlier how to define $0(F), 1(F)$, and $2(F)$, but Hilbert and Ackermann failed to add the required "etc." after these first three cases. A little pondering will lead to the insight that the ability to write a definition of $n(F)$ for any given n is based on an implicitly understood *recursive procedure*, the missing "etc" in Hilbert-Ackermann. Given $n(F)$, $n'(F)$ is formed by taking a fresh variable $x_{n'}$, then adding to the formula for $n(F)$ the parts $(Ex_{n'})$ and $F(x_{n'})$ and the negations $\overline{\equiv}(x_i, x_{n'})$ for $1 \leq i \leq n$, and so on in the appropriate places. When the inductive definition is supplied, the whole enterprise of a purely logical definition of the natural numbers collapses, for the properties of numbers are proved by an induction that corresponds to their buildup.

The second topic of interest in § 4 of chapter 4 in Hilbert-Ackermann is a discussion of three paradoxes that were instrumental in leading Gödel to discover the incompleteness of arithmetic, as detailed in the next chapter.

On the whole, with some hindsight, the topics beyond predicate logic with equality appear to be a rather strange choice. Why not present the elements of formal arithmetic? That was the almost exclusive topic beyond pure logic of the first volume of the *Grundlagen der Mathematik* to which Hilbert-Ackermann was meant to serve as a preparation, as stated in its preface. Perhaps the explanation is that Hilbert-Ackermann was confined to what was considered to be *pure logic*.

Hilbert-Ackermann had a second edition published in 1938, prepared by Ackermann. The main differences are that Gödel's completeness theorem for predicate logic is presented, results on the decision problem are added, and Church's proof of undecidability is mentioned. The part on type theory has been considerably shortened by leaving out the ramified version. There was a third edition in 1949, of which Haskell Curry, counted as a student of Hilbert's, wrote in a review (p. 265):

> Much water has spilled over the dam in the twenty-five years since this book was first published. This book is confined entirely to the classical logic; whereas we now know there are a variety of generalized systems, some with interesting applications, which have much the same relation to classical logic that various generalized geometries do to Euclidean. The new techniques developed by Gentzen are barely mentioned. Of course, it is quite true, as some insist, that one cannot do anything with those techniques that one cannot do with the older ones; just as one can go anywhere with a horse and buggy that one can reach with a Cadillac.

A fourth and final edition appeared in 1959, with an added section on intuitionistic propositional logic. An English translation of the second edition was published in 1950 with the title *Foundations of Mathematical Logic*.

The Hilbert-Ackermann book had a decisive effect on at least Gödel and Herbrand, both of whom studied it carefully, as we shall see in Chapter 8. It led Gödel to solve the completeness problem of predicate logic. The discussion of paradoxes was moreover a clear influence on the way to the incompleteness theorem. Carnap, on the other hand, was blind to the final formulation of the axioms and rules of classical predicate logic. In his widely read Königsberg address of 1930, on the logicist foundations of mathematics, he writes that there are in logic four propositional and two quantificational axioms, a rule of substitution, and a rule of implication, and he claims that "Hilbert and Ackermann have used these same axioms and rules of inference in their system"; not the whole truth, as that book was very explicit about the two rules of universal generalization and existential instantiation that must be added to the six axioms as in table 7.2.

(C) HILBERT AND BERNAYS: THREE BOOKS, THREE AUTHORS OR ONE?

Hilbert's lectures on logic after the impact of the *Principia* have been collected in the volume *David Hilbert's Lectures on the Foundations of Arithmetic and Logic 1917–1933*. Paul Bernays played a crucial role in this post-*Principia* era of Hilbert. He joined Hilbert in 1917 and soon developed a mastery of the formalism of the *Principia* that far superseded anything Russell and Whitehead had accomplished, with proofs of independence of the *Principia's* propositional axioms and then, obviously after a failed attempt to prove the independence of axiom IV, its derivation in the formal calculus. Here and there in Hilbert's papers and books on logic and foundations, the former always signed only by him, discoveries are credited to Bernays. A list, certainly just partial, is found in the next section. My general feeling after many years of study is that Bernays did not get the credit he deserved. He just got fired in April 1933, at the time the first volume of the big *Grundlagen der Mathematik* was in its final stages.

The textbook *Grundzüge der theoretischen Logik* was published under the name of David Hilbert and Wilhelm Ackermann in 1928. Reading Ewald and Sieg's introduction to the mentioned *Lectures* of Hilbert, we find that the book was in practice written by Bernays on the basis of Hilbert's 1917–18 lectures (Hilbert 2013, p. 49):

> But in fact virtually the whole book beginning with §10 of Chapter One is taken, often verbatim, from Part B of Bernays's 1917/18 typescript. §§1–9 of Chapter One are taken similarly from Bernays's typescript from the Winter Semester 1920.

Putting together §§1–9 and §10ff. gives "the whole book." Why is the second author Ackermann, not Bernays? As Ewald and Sieg note about Ackermann (p. 50), his "role seems to have been more that of textual editor than of co-author." The question of real authorship is left open, but an easy answer is at hand: Looking at the contents of the book, it presents the original *Principia*-style axiomatization of 1918, with disjunction and negation as primitive concepts. In the intervening ten years, the study of axiomatic logic was perfected by Bernays, in the form of separate groups of axioms for each of the connectives and quantifiers. Bernays' special aim was to single out the role of negation in logic, which had interesting side effects. As mentioned he succeeded in axiomatizing intuitionistic logic in 1925 by leaving out the double-negation axiom; nothing to publish by his standards, but the fact is shown by his complimentary letter to Heyting in 1930. At this stage of things, it would have been pointless for Bernays to publish a book with an antiquated approach to logic; therefore, enter Ackermann as Hilbert's textual editor.

Heyting had written his paper on the axiomatization of intuitionistic logic and offered it to the *Mathematische Annalen* in 1928, but he withdrew that plan when Hilbert threw Brouwer off the journal's editorial board. When the paper came out elsewhere in 1930, it became at once the standard presentation, well received by everyone. Whatever little Hilbert may have understood about intuitionistic logic, he certainly would not have welcomed such a brainchild of his archenemy Brouwer.

Five decades after the appearance of Hilbert-Ackermann, Bernays commented in interviews about the inadequacy of disjunction and negation as the basic notions in propositional logic in Hilbert's lecture course of 1917–18, as opposed to Frege's use of implication (cited in Hilbert 2013, p. 52):

> Yes, I discussed a lot with Hilbert right from the beginning—we were often of differing opinions. In formal respect, I went later back largely to Frege. That is even natural in a certain respect, if one wants to separate out the domain of pure logic of consequence [Folgerungslogik], for one cannot do it the way Russell does, namely so that he defines in advance implication through negation and disjunction.—Things stand so that the axioms that one finds in Russell, they have a more natural sense if one takes implication as a basic concept.

The ambivalence of Hilbert-Ackermann is implicitly noted even by Ewald and Sieg, who write on the one hand (Hilbert 2013, p. 2) that it "presents mathematical logic in its definitive modern form" and on the other hand (p. 23) that "this book does not reflect the proof-theoretic work of the 1920s."

The two volumes of the *Grundlagen* of 1934 and 1939 were written by Bernays. After the first volume was completed, proof theory was revolutionized by Gentzen's discovery of natural deduction and sequent calculus, by the proof of the consistency of arithmetic, and by the creation of ordinal proof theory. The second volume was as backward-looking in 1939 as Hilbert and Ackermann's book had been in 1928. It presents the first instance of what Haskell Curry described in 1963 as follows (p. 25):

> The German writers tend to shy away from the Gentzen technique and to devise ways of modifying the ordinary formulations so as to obtain its advantages without its formal machinery. This is much the same as if one attempted to develop group theory without introducing the abstract group operation.

The correspondence between Bernays and Ackermann shows how unwilling the latter was to rework the consistency proof of arithmetic in terms of the ε-calculus, and only repeated insistence on the part of Bernays led to the proof and to Ackermann's paper of 1940.

(D) PAUL BERNAYS: FROM ZURICH TO GÖTTINGEN AND BACK. Paul Bernays (1888–1977) studied in Berlin and Göttingen and obtained a doctorate under Edmund Landau with a thesis on number theory. Thereafter, he taught in Zurich, and became an assistant of Hilbert's after the latter had delivered a lecture on logic in Zurich in September 1917. Besides his work on proof theory, he is known for the Gödel-Bernays system of set theory and for his profound philosophical education and work that belongs to the *neo-Friesian* school of Leonard Nelson, an intriguing connection to a philosophy named after the fiercely anti-Semitic philosopher Jacob Fries.

After fifteen years at the University of Göttingen, Bernays was fired because he was a Jew, and in 1934 he moved to Zurich. His students in logic are even today unfairly counted as "Hilbert students," including Gentzen, Curry, Saunders MacLane, and several others. The difficult times Bernays had to face after 1933 are described in Menzler-Trott (2007, p. 101).

Personal merit seems not to have been an issue for Bernays. This aspect is at its most striking in the review of foundations of arithmetic Bernays wrote for the third volume of Hilbert's *Gesammelte Abhandlungen* in 1935, titled "Hilberts Untersuchungen über die Grundlagen der Arithmetik." It seems to have taken Bernays no effort to ascribe to Hilbert a great number of his own discoveries and results about the foundations of logic and arithmetic, beginning with "Hilbert's ε-substitution method." At most, he sometimes uses the passive mode, in unselfish, unlimited devotion.

There is at ETH-Zurich perhaps the most remarkable archive of documents of logic, the Bernays papers, including some five thousand letters and dozens of detailed notes for lecture courses from the late 1920s on. The latter are written in the obsolete Gabelsberger shorthand and are usually about 80 manuscript pages each. Two series about the foundations of mathematics were transcribed in the mid-1950s by Mr. Müller ("father of Prof. Dr. Gert H. Müller," as the catalogue states). Their contents can be described as follows. *Das mathematische Grundlagenproblem*, manuscript number Hs 973: 269, is a series of lectures from the Winter Semester 1942–43, consisting of 83 pages of transcription written in very light pencil. The discussion spans from Eudoxus in antiquity to Frege, with most of it dealing with the early developments. Another series of lectures with the same title was given in the Summer Semester of 1946 (Hs 973: 271), where now the emphasis is on the modern developments from Dedekind on.

The first results on logic after Bernays had joined Hilbert in Göttingen are contained in Bernays' *Habilitationsschrift* of 1918, and in elaborated versions of Hilbert's lectures, in fact often lectures that were given in tandem. These lectures have now been published in the mentioned volume that contains lectures from 1917 to 1933: first a series from 1917–18 with the title *Lectures on the Principles of Mathematics*; a second one, titled *Lectures on Logic*, from 1920; a third, titled *Lectures on Proof Theory*, from 1921–22 and 1923–24; and finally a fourth, titled *Lectures on the Infinite*, from 1924–25, 1931, and 1933.

Not all of Hilbert's lectures on logic and foundations have been preserved, and those that have been delivered to us come in different vestiges. Some are notes in Hilbert's hand, others are clear elaborations of the lectures, even typewritten, and prepared by Bernays. Still others are notes by participants in the courses. Such notes by Hellmuth Kneser are included in chapter 3 of the Hilbert lecture volume, and by comparing these to Bernays' elaborate version, one gets an impression of the original lectures.

As noted, Hilbert-Ackermann was in practice written by Bernays on the basis of Hilbert's 1917–18 lectures. The only direct comment about Hilbert-Ackermann on the part of Bernays that I have found is in the preface to the *Grundlagen*, where he writes that "the treatment differs from that of the book by Hilbert and Ackermann ... especially as regards the propositional calculus."

The expression *theoretische Logik* in the title of Hilbert-Ackermann strikes one as an oddity and is indeed nowhere to be found in Hilbert's lectures (using Amazon's "Search Inside This Book" function). In September 1927, some months before Hilbert signed the preface, Bernays delivered a talk in Göttingen with the title *Probleme der theoretischen Logik* (Bernays 1927a). The coincidence between its title and that or Hilbert-Ackermann, except for the first word, cannot be accidental. Bernays shows a witty sense of humor by starting his essay as follows:

> The theme of the talk as well as how it is named has been chosen in Hilbert's sense. As theoretical logic is indicated what is usually named symbolic logic, mathematical logic, algebra of logic, or logical calculus.

The talk and article were meant for the "Conference of German philologists and schoolmen," yet Bernays succeeds in introducing in a very accessible way the elements of logic. Propositional logic is explained through truth values, and then the rules of inference of logic are explained. As to them, Bernays makes the following memorable remark (p. 9):

> The rules of inference have to be chosen so that they eliminate logical thinking, for otherwise we would have to have logical rules anew for how the first rules are applied.

Axioms are given in separate groups for each of the connectives, with "the advantage that the separation of *positive logic* is made possible." When reading this, we should keep in mind what Bernays does not mention: that he had axiomatized intuitionistic logic by this method, five years before Heyting. I find that the message in the direction of Hilbert is subtle but clear: The purely classical Hilbert-Ackermann axioms present an outdated approach to logic.

A tentative list of Paul Bernays' formal discoveries from 1918 to 1933 includes:

1. A complete study of the propositional axioms of *Principia Mathematica* in the *Habilitationsschrift* of 1918, with partial publication in Bernays (1926). The possibility of an axiomatization with separate groups for each connective is introduced.

2. Formulation of principles of quantification with the τ-operation and its dual the ε-operation in 1922–23.

3. Axiomatizations of logic with separate groups for each connective.

4. Axiomatization of intuitionistic propositional logic in 1925.

5. Primitive recursive arithmetic with the uniqueness of recursively defined functions in place of the induction axiom in 1928.

6. The formal rule of existential elimination in 1928.

7. Axiomatization of classical predicate logic with just one operation, a generalization of the Sheffer stroke.

8. Provability of the Peano-Dedekind infinity axioms by suitable recursive definitions.

No similar list of Hilbert's formal discoveries in logic can be composed. One could perhaps mention the idea for a proof of the continuum hypothesis in 1925–26 in the paper on infinity, with its extension of recursive definition to constructive ordinals and the introduction of a hierarchy of functions of higher types. Even here, Hilbert ends his paper with thanks to Bernays, "for the valuable help, especially in what he has achieved with the proof of the continuum theorem." The extension to constructive ordinals is already seen in Ackermann (1924), Hilbert's doctoral student, whose real supervisor was Bernays. Hilbert thus appears to have been somewhat of an elderly "university baron" who as professor appropriated the discoveries of his younger associate; an in-house version of what is known as "Göttingen nostrification" as described in Courant (1981). It indeed appears odd that there is not a single article about logic and foundations that Hilbert coauthored with anyone. He made admissions, such as in the preface of Hilbert-Ackermann, that his series of articles on logic in the 1920s comes from "the closest collaboration with Bernays." Hilbert's first post-*Principia* logic paper of 1922 was of low quality by any standards, and light-years removed from the absolute precision of Bernays' *Habilitationsschrift*.

(E) **CRITICAL VOICES**. Not everyone admired Hilbert; here is what Thoralf Skolem wrote about the famous 1925–26 article Über das Unendliche (On the infinite):

> At the mathematical conference in Copenhagen 1925 it was spoken of as something epoch-making.... It surprises me greatly that mathematicians ...didn't understand right from the beginning how problematic—to use a mild expression—the essential contents of this paper were. Now it has become clear to everyone that it is not good. When in Vienna last autumn, I heard one mathematician use the expression "discrediting" about it. I mention this because there is a widespread confidence among mathematicians that it is in the first place elder people with a name in mathematics in the usual sense who can accomplish something in foundational research. It has turned out in reality that almost all advances in foundational research in recent times have come from younger people and essentially specialists.

This is my translation from the Norwegian original *Den matematiske grunnlagsforskning* that came out in the practically inaccessible *Nordisk matematisk tidskrift* in 1934. Skolem rather clearly declares Hilbert a dilettant without merit in logic and foundations—the more remarkable because Skolem had regular contact with Bernays, even visits in the late 1930s, and correspondence during the war.

In his 1923 paper on recursive arithmetic, Skolem had shown that the program initiated by Grassmann and Peano, of developing elementary arithmetic on a finitistic basis of recursive definitions of the arithmetic operations, could be carried through. Next to the development of pure logic as a basis, it was the first remarkable step to take in Hilbert's program of giving mathematics a finitistic foundation. Far from getting any recognition from Hilbert for his achievement, Skolem just had to watch his discoveries being used in papers such as those by Ackermann (1924, 1928), Hilbert's 1925–26 paper on the infinite, and elsewhere, perhaps with an occasional reference to Dedekind (1888). Hilbert's paper even had a French translation published right away in 1926, in the *Acta Mathematica*, the journal that had rejected the publication of Skolem's paper on recursive arithmetic in 1919.

The consistency of recursive arithmetic and related weak systems of arithmetic was one of the central themes of the first volume of

Grundlagen der Mathematik, published in 1934 but most of it finished by Bernays by 1930. Bernays made it clear in late interviews that Hilbert had not read any of the works of others but just knew about them because they were explained to him. Hilbert's "publicity management" resulted in wrong impressions given to everyone except an inner circle of those who knew who had done what, and a growing bitterness on the part of those whose discoveries were appropriated without recognition.

Without evidence to the contrary, one must still conclude that Bernays was content with the ways of Hilbert and his own role in Göttingen. Reinhard Kahle has suggested in conversations about this topic that the situation be looked at from the point of view of Bernays: What did he gain? Here is what Kahle wrote in February 2016 (my translation from German):

> You mention the problem of the authorship of the Hilbert-Bernays book. It is naturally objectively speaking beyond doubt that the book could have been published solely under the name of Bernays (leaving out the Nazism problem). But what would have been the consequence? Nobody would have read it! With Hilbert's name, it has achieved an authority, one that has influenced all of the modern developments.... Bernays had in this sense all the luck in the world, namely that Hilbert gave his name freely.

7.3. THE SITUATION IN FOUNDATIONAL RESEARCH AROUND 1930

(A) A SYNOPSIS OF THE GRUNDLAGEN. Bernays writes in the preface of volume 1 of the *Grundlagen der Mathematik*, dated March 1934, as follows about the presentation of Hilbert's proof theory:

> The carrying through of this project suffered an essential delay when, in a phase in which the presentation was close to an end, the appearance of the works of Herbrand and Gödel brought about a changed situation in proof theory.

Bernays decided to divide the project into two volumes. The "pre-Gödel" parts in them give an ample summary view of the situation of foundational research right before the appearance of the incompleteness theorems.

A first look at the first volume may give a rather unorganized impression, caused by the form of presentation. There are eight chapters, but in each, the text is continuous. Section and subsection headings and even smaller parts of text are indicated only in page headers and in the very detailed list of contents. Not a single theorem is singled out as such, nor any proofs or even definitions. To get an overall picture, one has to work one's way through the text and annotate it. The second volume of the Grundlagen, published in 1939, has clear chapter and section headings, but the text itself continues in the style of the first volume, with no definitions, theorems, or proofs singled out.

Here is a description of the overall structure, followed by details on the chapters on arithmetic and recursive definition, as a first orientation and for the benefit of those who don't read German.

The first volume begins with two introductory chapters:

§ 1 The problem of consistency of axiomatics as a logical decision problem.
§ 2 Elementary number theory—finitary reasoning and its limits.

They are followed by three chapters on pure logic:

§ 3 The formalization of logical inference I: the propositional calculus.
§ 4 The formalization of logical inference II: the predicate calculus.
§ 5 Addition of identity. Completeness of monadic predicate calculus.

The final three chapters deal with the following topics:

§ 6 Consistency in infinite domains. Beginnings of number theory.
§ 7 The recursive definitions.
§ 8 The concept of "the one such that" and its eliminability.

The 1939 second volume contains five chapters:

§ 1 Method of elimination of bound variables by Hilbert's ε-symbol.
§ 2 Proof theory of number theory by methods connected to the ε-symbol.
§ 3 Application of the ε-symbol to the investigation of the logical formalism.
§ 4 Arithmetization of metamathematics applied to the predicate calculus.
§ 5 Reasons for extending the methodological limits of proof theory.

There are in addition four supplements that deal with predicate logic, computability, purely implicational logic, and the formalization of analysis.

The first three chapters of the second volume belong to the pre-Gödel era, the fourth by its topic but not by its method.

In June 1934, in Geneva Bernays gave a lecture in French on "some essential points of metamathematics," published in 1935 in the Swiss mathematics journal *L'Enseignement Mathématique*, intended for mathematics teachers. It summarizes the central themes of the first volume of the *Grundlagen* in an accessible form. Bernays had spent the winter by his sister in Berlin, finished the *Grundlagen* in March, and was moving to Zurich. In the paper, the author is given as Paul Bernays (Zurich). Unfortunately, no English translation of this remarkable paper was made, contrary to his second Geneva lecture, the well-known "Sur le platonisme dans les mathématiques." The participants at the Geneva meeting included another fired Jewish professor from Göttingen, Paul Hertz, and Abraham Fraenkel, who had moved to Jerusalem four years earlier.

Bernays' paper has the following four sections, each explaining its topic with admirable clarity.

1. *An application of Herbrand's fundamental theorem to axiomatics* (pp. 70–75). Logical notation is introduced, and an example of a first-order axiomatization is given that emulates the strict order on natural numbers by the conditions of irreflexivity, transitivity, and extensibility. The consistency of the axiom system is proved through the method of Herbrand expansion.

2. *Proofs of consistency that relate to the ε-axiom* (pp. 75–80). The way from the standard first-order formulation of arithmetic to the ε-method is indicated, and results of Ackermann and von Neumann about the consistency of arithmetic with quantifier-free induction are then explained.

3. *Gödels's theorem about the proofs of consistency* (pp. 81–88). Gödel's argument is explained through Richard's paradox. The recursion scheme, arithmetic coding, and diagonalization of the provability relation are given.

4. *The relation between axiomatic number theory and intuitionistic arithmetic* (pp. 88–95). This part contains a remarkable anticipation of Gentzen's consistency proof of arithmetic by transfinite induction. It is followed by an explanation of the double-negation translation from classical to intuitionistic arithmetic.

Remarks that relate to these themes can be found in Bernays' 1935 article "Hilberts Untersuchungen über die Grundlagen der Arithmetik" in Hilbert's *Gesammelte Abhandlungen*, vol. 3, pp. 196–216.

It is easily seen from the layout of the *Grundlagen* that the ε-method was the main proof-theoretical tool for Bernays. The central "ε-theorem" is a result by which ε-terms can be eliminated from proofs of results of predicate logic that don't have ε-terms. Regarding the method, I share the opinion of Gentzen, who in a letter of 30 September 1937 to Bernays commented on the manuscript of the second volume of the Grundlagen, writing about the ε-theorem as follows (Gentzen 2016, p. 257):

> I would suggest to leave out this proof altogether; its result does not seem worth the effort to me. If I have understood right, this theorem is there only because of the ε-symbol that should be properly (p. 16 bottom) just an auxiliary for a more convenient treatment of questions that arise independently of the symbol.

The ε-method did not lead to success in the proof theory of arithmetic, but Gentzen's sequent calculus with its *Hauptsatz* (cut elimination theorem) did, as we shall see in chapters 9 and 10. This is what I think he alludes to by the mentioning of a "more convenient treatment." Bernays, though, was not ready to change his approach in 1937, and insisted with Ackermann until the latter produced a proof of consistency of arithmetic as an adaptation to the ε-method of Gentzen's proof in terms of sequent calculus. Today, the interest in the ε-method is historical, and too much effort would be needed to go into details here. Incidentally, even the proofs of consistency of first-order theories by Herbrand's theorem can be given a more convenient treatment by Gentzen's method of cut elimination. The axioms can be turned into additional rules of sequent calculus with cut elimination maintained, and consistency is usually an immediate corollary, as explained in Section 9.3(C).

As mentioned, Bernays (1935a) contains an incredibly clear anticipation of Gentzen's consistency proof of arithmetic. In his search of "intuitionistic proofs that cannot be formalized in $\mathfrak{N}$" (Hilbert's formalization of first-order arithmetic), he notes that the consistency of arithmetic is provable "in a formalism $\mathfrak{N}^*$ that one obtains from $\mathfrak{N}$ by the adjoining of certain non-elementary recursive definitions," and backs this up by an example. Then (p. 90):

> It follows that this sort of recursive definitions surpass the formalism $\mathfrak{N}$. On the other hand, such a recursive definition intervenes also in the formal deduction of the principle of transfinite induction applied to an order of the type of ordinal $\lim_n \alpha_n$, or
>
> $$\alpha_0 = 1 , \quad \alpha_{k+1} = \omega^{\alpha_k} (k = 0, 1....).$$
>
> This type of order can be realized for the integers by an order
>
> $$a \prec b$$
>
> that is definable by elementary recursions. And the said principle is expressed, for this order, by the formula
>
> $$(x)\{(y)(y \prec x \longrightarrow A(y)) \longrightarrow A(x)\} \longrightarrow (x)A(x)$$

Bernays suggests that this principle is not provable in the standard arithmetic formalism but is instead provable intuitionistically. Therefore (p. 91):

> It is apparent that the special case of the principle of transfinite induction considered is already an example of a theorem that is provable by intuitionistic mathematics, but not deducible in $\mathfrak{N}$.

> Consequently one would propose, in concordance with the theorem of Gödel, that one finds an intuitionistic proof of the consistency of the formalism $\mathfrak{N}$ in which the only part non-formalizable in $\mathfrak{N}$ is the application of the said principle of transfinite induction.

He adds that the proposal is "just one possibility at the moment." The specific example of recursive definition Bernays chose was not strong enough, as Skolem (1937, p. 439) pointed out. Bernays therefore modified his example in the second volume of the *Grundlagen* (p. 340).

(B) ELEMENTARY ARITHMETIC IN THE GRUNDLAGEN. Chapter 6 of volume 1 is dedicated to elementary number theory of a rather special kind, the only arithmetic operation in the whole long chapter being the successor. Sum, product, and so on appear only in chapter 7, through recursive definitions, the likely reason for this organization being that chapter 6 mainly deals with systems of arithmetic that don't use the induction axiom. Recursive definitions would not lead anywhere without the corresponding principle of proof.

Chapter 6 contains a presentation of what was known about the proof theory of arithmetic in the pre-Gödel era, with lots of detailed methods and results that may be somewhat difficult to discern. It begins with materials that prepare the reader for a proof of consistency of arithmetic without the induction axiom (p. 209). First a treatment inspired by Dedekind's is explained, written in detail in the language of predicate logic. There is one basic predicate $P(a,b)$, *the successor of a is b*, and the four axioms (p. 213)

$$(D) \quad \begin{cases} (x)(Ey)P(x,y), \\ (Ex)(y)\overline{P(y,x)}, \\ (x)(y)(z)(P(x,y)\,\&\,P(x,z) \to y=z), \\ (x)(y)(z)(P(x,z)\,\&\,P(y,z) \to x=y). \end{cases}$$

Bernays replaces this two-place predicate by a function, the "strich-symbol" of Schröder that even Dedekind used for successor. This only primitive function is explained as follows (p. 219):

> It is customary in mathematics to give the process of continuation [of the number series] through "+1." This kind of notation has, however, the defect that the conceptual distinction between taking "$a + 1$" as the number that follows a and on the other hand as the sum of a and 1 fails to be represented.

The axioms with the successor notation now read (p. 215):

$$(D_0) \quad \begin{cases} (x)(Ey)(x' = y), \\ (Ex)(y)(y' \neq x), \\ (x)(y)(z)\,(x' = y\,\&\,x' = z \to y=z), \\ (x)(y)(z)\,(x' = z\,\&\,y' = z \to x=y). \end{cases}$$

These axioms and those for equality are further modified into the following quantifier-free axioms (p. 220):

(I_1) $a = a$,
(I_2) $a = b \rightarrow (A(a) \rightarrow A(b))$,
$(<_1)$ $\overline{a < a}$,
$(<_2)$ $a < b \,\&\, b < c \rightarrow a < c$,
$(<_3)$ $a < a'$,
(P_1) $a' \neq a$,
(P_2) $a' = b' \rightarrow a = b$.

These are the axioms of *identity*, *strict order*, and *infinity*.

Pages 221–248 give a proof that the formula $0 \neq 0$ is not derivable in the axiom system. The proof contains two aspects that are similar to Gentzen's methods as detailed in the next chapter. Bernays resolves his axiomatic linear derivations into "proof threads" that display the deductive dependences of formulas in assumption-conclusion sequences. From these, tree diagrams are constructed (p. 223) that are quite similar to the tree diagram we saw in connection with Skolem's lattice-theoretic proof toward the end of Section 3.3. In hindsight, Gentzen's step to consider such a tree structure directly as being the derivation itself seems a little one, but Bernays does not take it.

Another method in the long proof also found in Gentzen is to eliminate free variables. As the assumed conclusion is a numerical formula, namely $0' \neq 0$, the taking of instances of the axioms can be permuted up in a derivation so that, in the end, top formulas become axiom instances. All of these are decidable numerical formulas, each of them true by a finitary contentful consideration (Bernays says "verifiable," p. 238), and with verifiability preserved under the only rule of inference, implication elimination.

After the consistency theorem, Bernays gives an example of a verifiable formula and shows that it is not derivable (pp. 249–252):

$$a < b \rightarrow a' = b \vee a' < b$$

A condition is given by which verifiable formulas become derivable (p. 253), and then it is shown that the condition is satisfied (pp. 255–263). The upshot of this lengthy discussion is the axiom system denoted by (A), with what later came to be called *Robinson's axiom* in place of the induction principle (p. 263):

$$(A) \quad \begin{cases} a = b \rightarrow (A(a) \rightarrow A(b)), \\ a < b \,\&\, b < c \rightarrow a < c, \\ a \neq b \rightarrow a < b \vee b < a, \\ \qquad a < a', \\ \quad a < b \rightarrow \overline{b < a'}, \\ \qquad \overline{a < 0}, \\ a \neq 0 \rightarrow (Ex)(x' = a). \end{cases}$$

Robinson's axiom is a classically equivalent formulation of the last one, $a = 0 \vee (Ex)(x' = a)$. Note that reflexivity of equality is a theorem, easily derived from the instance $a \neq a \rightarrow a < a \vee a < a$ of the third axiom together with the fifth, $a < a \rightarrow \overline{a < a'}$. The consequent contradicts the fourth axiom $a < a'$, by which $\overline{a \neq a}$ follows.

The overall result can now be given in terms of system (A), as follows in (p. 264):

The domain of derivable formulas that do not contain formula variables coincides with that of the verifiable formulas.

Given a prenex formula without free variables, its derivability coincides with its "contentful finitary interpretability," explained as follows through a representative example (p. 265):

A formula such as

$$(x)(Ey)(z)(Eu)A(x, y, z, u)$$

in which $A(x, y, z, u)$ has no variables besides x, y, z, u is derivable if and only if we can determine for each numeral $\mathfrak{x}_1$ a numeral $\mathfrak{x}_2$ such that for each numeral $\mathfrak{x}_3$ a numeral $\mathfrak{x}_4$ can be determined such that the numerical formula

$$A(\mathfrak{x}_1, \mathfrak{x}_2, \mathfrak{x}_3, \mathfrak{x}_4)$$

is true.

Here is a fact that gives clear expression to the complete correspondence between the formalism and the contentful conception [inhaltliche Fassung] in the domain fixed by axioms (A).

Bernays dedicates the following pages to showing that the standard induction axiom is conservative over the system (A) for formulas without schematic formula variables (pp. 265–270). Then follows yet another system (B), in which the induction axiom finally appears (p. 273):

$$(B) \begin{cases} a = b \rightarrow (A(a) \rightarrow A(b)), \\ \overline{a < a}, \\ a < b \,\&\, b < c \rightarrow a < c, \\ a < a', \\ a < b \rightarrow a' = b \lor a' < b \\ 0 = 0, \\ A(0) \,\&\, (x)(A(x) \rightarrow A(x')) \rightarrow A(a) \end{cases}$$

It is shown that this system is stronger than system (A) by a proof that the induction axiom is not derivable in (A) (pp. 273–276).

The final two topics in chapter 6 are first a proof of the independence of the seven axioms of system (B) (pp. 277–279), and similarly for system (A) (pp. 279–283), and secondly a proof of the equivalence of the induction axiom in (B) with the *least number principle* (pp. 283–284):

$$A(a) \rightarrow (Ex)(A(x) \,\&\, (y)(A(y) \rightarrow x = y \lor x < y))$$

The proofs of independence show that Bernays had looked deep inside the formalisms presented and knew to perfection how they worked.

(C) COMPUTATION TURNED INTO DEDUCTION. The close to a 100-page chapter 7 on recursive arithmetic in volume 1 of the *Grundlagen* was the first systematic presentation of the topic. It owes a lot to Skolem's 1923 paper but refers even to Peano's *Formulario* in its last 1908 edition, to Peano's original work of 1889, and to Dedekind (1888). The scheme of primitive recursion first appears in a general form in Gödel (1931).

In the beginning pages of the chapter, there is the wonderful, clear insight that the recursive definition of basic arithmetic operations turns computation into deduction, repeated several times (pp. 288, 290, 292):

> The formalization of the intuitive procedure of a recursive definition through the recursion scheme depends on obtaining, for a recursively

introduced function sign ... , a *derivable equation* ... with the help of the
recursion equations and the axioms of equality.

$$\vdots$$

If c is a numeral, the computation of the value of $\mathfrak{f}(\mathfrak{a}, \ldots, \mathfrak{t}, \mathfrak{z})$ is already
completely formalized through the derivation of the equation

$$\mathfrak{f}(\mathfrak{a}, \ldots, \mathfrak{t}, \mathfrak{z}) = \mathfrak{c} \ldots$$

$$\vdots$$

We can reconstruct completely the recursive procedure of computation of
finitary number theory in our formalism, through the deductive application
of the recursion equations.

In the consistency proofs of Chapter 6, all equations were reduced
to numerical verifiable equations by the "substitution method" (Erset-
zungsmethode). The method does not work for recursive definitions in
the presence of bound variables (p. 294). Therefore, Bernays considers
the system of quantifier-free axioms for identity, strict order, and
infinity as detailed previously, and systems (A) and (B) with the last
axiom left out in both. The consistency result is expressed as follows
(p. 298):

> If the elementary calculus with free variables is extended by the addition
> of recursive definitions, of verifiable axioms ... and equality axiom (I_2),
> then each formula derivable in the system and without formula variables is
> verifiable.

Bernays adds that the result applies even for quantifier-free uses of the
induction axiom with the consistency of quantifier-free arithmetic as a
corollary (p. 299).

An example is given that shows how the consistency result can
become lost under the addition of recursive definitions, with the more
general conclusion (p. 301):

> The situation found here can be set straight also from the intuitive point of
> view. A system of two equations of the form

$$\mathfrak{f}(0) = \mathfrak{a},$$
$$\mathfrak{f}(n') = \mathfrak{b}(n, \mathfrak{f}(n))$$

puts a requirement on the function $\mathfrak{f}(n)$. Whether it is satisfiable is not revealed by the structure of the recursion equations themselves but depends instead *on the characteristic properties of the successor function*, namely that this function never leads to the value 0, and that two different argument values always correspond to two different values of the successor function.

Therefore the admission of recursive definitions amounts to an *implicit characterization of the successor function.*

This insight leads Bernays "to assume that Peano's axioms $(P_1), (P_2)$ can be derived by recursive definitions" (pp. 302–303). The first infinity axiom $a' \neq 0$ can indeed be derived from $0' \neq 0$ and the function Bernays designates by α (p. 302):

$$\alpha(0) = 0$$

$$\alpha(n') = 0'$$

The equality axiom gives $a' = 0 \rightarrow \alpha(a') = \alpha(0)$ and then the recursion equations $0' = 0$ against the axiom $0' \neq 0$, so that $a' \neq 0$ follows.

The second infinity axiom, $a' = b' \rightarrow a = b$, is derived by the *predecessor function* δ in the now standard way (ibid.):

$$\delta(0) = 0$$

$$\delta(n') = n$$

The equality axiom gives again $a' = b' \rightarrow \delta(a') = \delta(b')$ and the second recursion equation then $a = b$.

The derivation of the two Peano axioms is announced in Bernays' long letter to Gödel of 18 January 1931 in response to Gödel's incompleteness article that he had received four days earlier in the form of galley proofs (as on p. 82 of Gödel 2003). It is worth comparing Bernays' interpretation with that of Dedekind. In the latter, the first infinity axiom shows that the image of N under the successor function is a proper subset of N, and the second axiom gives a one-to-one correspondence between this subset and N, so Dedekind's two clauses for the infinity of a set are satisfied. In Bernays, instead, the second axiom guarantees that when in a recursive call the step back from a successor n' is taken, the new argument n of the recursive function is unique. Similarly, the first axiom just dictates that the recursive call ends when 0 is reached.

Now the system of recursive arithmetic consists of the two equality axioms $(I_1), (I_2)$, the axiom $0' \neq 0$, and the scheme of recursive definition. Even the first equality axiom can be dropped when sum and product are given their recursive definitions. Bernays just left out the axiom and somehow missed giving the justification, by oversight I think. Reflexivity can be derived as follows: By the first recursion equation for sum, we have $a + 0 = a$. The replacement scheme has the instance

$$a + 0 = a \rightarrow (a + 0 = a \rightarrow a = a)$$

Here the predicate $A(x)$ is $x = a$.

The development of recursive arithmetic is detailed as follows.

1. *Definition of sum and product.* First, the recursive definitions of sum and product are given (pp. 307–309), with the same somewhat strange order of results as in Dedekind, detailed in Section 2.2(B).

2. *A logic-free formulation.* Bernays shows how the operations of propositional logic can be interpreted in purely arithmetic terms (pp. 309–312):

Each variable-free formula can be transformed into an equation of the form t = 0.

Point 2 was met already in Section 6.2(C): an equation $a = b$ is transformed into the form $t = 0$ by the symmetric truncated difference of a and b. First, form the simple truncated difference (p. 303):

$$\delta(a, 0) = a$$
$$\delta(a, n') = \delta(\delta(a, n))$$

The value of $\delta(a, n')$ is the predecessor of $\delta(a, n)$, rather peculiarly the same symbol used for two functions in one formula. (In Bernays (1951), the notation of Curry (1941) is used: $a \dot- b$.) One subtracts from a until 0 is reached or there is nothing to subtract. The symmetric difference $\delta(a, b) + \delta(b, a)$ has value 0 if and only if $a = b$, so we know how to translate any equation into the form $t = 0$.

To translate negation as in $a \neq b$, first translate it into the form $t \neq 0$. In general, $a \neq 0$ if and only if $a = \delta(a)'$, i.e., if the successor of the predecessor of a is equal to a. This leads now to an equation of the form $t = 0$ for $a \neq b$. Conjunction and disjunction are handled by the equivalences (p. 309):

$$a + b = 0 \text{ if and only if } a = 0 \& b = 0$$
$$a \cdot b = 0 \text{ if and only if } a = 0 \vee b = 0$$

If formulas are in conjunctive normal form, the preceding is sufficient for the translation of any propositional formula. The same method was used by Curry (1941) and also invented by Goodstein, who by his own admission had by 1940 never opened the *Grundlagen*. As mentioned in Section 6.2(C), the origins of the idea can be traced back to Gödel's incompleteness article, in the result by which the operations of propositional logic maintain the "numeralwise representability" of primitive recursive relations. A similar reduction is seen in Gödel's (1958) interpretation of intuitionistic arithmetic in terms of primitive recursive functionals.

3. *Bounded quantification.* The third stage in recursive arithmetic is to define the *bounded quantifiers*, a notion perhaps inspired by Skolem's 1923 discussion of arithmetic translations of formulas of the type (p. 313):

$$(x)(x \leq n \rightarrow A(x))$$
$$(Ex)(x \leq n \& A(x))$$

4. *Development of recursive number theory.* There now follow some of the standard topics of recursive number theory, namely the definitions of *divisibility* (p. 316), *prime factorization* (p. 319), and *greatest common divisor* (p. 323).

5. *Various forms of recursion.* The next item is a presentation of variants of recursive definition: first the scheme of *primitive recursion* (p. 326), then *course-of-values recursion* (p. 326), and last *simultaneous recursion* (p. 328).

6. *Beyond primitive recursion.* There are two ways of showing that the scheme of primitive recursion can be surpassed. The first one is the *diagonal method.* Assume all primitive recursive functions of one argument are listed one after the other, and let $\chi(n, m)$ give the value of the mth function for argument n in the list. The expression $\chi(n, n) + 1$ defines a one-place function, but it is different from all the primitive recursive functions in the list (p. 330).

A direct proof of the existence of functions that are not primitive recursive was given by Ackermann (1928). The idea is to construct a function that "grows faster than any function definable in a primitive recursive way" (p. 330). Skolem must have considered himself the creator of recursive arithmetic to the extent that he commented on Ackermann's discovery in the following terms (1947, p. 502):

> Indeed, it is a fact that before the beginning of the recent investigations concerning recursive functions no mathematician has ever defined a computable function which is not primitive recursive.

Ackermann's example was simplified considerably by Rosa Politzer, a Hungarian and one of the very few women to contribute to the development of the foundations of mathematics early on in a decisive way. She presented a contributed paper on primitive recursive functions at the International Congress of Mathematicians that took place in Zurich in September 1932. An abstract with the title "Rekursive Funktionen" was published in the second volume of the proceedings, pp. 336–337, followed by one by László Kalmár on the decision problem of predicate logic. The latter had learned the basics of recursive arithmetic from Bernays in Göttingen, and this led to Politzer's work. The sources she mentions are Hilbert (1925–26), Ackermann (1928), and Gödel's (1931) incompleteness paper. Gödel's paper gives the scheme of definition by primitive recursion in a general form, independently of any particular examples such as addition.

Politzer's paper was the first one in which recursive functions were studied on their own as a separate topic. The results announced in the summary presentation were detailed in two papers published in the *Mathematische Annalen* in 1934 and 1935 through the mediation of Bernays, with a third follow-up in 1936. Through these results, Rosa Politzer can be considered the one who established the theory of recursive functions as a chapter of mathematics of independent interest. Her role "as the leading contributor to the special theory of recursive functions" is well recognized by Stephen Kleene in his review of Politzer's 1951 book *Rekursive Funktionen*, the first book devoted to the topic. After the publication of the summary, Politzer changed in January 1934 her Jewish last name to Péter, likely suggested by the name of her university, in common with many Hungarian Jews who in this way sought to evade discrimination.

Bernays also took part in the Zurich conference, and the two worked together one year later in the summer of 1933 in Göttingen, as is shown by their extensive correspondence preserved in part in Zurich. Politzer's formulation of the basics of recursive function theory played an important role in the 100-page chapter 7 devoted to the topic in volume 1 of the *Grundlagen der Mathematik*. What Politzer, following Gödel, calls recursive functions in her summary later came to be called primitive recursive functions. The results on various forms of recursion come mainly from her 1934 paper in which the author is given as "Rózsa Péter (Politzer) in Budapest." Her next paper contained the mentioned diagonal argument and an improvement of Ackermann's example of a function that is not primitive recursive, published in 1935, with her Jewish name dropped. Bernays barely had time to simplify accordingly his treatment of the example in the *Grundlagen;* even so, it takes a full ten pages in Bernays' presentation (pp. 332–341). A third paper on multiple recursion appeared in 1936, with the 1934 and 1935 papers accurately referred to as authored by "Rózsa Péter (Politzer)" and "Rózsa Péter," respectively.

Politzer defended her doctoral thesis in Budapest in 1935. Had it not been for the situation in Germany, one can safely presume that she would have been another of Bernays' students in Göttingen. – Rosa Politzer survived the war, tagged with a yellow star and spending the last months of the Nazi terror hiding in the attic of an exposed building, a place to which the deportation troops did not dare enter for fear of shooting and shells from exploding bombs. Afterward, she received an academic position that became a professorship she held until the 1970s.

Gabriel Sudan was a student in Göttingen, who in 1925 defended his thesis *Über die geordneten Mengen* (On the ordered sets). In 1927, he published a paper in a mathematics journal in Romania, his home country, with the title *Sur le nombre transfini ω^ω*. The paper takes up the hierarchy of higher-type recursive functions of Hilbert's infinity paper and refers to Ackermann's example that goes beyond primitive recursion. The main result is the determination of what is today called the ordinal of primitive recursive arithmetic: Each ordinal below ω^ω can be defined by primitive recursion on natural numbers, and each function definable by primitive recursion corresponds to an order type below that ordinal. Sudan's result, though, is not mentioned in the *Grundlagen*, nor is there a single letter left in the Bernays archive, but Ackermann (1928) mentions Sudan's paper.

Returning to the *Grundlagen*, there follow in chapter 7 three systems of arithmetic.

1. *Presburger arithmetic.* The system of arithmetic with just addition is named after Mojzesz Presburger (1904–1943?), who published just one article, in 1930, and perished during the Nazi era. Presburger showed his arithmetic to be consistent, syntactically complete, and decidable. Bernays outlines these results, as proved by his methods for systems (A) and (B), in a lengthy discussion (pp. 357–368). He then shows that product cannot be incorporated into the system of Presburger arithmetic without losing the preceding results.

Incidentally, and as a kind of dual to Presburger arithmetic, Skolem (1930) showed that arithmetic with just product behaves in a way similar to Presburger's.

2. *First-order arithmetic with sum and product.* Bernays defines an axiom system denoted (Z) that has the equality axioms, the two infinity axioms, recursion equations for sum and product, and the schematic induction axiom. This can be considered the standard system of first-order arithmetic with equality and the functions of successor, sum, and product as primitives (p. 371):

$$
(Z) \quad
\begin{cases}
a = a, \\
a = b \rightarrow (A(a) \rightarrow A(b)), \\
a' \neq 0, \\
a' = b' \rightarrow a = b, \\
a + 0 = a, \\
a + b' = (a + b)', \\
a \cdot 0 = 0, \\
a \cdot b' = a \cdot b + a, \\
A(0) \& (x)(A(x) \rightarrow A(x')) \rightarrow A(a).
\end{cases}
$$

Bernays notes that the method of reduction, substitution of free variables, is not applicable. He gives examples of formulas that express, in turn, famous open problems in number theory such as the Goldbach and twin prime conjectures (p. 371), Fermat's theorem, and the solvability of Diophantine equations. It is then clear that (Z) *"gives a formalization of the entire theory of numbers"* (p. 372). The applicability of the methods of Chapter 6 would lead to decidability, something Bernays clearly takes not to be possible. Thus, the known methods of consistency proof need to be extended if proof theory is to progress to full first-order arithmetic.

3. *Arithmetic without the replacement scheme.* The replacement scheme can be reduced to nonschematic replacement axioms. In the end, this result depends on the possibility of deriving replacement for compound formulas from that for atomic formulas, with reflexivity restored as an axiom. Bernays gives as examples the following axioms (p. 374):

$$a = a,$$

$$a = b \rightarrow (b = c \rightarrow a = c),$$

$$a = b \rightarrow a' = b',$$

$$a = b \rightarrow (a < c \rightarrow b < c),$$

$$a = b \rightarrow (c < a \rightarrow c < b).$$

Next, he shows how compound formulas inherit the replacement property (pp. 374–375). When arithmetic operations are added, the respective replacement axioms have to be added, say as the one for sum $a = b \rightarrow a + c = b + c$ (p. 377). In the end, we have (p. 378):

The use of formula variables in the derivation of number-theoretic formulas can be be avoided in general.

(D) THE "TEMPORARY FIASCO" OF PROOF THEORY. In the preface of the second volume of the *Grundlagen*, Bernays writes that "the temporary fiasco of proof theory was caused by overly tight methodological requirements." The fiasco was, of course, Gödel's incompleteness theorem. There has been a rather wide belief that the result was not well received in Hilbert's Göttingen. For example, Saunders Mac Lane, who studied with Bernays there at the time, writes about Gödel's result (2005, p. 51): "I wonder whether his work was suppressed in Göttingen because it was a blow to the work of Hilbert and others, or whether I was just so busy that I ignored the excitement." Bernays' long letter to Gödel only four days after the reception of his article contains the following (Gödel 2003, p. 82): "What you have done is really a remarkable step ahead in the research on foundational problems."

The doctrines of Hilbert's archenemy, the intuitionist Brouwer, showed the way out. With Gödel's results, a finitary, "absolutely reliable" consistency proof of the kind envisaged by Hilbert would not be possible. Some, von Neumann foremost, declared the foundational enterprise dead: The consistency of mathematics would remain forever unprovable in some absolute sense. Bernays instead sought a way out through intuitionism that he took to go beyond the Hilbertian finitism. There is no lack of clarity in Bernays' assessment: Brouwer had it right on his critical point, namely that the law of excluded middle has found no justification beyond finitary situations (*Grundlagen*, vol. 1, p. 34):

> The complicated situation that the finitary standpoint faces here in respect of the denial of judgments corresponds to Brouwer's thesis about the invalidity of the law of excluded middle for infinite totalities. This invalidity obtains in fact in the finitary standpoint in the sense that one does not succeed in finding, for existential as well as universal judgments, a negation with a finitary meaning that satisfies the law of excluded middle.

Bernays describes finitism as a categorical buildup of mathematics, in the sense that nothing is assumed but everything is built up finitistically from decidable concepts. Brouwer's intuitionism brings to this picture the new element that hypothetical proofs also are considered, as well as mathematical constructions made on top of such assumed proofs. Bernays wrote (p. 43):

> The methodological point of "intuitionism" that is at the basis of Brouwer is formed by a certain *extension of the finitary position* [Erweiterung der finiten Einstellung], namely, an extension in so far as Brouwer allows the introduction of an assumption about the presence of a consequence, resp. of

a proof, even if such a consequence, resp. proof, is not determined in respect of its visualizable constitution [nicht... nach anschaulicher Beschaffenheit bestimmt]. For example, from Brouwer's point of view, propositions of the following forms are allowed: "If proposition B holds under assumption A, also C holds," and also "The assumption that A is refutable leads to a contradiction," or in Brouwer's mode of expression, "the absurdity of A is absurd."

The essence of intuitionism as given here is that it is allowed to assume conditionals and, even more simply, the presence of a hypothetical proof. One would think that this was no novelty in principle, for what are mathematical axioms if not conditionals that are assumed? Bernays thinks instead that there is no hypothetical element in the practice of logicism or formalism. In this light, Gentzen's departure from these traditions in his setting up of natural deduction in 1932 is the more remarkable because the most central idea in natural deduction is to consider hypothetical inferences.

Bernays proceeds with the discussion in very general terms, the problem being always how to extend the finitary standpoint, and ends with the conclusion that we are still far away from even a solution to the consistency problem of arithmetic (p. 44). The solution was instead much closer than he could imagine, for Gentzen had it by the end of 1934.

∗ ∗ ∗

The experts most useful for me in the study of the development of logic in the 1920s have been Reinhard Kahle, Paolo Mancosu, and Richard Zach. Most of the work of Mancosu is found in the hefty collection, *The Adventure of Reason: Interplay between Philosophy of Mathematics and Mathematical Logic, 1900–1940* (Mancosu 2010). Original papers from the time of the foundational debate are found in Mancosu (1998). Newer original material is contained in the 1000-page volume of Hilbert's lectures on logic and foundations of arithmetic between 1917 and 1933 (Hilbert 2013), though the lengthy introductions must be read with some caution: The unlimited admiration of Hilbert's achievement gives a distorted view of who actually did what.

I have treated what is called Hilbert's ε-method very briefly, mainly because I share the opinion of Gentzen that "it's not worth the effort"; the same results and more are achieved by more forceful methods of analysis of the structure of formal proofs, as in Chapter 9. Zach (2003) gives accounts of Ackermann's work on the ε-method. Detailed original systematic presentations of the method by its real inventor, Bernays, can be found in the two Hilbert-Bernays Grundlagen volumes.

8

GÖDEL'S THEOREM: AN END
AND A BEGINNING

8.1. HOW GÖDEL FOUND HIS THEOREM

(A) GÖDEL AND HIS THEOREM. Kurt Gödel was born in 1906 in Brünn, a city that at that time belonged to the Austro-Hungarian empire. He started to study physics at the University of Vienna in 1924, then changed to mathematics in 1926. That same year, he started attending the meetings of the "Vienna Circle." These were weekly gatherings on philosophical topics that were headed by the philosopher Moritz Schlick. The philosophy of the circle came to be known as logical empiricism and had an enormous effect on the world of philosophy. Gödel later wanted to emphasize that he by no means shared all of the philosophical ideas of the circle. His professor of mathematics Hans Hahn was a regular member of the circle. In the meetings, Gödel came to know the philosopher Rudolf Carnap and the mathematician Karl Menger, in whose mathematical colloquium he later presented many of his results.

Gödel's best-known discoveries are the completeness proof of predicate logic, the incompleteness theorems, and the proof of consistency of the continuum hypothesis.

1. *Completeness of predicate logic.* Gödel's first result was the proof of completeness of the predicate calculus that he found in 1929. The first volume of his *Collected Works* (1986) revealed a surprise: Gödel had submitted his proof as a doctoral thesis to the University of Vienna in 1929, with the title *Über die Vollständigkeit des Logikkalküls*. The publication of the original thesis text in 1986 brought to light his insight

that a constructive proof of completeness would amount to a decision method for predicate logic, an insight of which there is no trace in the published paper of 1930, "Die Vollständigkeit der Axiome des logischen Funktionenkalküls." Apparently, one is to conclude that in Gödel's mind the decidability of predicate logic was unlikely, contrary to, say, Jacques Herbrand.

There has been some discussion of the extent to which Skolem in his 1922 paper may have anticipated Gödel's completeness result. Gödel himself wrote to Van Heijenoort about the matter, in connection with the preparation of the collection *From Frege to Gödel*; all of this correspondence can now be read from the two-volume selected correspondence of Gödel that forms the fourth and fifth volumes of his *Collected Works* (2003). The first mention of Skolem's 1922 paper is in a letter of Gödel's of 4 October 1963:

> I think I first read it about the time when I published my completeness paper. That I did not quote it must be due to the fact that either the quotations were taken over from my dissertation or that I did not see the paper before the publication of my work.

Gödel's interpretation of Skolem, from a letter of 14 August 1964, is that Skolem implicitly proved:

> 'Either A is provable or $\overline{A}$ is satisfiable' ('provable' taken in an informal sense). However, since he did not clearly formulate this result (nor, apparently, had made it clear to himself), it seems to have remained completely unknown.

On August 27, Gödel writes, more sharply:

> Skolem advances some syntactic considerations, but they do not establish (nor, I think, are intended to establish): 'If A is not provable, A is satisfiable.'

(The unnegated second A is a slip of Gödel's or a misprint.) On 18 September, Gödel wrote that Skolem "had no clear ideas in these matters. However, anyone who has can, without any further mathematical inferences, immediately draw the conclusion I stated in my previous letter."

In Gödel's understanding, from the letter of December 1967 to Wang, Skolem's viewpoint in logic and mathematics was that "meaning is attributed solely to propositions which speak of concrete and finite objects, such as combinations of symbols," and this viewpoint prevented him from taking the necessary nonfinitary step needed for a full statement of the completeness theorem. Gödel's view is in accordance with what Skolem's 1920 paper suggests. It is further confirmed by Skolem's paper on primitive recursive arithmetic that was published belatedly in 1923. Its writing was prompted, as Skolem repeatedly emphasized, by his reading of the *Principia Mathematica* in 1919. Each instance of a free-variable derivation has a finitistic sense, whereas quantification over an infinite domain remains problematic. Thus, this work displays precisely the finitistic, combinatorial conception of logic and mathematics Gödel took to be characteristic of Skolem's work. One aspect that seems not to have been mentioned in discussions of Gödel's completeness proof is that the undecidability of predicate logic was not established until 1936. Before that date, no one was in a position to claim that a nonfinitary step in proofs of completeness is unavoidable. From this point of view, Gödel's charge, in the December 1967 letter to Wang, that Skolem had demonstrated a surprising "blindness (or prejudice, or whatever you may call it)," carries with it a touch of hindsight.

2. *The incompleteness theorems.* By the summer of 1930, Gödel had found what is now known as his first incompleteness theorem. It states that any sufficiently expressive formal system of arithmetic contains arithmetic statements that are true but unprovable in it, or else the system is contradictory. The result became public knowledge in a conference in Königsberg, East Prussia, in September 1930, and immediately aroused great interest from such leading young figures as Johann von Neumann and Karl Menger.

Gödel's paper on incompleteness was published in early 1931. It contains two incompleteness theorems; the first was given in the previous paragraph. The second one says that the unprovability of a contradiction, that is, the statement of consistency of the system itself, is among such unprovable propositions. Both theorems were great surprises, and the second one especially completely changed ideas about the foundations of mathematics. Today we can see that Gödel's incompleteness theorems and the methods he invented for their proof belong among the greatest discoveries of the twentieth century. Their importance in logic and foundations of mathematics is not unlike the role that, say, special relativity and quantum mechanics have had in physics.

The incompleteness paper was also important for the development of the notions of formal system and computability: It contains a general scheme for the definition of primitive recursive functions that are used in the arithmetic coding (Gödel numbering) of formal expressions and of proofs of arithmetic and in a result by which not only the system of *Principia Mathematica* but already the first-order theory of arithmetic is incomplete.

Gödel's other central discoveries include the following.

3. *Interpretation of classical arithmetic in intuitionistic arithmetic.* In 1932, Gödel found a translation method for arithmetic formulas by which the problem of the consistency of classical arithmetic is reduced to the corresponding problem for a constructive version of arithmetic (Gödel 1933a).

4. *The definition of general recursive functions.* Gödel gave the "Herbrand-Gödel" definition of general recursive functions, the first definition of such functions, in a series of lectures in 1934.

5. *Consistency of the continuum hypothesis.* Gödel gave a proof that no contradiction follows from the addition of the axiom of choice and of Cantor's continuum hypothesis to the axioms of set theory (1938a). Twenty-five years later, Paul Cohen proved what Gödel believed but did not succeed in proving himself, that also the negations of these can be consistently added to the other axioms of set theory. It follows that known axioms of set theory do not decide the continuum hypothesis.

6. *The theory of primitive recursive functionals.* From 1940 on, Gödel worked on a "logic-free" interpretation of intuitionistic arithmetic in a constructive functional hierarchy, with the aim of giving proofs of the consistency of arithmetic and analysis (Gödel 1941, with a separate short published article 1958).

7. *The field equations of general relativity.* Gödel found solutions to the field equations of Einstein's general theory of relativity, among the first such solutions altogether (Gödel 1949). A cyclic time coordinate is possible in some of these solutions, which led to speculations about "time travel."

Gödel visited the newly established Institute for Advanced Study in Princeton, close to New York, several times during the 1930s, and stayed there permanently from 1940 until his death in 1978.

Serious work on Hilbert's program started in the early 1920s in collaboration with Hilbert's assistant Paul Bernays, his student Wilhelm Ackermann, and others such as Johann von Neumann among the foremost. In Hilbert (1928), the completeness of predicate logic is given as an important open problem. Predicate logic is the logical language that forms the basis of the formalization of such parts of mathematics as arithmetic and elementary geometry. Gödel's doctoral thesis settled the completeness problem.

In pursuing the problems of consistency and completeness of Hilbert's program, Gödel came to think of the following problem: to show that analysis, i.e., the theory of real numbers, is *conservative* over arithmetic. Consider the language of arithmetic LA and the language of real numbers LR. Each statement of LA is also a statement of LR but not the other way around. Let TA be the axiomatic theory of arithmetic and TR that of the real numbers. Conservativity of TR relative to TA is defined as: If a statement A of LA is provable in TR, it is already provable in TA. In other words, the use of real numbers and their axioms would not in principle be needed in proofs about the natural numbers.

Assume, for the sake of argument, that TR is not conservative over TA. If TR were inconsistent but TA instead consistent, TR would trivially not be conservative. Therefore, we assume TR is consistent. It follows also that the subtheory TA is consistent. There is then, by assumption, some statement A in LA that is provable in TR but not in TA. Therefore, by the consistency of TR, A must be true, and we have an arithmetic statement that is true but not provable in the formalized arithmetic theory TA.

Gödel's amazing discovery is, of course, that the preceding train of thought is not "for the sake of argument," namely to prove conservativity through an impossibility, but rather the contrary: There do exist exactly the kind of statements hypothesized in the argument. This being so, one might think: "Why, then, not just use the theory TR for proving theorems of arithmetic, instead of the incomplete theory TA?" The answer is that even TR would be incomplete as an axiomatization of arithmetic. No formal system that contains arithmetic as a part can be a complete axiomatization of arithmetic.

(B) FROM LOGICAL PARADOXES TO INCOMPLETENESS. The 1928 book by Hilbert and Ackermann was Gödel's main source on logic and foundations; one can say that without it, he would not have come to the completeness and incompleteness theorems. The former is a clearly stated open problem in Hilbert-Ackermann. Inspiration for the latter comes from the discussion of the logical paradoxes in section 4 of the fourth chapter of Hilbert-Ackermann. The authors explain three paradoxes in turn, each involving higher-order logic (pp. 92–98).

1. *Paradox of self-predication.* Let $P(F)$ be a second-order predicate. Then in particular, it can happen that P can be "predicated of itself," $P(P)$. Next, let $Pd(P)$ be the property that a predicate P can be predicated of itself. Then even the negation $\overline{Pd}$ is a higher-order predicate of predicates. Therefore, we can form $Pd(\overline{Pd})$ and $\overline{Pd}(\overline{Pd})$. If $\overline{Pd}(\overline{Pd})$, then $Pd(\overline{Pd})$, and if not $\overline{Pd}(\overline{Pd})$, then $\overline{Pd}(\overline{Pd})$, a contradiction just as in Russell's paradox (p. 93).

2. *Paradox of the liar.* Let $Bh(X)$ state that a person $\mathfrak{P}$ asserts (behaupten) X within given time limits t. Let $\mathfrak{P}$ state: "All things $\mathfrak{P}$ asserts within time limits t are false." This proposition, denoted $\mathfrak{A}$, can be expressed formally by a second-order quantifier as (ibid.)

$$(X)(Bh(X) \to \overline{X})$$

Formula $\mathfrak{A}$ implies any of its instances, in particular the one with $\mathfrak{A}$ in place of X, so we have $\mathfrak{A} \to (Bh(\mathfrak{A}) \to \overline{\mathfrak{A}})$. Switching the conditions gives $Bh(\mathfrak{A}) \to (\mathfrak{A} \to \overline{\mathfrak{A}})$ that together with $Bh(\mathfrak{A})$ gives $\mathfrak{A} \to \overline{\mathfrak{A}}$. The converse implication is derived similarly (p. 95).

3. *Least undefined number paradox.* The third paradox in Hilbert-Ackermann is the most involved one. Briefly stated, the notion of "least number not to become defined in the twentieth century in our logical symbolism" leads to a contradiction because the said statement can be expressed in the symbolism of Hilbert-Ackermann.

Gödel himself later stressed the importance of distinguishing between:

1. The notion of *truth.*
2. The notion of *provability* within a formal system.

Reading Gödel's own proof of 1931, we find such detailed constructions in it that they could not have been the way to the theorem. Rather, intuitive considerations as suggested by the discussion of the paradoxes in Hilbert-Ackermann revealed the possibility of incompleteness. Gödel's paper begins, indeed, with a description of the intuitive argument of the paradox of the Liar: "This statement of mine is false." Well, if it is false, it was true for the Liar to say it. If it was true, it was false. By replacing falsity with unprovability, one escapes the paradoxical consequences of such "self-referential" statements as the liar's: Consider

(8.1)   Statement (8.1) is unprovable.

Let us assume, as we always have to when discussing Gödel's theorem, that our system of proof is consistent. Then

(8.2)   If (8.1) is true, it is unprovable.
(8.3)   If (8.1) is false, the contrary of (8.1) is true, so (8.1) is provable.

The assumption in (8.3), the falsity of (8.1), leads to the provability of (8.1), which by consistency in turn gives that (8.1) is not provable. Therefore, the assumption of the falsity of (8.1) is impossible, and we have found a true but by (8.2) unprovable statement.

The preceding argument has an air of artificiality in it; especially the self-reference in (8.1) could be challenged as meaningless. In his paper, Gödel succeeds in constructing self-referential propositions in arithmetic that state their own unprovability, not unlike the construction of a logical formula in Hilbert-Ackermann that defines the least undefined number. In the formal development of his article, Gödel shows how such self-reference is effected by a method known as *Gödel numbering*. Arithmetic expressions are interpreted as sequences of symbols built according to the rules of the formal language of arithmetic. Each possible finite sequence of symbols is given a unique natural number as a code. Some sequences of symbols are correctly constructed formulas, and the corresponding Gödel numbers have the property that they code a correct formula. Let $gnf(A)$ ("Gödel number of formula A") be the code of the sequence of symbols A. We can now introduce an arithmetic one-place predicate $CF(n)$, which states that n is the Gödel number of a correctly formed formula. Thus, $CF(n)$ is a true arithmetic statement whenever n codes a correct formula. Next we code sequences of sequences of symbols. Some such sequences of sequences are correct derivations of the formulas that appear as the last sequence of symbols. Let $gnd(\mathcal{D})$ ("Gödel number of derivation $\mathcal{D}$") code such a sequence of sequences of symbols, and let $gnf(A)$ code the last sequence. Further, let $CD(m)$ express that m codes a correct derivation. Next, we introduce a two-place arithmetic *provability relation* $PR(n,m)$ such that $PR(n,m)$ is true whenever n codes a correct formula A, so that $gnf(A) = n$ and $CF(n)$ is true, and m codes a correct derivation $\mathcal{D}$ of A, so that $gnd(\mathcal{D}) = m$ and $CD(m)$ is true. We can now express provability of a formula A as the existence of a number that codes a derivation of A: $\exists x PR(gnf(A), x)$. This formula defines a one-place arithmetic predicate $\exists x PR(n, x)$ that has, for each n, a Gödel number $gnf(\exists x PR(n, x)) = k$. The crucial point of Gödel's proof is that there is a "diagonal number" in this construction, i.e., some number l such that $gnf(\exists x PR(l, x)) = l$. The formula with Gödel number l is nothing but $\exists x PR(l, x)$, so we have a formula that states its own provability. By treating the formula $\sim \exists x PR(n, x)$ in an analogous way, we find a formula in the language of arithmetic that states its own unprovability.

The possibility to express provability in arithmetic inside the language of arithmetic gives as a special case an arithmetic expression for the consistency of arithmetic. Axiomatizations of arithmetic contain, or, depending on the formulation, easily imply, the negation of $0 = 1$. Therefore, if $0 = 1$ also were provable, such an axiomatization of arithmetic would be inconsistent. The contrary is given by

$$(8.4) \qquad \sim \exists x PR(gnf(0 = 1), x)$$

(C) GÖDEL'S PAPER: THE PERFECTION OF A NOVICE. Gödel's 1931 paper is in four sections, with a grand opening that is fully justified by what the work accomplished:

> The development of mathematics in the direction of greater exactness has led, as is well known, to the formalization of large parts of it in a way in which proving can be conducted according to a few mechanical rules. The most comprehensive formal systems so far are the system of Principia Mathematica (PM) on the one side, the system of axioms of set theory of *Zermelo-Fraenkel* on the other (as further developed by J. v. Neumann).

Both of these systems are so broad that all proof methods used in today's mathematics are formalized in them, i.e., reduced back to a few axioms and rules of inference. Therefore, the conjecture lies close at hand that these axioms and rules of inference are even sufficient for the decision of *all* mathematical questions that can be formally expressed at all in the systems in question. It will be shown in the following that this is not the case, and that there are instead in both of the mentioned systems problems from the theory of the usual entire numbers, even relatively simple ones, that cannot be decided from the axioms.

Gödel's own intuitive argument for incompleteness in his section 1 is based on the possibility of coding such self-referential statements as those just given. We follow his argument from this section of his paper, with just some inessential changes in notation. He considers properties of natural numbers; that is, one-place arithmetic predicates. These can be ordered lexicographically in a sequence $R_1, R_2, \dots$. Thus, $R_n(m)$ states that the number m has the nth property in the sequence of all arithmetic properties. Next, let $Bew(R_n(m))$ (for the German "Beweisbar") express that $R_n(m)$ is provable, and consider the negation $\sim Bew(R_n(n))$ of $Bew(R_n(n))$. This negation is a one-place predicate, i.e., there is some k such that

(8.5) $\sim Bew(R_n(n)) = R_k(n)$

Gödel now proves that neither $R_k(k)$ nor $\sim R_k(k)$ is provable. (All of this assumes, as stated, that arithmetic is consistent.) If $R_k(k)$ is provable, it is true, and by (8.5) $\sim Bew(R_k(k))$ holds, contrary to the assumption that $R_k(k)$ is provable. If on the other hand $\sim R_k(k)$ is provable, then by (8.5) the contrary to $\sim Bew(R_k(k))$ would hold, namely $Bew(R_k(k))$ would hold. Then $R_k(k)$ is provable, which is against the assumption.

Gödel mentions that the preceding argument bears a resemblance to *Richard's paradox*: The expressions of the (French) alphabet can be put in a denumerably infinite sequence in lexicographical order. Next, those expressions that don't define a number are crossed out, and the ensuing denumerably infinite sequence diagonalized, to produce a number defined by an expression but not in the list.

In the main part of Gödel's article, section 2, which spans pages 176 to 191, the details are worked out, in particular the formal language and system of proof denoted by P (for the *Principia*) and the Gödel numbering of formal expressions and proofs, and secondly the expressibility inside the language of arithmetic of the properties of the syntax of arithmetic such as the notions of a formula, an arithmetic predicate, a derivation, and provability.

The preliminaries in section 2 include a presentation of primitive recursive functions and relations. It was a central source of inspiration for Rosa Politzer and others in the early 1930s.

The scheme of primitive recursion in Gödel, just called recursion at the time, is as follows (p. 179):

A number-theoretic function $\varphi(x_1, x_2 \ldots x_n)$ is *recursively defined* from the number-theoretic functions $\psi(x_1, x_2 \ldots x_{n-1})$ and $\mu(x_1, x_2 \ldots x_{n+1})$ when the following holds for all $x_2 \ldots x_n, k$:

$$\varphi(0, x_2 \ldots x_n) = \psi(x_2 \ldots x_n)$$

$$\varphi(k+1, x_2 \ldots x_n) = \mu(k, \varphi(k, x_2 \ldots x_n), x_2 \ldots x_n)$$

A number-theoretic function φ is *recursive* if there is a finite sequence of number-theoretic functions $\varphi_1, \varphi_2 \ldots \varphi_n$ that ends with φ and has the property that every function φ_k of the sequence is either defined recursively by two previous ones in the sequence or arises through substitution from any one of the preceding or is, finally, a constant or the successor function $x + 1$.

Next, a relation $R(x_1, x_2 \ldots x_n)$ is defined to be recursive if there is a recursive function $\varphi(x_1, x_2 \ldots x_n)$ such that $R(x_1, x_2 \ldots x_n)$ is equivalent to $\varphi(x_1, x_2 \ldots x_n) = 0$. It follows that recursive relations are closed with respect to the operations of propositional logic and also with respect to bounded quantification. The former observation is at the basis of the "logic-free" arithmetic of Bernays in the second volume of the *Grundlagen*, that of Curry (1941), and that of Goodstein (1945). Gödel himself gave a similar logic-free variant of arithmetic, the *Dialectica interpretation* (Gödel 1958).

Gödel's aim with the definitions and lemmas is to show that the functions and relations needed for his theorem, such as the provability relation $PR(n, m)$ in the previous subsection, can be expressed as recursive functions and relations in arithmetic. It follows that the functions are computable and the predicates decidable. Gödel gives a list of 45 such functions and relations (pp. 182–186) that ends with the provability relation, $x\,B\,y$ in Gödel's notation—in a few years, the notion of a formal system in general would be specified in these terms. Now the provability of a formula with Gödel number x can be expressed by Gödel's formula 46, $Bew(x) \equiv (Ey)\,y\,B\,x$.

Section 2 ends with the main theorem VI and its proof, "the general result about the existence of undecidable propositions," formulated in the special language Gödel had created, the details of which are not reproduced here. Section 3 presents consequences of the main result for various formal systems. First of all, theorem VI is used to give, as promised in the title of the paper, the incompleteness of the system of *Principia Mathematica* (p. 193):

> Theorem VIII: *In each of the formal systems mentioned in theorem VI, there are undecidable arithmetic propositions.*

It is at once noted that the same holds even for systems of set theory, and, in the other direction so to say:

> Theorem IX: *In all formal systems mentioned in theorem VI, there are undecidable problems of the first-order logical calculus.*

This theorem follows from the next result (p. 194):

> Theorem X: *Each problem of the form $(x)F(x)$ (with F recursive) can be reduced back to the question of satisfiability of a formula in the first-order logical calculus.*

By the correspondence between recursive relations and functions, the last problem can be expressed equivalently in the following form: With f recursive, is $f(x) = 0$ for all x? Kleene (1936) showed this question to be undecidable but, as noted by Davis (2013), failed to notice that Gödel's theorem X would have given at once the negative solution of the *Entscheidungsproblem* of predicate logic.

A short final section of Gödel's paper, section 4, briefly explains the second incompleteness theorem, numbered XI (p. 196):

> Let κ be an arbitrary recursive consistent class of formulas, then we have: The formula that states the consistency of κ is not κ-provable; in particular, the consistency of P is unprovable in P under the condition that P is consistent.

One who takes the trouble of going through the whole paper can only wonder how Gödel, with very short experience in research, was able to figure out the right formal machinery for his result and to execute the necessary steps with no errors whatsoever.

A comment on the title of Gödel's paper. In the title, the notion of "formal undecidability" of a proposition means that neither the proposition nor its negation is provable within the system. This notion has to be kept separate from the standard meaning of decidability and undecidability: A theory is decidable if there is an algorithm for deciding, for any formula of the theory, if it is derivable. If there is no such algorithm, the theory is undecidable. Gödel's notion is that of *independence*, sometimes rendered as "unsolvability" to distinguish it from the standard notion of undecidability. The Roman numeral I in the title indicates the plan to publish a continuation, namely a detailed proof of the second incompleteness theorem. Because of the rapid acceptance of his results, Gödel changed his plans. A proof of the second theorem along the lines he had planned can be found in the second volume of the *Grundlagen der Mathematik*, 1939.

Three years after the incompleteness paper, in 1934, Gödel gave the influential Princeton lectures on incompleteness, on which notes were prepared by Stephen Kleene and J. Rosser, with the title "On Undecidable Propositions of Formal Mathematical Systems." These notes contained what is now called the Herbrand-Gödel definition of general recursive functions. Just two years later, people in Princeton, including Church and Kleene, came up with apparently different proposals for a notion of computability that very soon were proved to be equivalent. The Gödel notes were not made generally available until some thirty years later, in Martin Davis' collection *The Undecidable* (1965), together with Kleene's work that contained the definition of *general recursive functions*, Church's λ-*definable functions*, Turing's *computable functions*, and Post's *combinatory processes*. All of these appeared in 1936 and turned out to define the same class of functions that the definition of Herbrand and Gödel does.

Gödel saw at once the significance of Turing's definition of "mechanical computability," as is shown by an untitled manuscript dated to 1938 (Gödel 1995, p. 168):

He has shown that the computable functions defined in this way are exactly those for which you can construct a machine with a finite number of parts which will do the following thing. If you write down any number $n_1, \ldots, n_r$ on a slip of paper and put the slip into the machine and turn the crank, then after a finite number of turns the machine will stop and the value of the function for the argument $n_1, \ldots, n_r$ will be printed on the paper.

There has been a very close to universal consensus that the various equivalent definitions of recursiveness correspond to an informal notion of computability. I know two exceptions. The first one was hinted at in Gentzen's review of Emil Post's 1936 formulation of the notion of a recursive function in the paper "Finite Combinatory Processes." He mentions other definitions of recursiveness by Church, Gödel and Herbrand, and Kleene, and then notes that a process is finite if it ends in a finite number of steps. In this connection, he points at the problematic aspect of the notion of general recursiveness by putting in italics his final comment (in Mentzler-Trott 2007, p. 76):

Of what kind the proof of finiteness of the process should be is, as also in Church, not specified.

A second critical voice is that of Rosa Politzer in her 1951 book and very explicitly in her 1959 article "Rekursivität und Konstruktivität." Briefly, the argument is that general recursiveness of a function f is a condition of the form $\forall x \exists y f(x) = y$. With Kleene, the proof of existence of the function value need not be constructive, and the notion therefore becomes too broad. If the existence proof is required to be constructive in some specific sense, the definition is circular because the aim was to capture the notion of constructivity through general recursiveness (p. 228).

Despite the impossibility of a definitive constructive notion of computability, something is gained in any case by general recursiveness: If it is agreed that all computable functions are at least *contained* in the general recursive ones, negative results about uncomputability and undecidability can be proved: Anything beyond the class of general recursive functions is even beyond the class of computable functions, however hazy in its borders the latter notion may be.

8.2. CONSEQUENCES OF GÖDEL'S THEOREM

(A) THE UNPROVABILITY OF CONSISTENCY. The foundational problems of Hilbert's program, such as consistency and completeness, are to be solved by mathematical means. The question naturally arises as to what methods can be used in the study of the foundational problems. Because the consistency of mathematics was at stake, Hilbert had required that the allowed methods be "absolutely reliable." He then identified such methods as those that can be seen as "symbolic manipulations of concrete signs according to specific mechanical rules." The basic acts in such manipulation consist in identifying two sequences of symbols as equal, in identifying a sequence of symbols as having a specific form, and in identifying a number of given sequences of symbols as an instance of a rule for the construction of a proof. (Say we have three sequences, the first two having the form of premises of a rule, and the last the form of a conclusion.)

Hilbert's thought was that the representation of mathematics as a mechanical manipulation of concrete symbol sequences according to finitary rules carries with it absolute evidence. By this it is meant not that errors are not possible but rather that any errors can be pointed out concretely, in the same way as typing errors. Those parts of mathematics that do not conform to the finitary restrictions must be "saved" by showing that they would never lead to a contradiction. Moreover, if a finitary statement with a concrete meaning is provable by the use of infinitistic notions and methods, it should also be provable finitistically. To put it as a slogan, *the infinite should be conservative over the finite.*

The specific "transfinite" form of arithmetic proof that worried Hilbert is as follows. Consider some property $A(n)$ of natural numbers such that the assumption that for all numbers n the negation of $A(n)$ is the case leads to an impossibility. Then, since it is not the case that all numbers have the negative property, there is some number x that has the contrary property $A(x)$. If we start running through the natural number series $0, 1, 2, \ldots$, the mere proof of existence of some x such that $A(x)$ is the case does not tell us anything about how far in the series we may need to go.

We can put the preceding as follows. Assume that the existence of a number with the property A is impossible, derive a contradiction, and conclude that there is a number with the property. This "transfinite" existence proof is an instance of the principle of indirect proof, the general form of which is: Assume not-A to be the case, derive a contradiction, and conclude that A is the case. One way to express this principle is to admit as true any instance of the "law of excluded middle" that says that for any statement A, either A or its contrary not-A is the case. Hilbert's specific aim was to justify the use of this law in cases in which an infinity of objects is considered, such as the infinity of the natural number series.

Gödel's theorem showed that the central aim of Hilbert's program is an impossibility. Gödel himself did not first make this conclusion in any categorical way, at least not in public, but Johann von Neumann did, after he had found that the consistency of arithmetic is among the un-provable Gödelian propositions. He conjectured that the consistency of those parts of mathematics that use infinitistic principles is unprovable in an absolute sense. However, in 1932, a way out was found through the translation from a classical to a constructive or intuitionistic system of arithmetic, mentioned among Gödel's achievements. It showed two things:

1. The use of indirect proofs in arithmetic does not lead to contradictions if other parts of arithmetic don't.

2. Constructive methods surpass finitistic methods.

Gentzen was led to develop a proof theory of arithmetic by which he had succeeded in proving its consistency by the end of 1934. Much later, in interviews Sue Toledo conducted with Gödel in 1974, she recorded Gödel's view as follows (Toledo 2011, p. 203):

> Hilbert's program was completely refuted, but not by Gödel's results alone. That Hilbert's goal was impossible became clear after Gentzen's method of extending finitary mathematics to its utmost limits.

(B) DIFFERENT APPROACHES TO GÖDEL'S THEOREM. There are at least four principal ways of approaching Gödel's two incompleteness theorems. Once the consistency of arithmetic was found to be among the unprovable propositions, approaches less intricate than Gödel's original one were found. We now discuss Gödel's approach and some alternatives.

1. *Coding of formulas and proofs*. Gödel's first approach was through the coding of the provability predicate in arithmetic and a diagonalization.

2. *Analysis of formal derivations in arithmetic*. The first alternative to Gödel's approach was Gentzen's direct proof of the underivability of a contradiction in a formal proof system for arithmetic, found in 1934 and published in 1936. Gentzen first developed general logical calculi called "natural deduction" and "sequent calculus," to be used in the analysis of the structure of mathematical proofs. The basic idea of proof analysis was to define simplifications of derivations (formal proofs) in natural deduction and sequent calculus and to show that these simplifications terminate in a predictable number of steps. Next, assuming that there is a derivation of a contradiction, simplifications are applied to the derivation until it assumes a particular form. Then it is shown that there cannot be any such form and thus no derivation of a contradiction. The crucial point of the method is the ordering of derivations according to their simplicity. It would not be useful to order derivations such that first all derivations that consist of one step of inference would be listed, then those with two steps, and so on, because there is already an infinity of one-step derivations. If we take any derivation and locate its place in the ordering, then there should be only a finite, bounded number of derivations in the ordering before the one we took. The principle that guarantees this termination in the case of arithmetic is called "transfinite induction up to the first ε-number." It is a principle that can be expressed in arithmetic but not proved; call it *Gentzen induction*. Here, an arithmetic principle was found that is independent of the standard axioms of arithmetic yet not a product of Gödel's method of coding. In another work, conducted in 1939 and published in 1943, Gentzen showed that all induction principles strictly weaker than Gentzen induction are in fact equivalent to the standard induction principle of arithmetic, and that Gentzen induction itself is unprovable in arithmetic. By this unprovability, the incompleteness of arithmetic was shown in a direct and natural way, in contrast to Gödel's original proof, which used an arithmetic coding of the provability

relation. Gentzen induction is a measure of the deductive strength of the axioms of arithmetic. With these results, Gentzen completely clarified the situation with Gödel's incompleteness theorems and initiated a field known as ordinal proof theory. It is the most beautiful but also the most difficult way of approaching Gödel's incompleteness theorems.

3. *The modal logic of provability.* The addition of provability as a primitive to logical language leads to an abstract logical formulation of incompleteness. The approach was suggested first by von Neumann in a letter to Gödel in January 1931 (see Gödel's *Collected Works*, vol. IV). In 1933, Gödel gave an interpretation of intuitionistic logic as a logic of provability. Toward the late 1930s, he further developed provability logic. In provability logic, one adds to standard propositional logic a logical operation by which from a formula A one can form a formula $\Box A$, to be read as "A is provable," or more briefly, "box A." Such an operation is easily suggested from the provability predicate in Gödel's own article: the predicate in Gödel's formula 46, $Bew(x) \equiv (Ey)y\,Bx$. In provability logic, propositional logic is extended with the provability operator and its axioms. It gives a neat, abstract approach to Gödel's theorem.

4. *Computable functionals.* Gödel's *Dialectica* interpretation uses a hierarchy of primitive recursive functionals to interpret arithmetic, which Hilbert introduced in 1925–26 in his attempt at the continuum hypothesis. Classical logic allows the use of the law of excluded middle, and therefore also indirect existence proofs. These are excluded in intuitionistic logic, and it was thought that a formal system that uses intuitionistic logic instead of classical logic could form the basis for a proof of the consistency of arithmetic. The Gödel-Gentzen translation of 1932–33, mentioned in Section 8.1, showed this to be a feasible plan, but Gödel was still not satisfied with intuitionistic logic itself. Through computable functionals he found an alternative "logic-free" approach to the foundations of arithmetic.

(C) INFORMAL STATEMENTS OF GÖDEL'S THEOREM. There have been many attempts at generally comprehensible explanations of Gödel's theorem. The publication of Gödel's correspondence in two volumes in 2003 brought to light three such brief statements by Gödel himself. In the passages that will be quoted, "decidable" and "undecidable" are meant in the sense of "solvable" and "unsolvable." The first, in chronological order, is from 1962 (*Collected Works* V, p. 176):

> My theorems show only that the *mechanization* of mathematics, i.e., the elimination of the *mind* and of *abstract* entities, is impossible, if one wants to have a satisfactory foundation and system of mathematics.
>
> I have not proved that there are mathematical questions undecidable for the human mind, but only that there is no *machine* (or *blind formalization*) that can decide all number theoretic questions (even of a certain very special kind).

A second statement, from 1967, reads (*Collected Works* V, p. 162):

> Whether every arithmetical yes or no question can be decided with the help of some chain of mathematical intuitions is not known. At any rate it has not been proved that there are arithmetical questions undecidable by the human mind. Rather what has been proved is only this: Either there are such questions or the human mind is more than a machine. In my opinion the second alternative is much more likely.

A third statement, from 1969, is (*Collected Works* IV, p. 330):

> A few immediately evident axioms from which *all* of contemporary mathematics can be derived do not suffice for answering all Diophantine yes or no questions of a certain well-defined simple kind.
>
> Rather, for answering all these questions, infinitely many new axioms are necessary, whose truth can (if at all) be apprehended only by constantly renewed appeals to mathematical intuition, which is actualized in the course of the development of mathematics. Such an intuition appears, e.g., in the axioms of infinity of set theory.

In volume III of the *Collected Works*, a finished 20-page lecture from 1951 was published for the first time. Its title is "Some basic theorems on the foundations of mathematics and their implications." Gödel put into italics the following passage (p. 310):

> *Either mathematics is incompletable in the sense, that its evident axioms can never be comprised in a finite rule, that is to say, the human mind (even within the realm of pure mathematics) infinitely surpasses the powers of*

any finite machine, or else there exist absolutely unsolvable diophantine problems of the type specified.

The Diophantine problems are questions regarding whether a Diophantine equation has a solution. Such an equation is any equation of the form $ax^n + bx^{n-1} + \cdots + cx + d = 0$ in which the coefficients $a, b, \ldots, c, d$ are integers and the solution also has to be an integer.

Gödel's "blind formalization," in the first quotation, would consist of a system in which the proofs of all theorems can be generated, one after the other, and the possibility to "decide" any number-theoretic question means that the list of theorems contains one that answers it. Even if there were such a formalization, it would decide number-theoretic questions only in a very weak sense. Abstractly speaking, if there is a solution, it will eventually be found. For us, instead, a question remains undecided as long as a solution has not been found. Things would be different if we could determine, on the basis of the question posed, an upper limit to how far in the list of theorems we have to go in order to find an answer or to conclude that there is no answer.

8.3. TWO "BERLINERS"

In 1925, at the premature age of twenty-two, Johann von Neumann finished a long paper with the title *Zur Hilbertschen Beweistheorie*, which would be published two years later. The arguments in the paper are complicated and were studied by very few. A main result was a consistency proof of arithmetic with the principle of inductive proof restricted to quantifier-free formulas. Another important observation was that Ackermann's (1924) purported proof of the consistency of arithmetic and analysis contained a fatal flaw. In these years, von Neumann was enormously active and produced one mathematical invention after another, including also themes at the borderline of pure mathematics such as the Hilbert space formulation of quantum mechanics (in 1926) and game theory (in 1928). Thus, logic and foundations were just one topic among many others.

Von Neumann lectured on proof theory in 1928–29 and 1930–31 in Berlin. The latter lecture course turned into an event without precedent, for von Neumann became solely responsible for the immediate acceptance of Gödel's incompleteness theorem. His first reaction to Gödel's explanation of incompleteness, at the Königsberg meeting of September 1930, was to point out the existence of undecidable polynomials. Contemporary accounts tell of the tremendous excitement that these developments aroused among the Berlin mathematicians. Carl Hempel, later a very famous philosopher, was one of the participants, and his recollection (Hempel 2000, pp. 13–14), even evidenced by contemporary correspondence (see Mancosu 1999), is as follows:

> I took a course there with von Neumann which dealt with Hilbert's attempt to prove the consistency of classical mathematics by finitary means. I recall that in the middle of the course von Neumann came in one day and announced that he had just received a paper from ... Kurt Gödel who showed that the objectives which Hilbert had in mind and on which I had heard Hilbert's course in Göttingen could not be achieved at all. Von Neumann, therefore, dropped the pursuit of this subject and devoted the rest of the course to the presentation of Gödel's results. The finding evoked an enormous excitement.

These are later recollections; for example, it is known that von Neumann got the proofs of Gödel's paper around the tenth of January 1931. The lectures before that date were based on what Gödel had told him and what he had found out himself.

Jacques Herbrand was born in 1908 and received his education at the prestigious Ecole normale superieure of Paris. He finished his thesis *Recherches sur la théorie de la démonstration* at the precocious age of 21 in the spring of 1929. The thesis contains what Herbrand is remembered for, *Herbrand's theorem*, by which he tried to reduce the decidability of predicate logic to that of propositional logic.

What little is left of Herbrand's papers is kept at the library of the Ecole normale under the name "Dossier Vessiot: VES 50 Herbrand," from which I have drawn most of the details that follow.

Herbrand's knowledge of logic and foundations stems mainly from Hilbert and Ackermann (1928) and from the *Principia Mathematica*. For example, as we have seen, Hilbert-Ackermann practically call for the deduction theorem as a means to show that any derivation of a theorem from mathematical axioms can be converted into a derivation of an implication in pure logic. Herbrand's thesis is the first place in which this deduction theorem is proved (1930, p. 107). A popular essay of his on "Hilbert's logic" contains the following informal statement (1930a, p. 213):

> It can easily be shown that if the proof of a proposition P in a theory makes use of only hypotheses $H_1, H_2, \ldots, H_n$, then $H_1 \& H_2 \& \ldots \& H_n. \supset P$ is an identity, and conversely if such a proposition is an identity, then P is true in the theory under consideration.

The title of the printed version is *Les bases de la logique hilbertienne*, changed from *Une nouvelle logique mathématique*, as in the 19-page typewritten manuscript in the Dossier Vessiot.

After a year of military service and a defense of the thesis, Herbrand went to stay for the next academic year in Germany, first Berlin from October 1930 to late spring 1931, and from then until July in Hamburg and Göttingen. These stays were in part prompted by his work on algebra, where Emil Artin in Hamburg and Emmy Noether in Göttingen were the leading figures.

Crucial contemporary evidence of von Neumann's reaction to Gödel, in his lectures titled *Hilbertsche Beweistheorie* in the Winter Semester 1930–31 and otherwise, is given witness through the presence of Jacques Herbrand in Berlin. There is a letter of Herbrand's of 28 November 1930 to the director of the Ecole normale, Ernest Vessiot, in which he tells of the "extremely curious results of a young Austrian mathematician who succeeded in constructing arithmetic functions Pn with the following properties: one calculates Pa for each number a and finds $Pa = 0$, but it is impossible to prove that Pn is always zero."

Herbrand's letter to his friend Claude Chevalley five days later, on 3 December, has the looks of a rough diamond: hard to take for what it is without a close look, because of the worst handwriting imaginable, but full of sparkling ideas that seem to spring from nothing. It shows that von Neumann had come to the second incompleteness theorem through what must have been Gödel's informal explanation in Königsberg, very similar to the introduction of Gödel's 1931 paper with its self-referential diagonalization through arithmetization. In the letter, Herbrand explains von Neumann's argument as follows:

> Let T be a theory that contains arithmetic. Let us enumerate all the demonstrations in T; let us enumerate all the propositions Qx; and let us construct a function $Pxyz$ that is zero if and only if demonstration number x demonstrates Qy, Q being proposition number z.

We find that $Pxyz$ is an effective function that one can construct with arithmetic functions that are easily definable.

> Let β be the number of the proposition $(x) \sim Pxyy$ ($\sim$ means: not); let Ax be the proposition $\sim Px\beta\beta$
> $\qquad A$ the proposition $(x).Ax$ (Ax is always true)
> Ax, equivalent to: demonstration x does not demonstrate the proposition β; so

$Ax. \equiv .$ demonstration x does not demonstrate A

Let us enunciate:

$Ax. \equiv . \sim D(x, A)$

1) Ax is true (for each cipher x); without it $D(x, A)$ would be true; therefore A; therefore Ax; therefore $\sim D(x, A)$.
2) A cannot be demonstrated for if one demonstrates A, Ax would be false; contradiction.
Therefore: $A0, A1, A2\ldots$ are true
$(x)Ax$ cannot be demonstrated <u>in T</u>

Next in Herbrand's letter comes von Neumann's striking addition to Gödel's first theorem: with $D(x, A)$ standing as above for: proof number x demonstrates proposition A, Herbrand writes in the letter the magic formulas

> 3)$\qquad \sim A \rightarrow D(x, A)$ et $D(z, \sim A)$
> therefore: $\sim (D(x, A)$ et $D(z, \sim A)) \rightarrow A$

The conclusion, for the unprovable proposition A, is that "if one proves consistency, one proves A." Consistency requires that for any proposition A, there do not exist proofs of A and $\sim A$; that is, $\sim \exists x \exists z (D(x, A)$ et $D(z, \sim A))$, or in a free-variable formulation, for each x and z, $\sim (D(x, A)$ et $D(z, \sim A))$.

It is surely the case that Gödel had not found the second incompleteness theorem by September 1930. Had that been the case, von Neumann would not have had any independent discovery of the second incompleteness theorem to report.

The two-volume correspondence of Gödel (2003) contains exchanges with von Neumann that bring to light the circumstances of Gödel's greatest discovery. Von Neumann explained the unprovability of the consistency of arithmetic to Gödel in a letter in November 1931 but wanted to delay publication until Gödel himself was finished. Gödel responded that he, too, had discovered what is now called Gödel's second incompleteness theorem. Von Neumann withdrew his plans to publish but continued to work on the topic. A letter from January 1932 shows that he had arrived at the idea of provability logic, with the specific result that a propositional logic to which a provability operator has been added is decidable. He also offered to "write it up" if Gödel is interested or "has not thought of it himself." Apparently, Gödel did not catch on to this at that moment, and consequently we do not have any von Neumann version of provability logic, unless, of course, some forgotten notes pop up from some source. A hint is contained in Herbrand's letter, the magic formula with the relation $D(x, A)$, "proof number x demonstrates proposition A." Rather startlingly, a bit over two years after von Neumann's offer to explain the details of a provability logic, we find Gödel writing in a letter to von Neumann on 14 March 1933 about "an interpretation of Heyting's calculus for sentential logic by means of the concept 'provable'," as if this were an idea of his, not von Neumann's.

The "Berlin section" of proof theorists came to an abrupt end, though not, contrary to what has often been told, through von Neumann's abandoning foundational study after Gödel's result. Von Neumann conjectured on the basis of his independent proof of the second incompleteness theorem that "the consistency of mathematics will remain forever unprovable," which could undermine the interest in foundations, but as seen, he also continued with the idea of a provability logic. The conjecture is even mentioned in Bernays (1932), a brief review of the foundational situation after Gödel's theorem. In 1931, von Neumann left for Princeton, not having won a professorship in Germany. The matter in itself is hard to believe, if one compares how he tried and how most everyone else found a position, even those with only a fraction of his scientific production. This failure, and the subsequent loss of von Neumann to the United States, was caused by the anti-semitism of the German academic culture, as reported in detail in Hashagen (2008).

The second Berliner Jacques Herbrand stayed in Berlin until May 1931. Three lectures of his on proof theory are recorded in the yearly accounts of activities of mathematical societies as found in the *Jahresberichte der Deutschen Mathematiker-Vereinigung*: "Zur Beweistheorie" (On proof theory), 21 April 1931, Berlin; "Hilbertsche Metamathematik und die Widerspruchsfreiheit der Arithmetik" (Hilbertian metamathematics and the consistency of arithmetic), 22 April 1931, Halle; and "Über einige metamathematische Fragen" (On some metamathematical questions), 14 July 1931, Göttingen.

In Göttingen, if not earlier, Herbrand must have realized that his idea of who had done what in logic and foundational research was biased. His *Logical Writings*, edited by Warren Goldfarb in 1971 (French first edition 1968), reveals altogether five references to Bernays and five times as many to Hilbert, one of them ascribing even the Ackermann function to Hilbert (p. 291). Of the references to Bernays, two (pp. 95, 97) are to the article by Bernays and Schönfinkel (1928) on the decision problem, the third (p. 120) to Bernays (1927b), a four-page "Addition to Hilbert's talk on *The Foundations of Mathematics*." An "Unsigned note on Herbrand's thesis written by Herbrand himself" tells that when Herbrand began his work, the state of questions about decidability and consistency was as follows (pp. 274–275):

> Hilbert had limited himself to giving schemata for proofs, nearly all of which were subsequently seen to be false. Only his proof of the consistency of the simplest axioms of arithmetic could be sketched in a somewhat complete manner by his pupil Bernays. The only important contribution to the theory had been furnished by von Neumann, who proved in a complete manner the consistency of a fragment of arithmetic.

That was the fourth mention of Bernays, and only the stay in Göttingen brought a change in Herbrand's vision, seen from a reference to discussions with Bernays in a note he added to his last paper finished there, in July 1931 (Van Heijenoort 1967, p. 294). The paper clearly shows the influence of von Neumann's conjecture about the unprovability of consistency.

As mentioned, Herbrand had worked on algebra with Artin and Noether toward the end of his German stay. He had at that time already decided to leave logical studies behind and to concentrate on algebra, as can be seen from his application for a Rockefeller scholarship to be spent in Princeton.

Shortly after leaving Göttingen, Herbrand died in a mountaineering accident on 27 July when he was only twenty-three years old. Documents at the Ecole normale in Paris show that the accident was a trivial fall on easy ground during the descent. There is in Paris even a letter of condolence to Herbrand's father in which "Jean de Neumann" describes Herbrand as a "collaborateur et cointeressé tres precieu a Berlin." The letter was sent from Hungary, but von Neumann was already preparing to become professor of mathematical physics in Princeton, and that move left no one in Berlin to continue work on proof theory.

* * *

There are fewer formal details in this chapter than in most of the others. In part, this is because approaches to incompleteness have changed during the more than eighty years since its discovery, so it isn't worth going into all the details. Also, there are other sources that discuss the details of Gödel's paper, in particular the introduction to the first volume of his *Collected Works* (Gödel 1986). John Dawson's biography *Logical Dilemmas: The Life and Work of Kurt Gödel* is the first place to start exploring further.

The passages from Gödel's 1931 paper are my own translations straight from the original German text, unaffected by the interpretations in Van Heijenoort's *From Frege to Gödel*, namely from an offprint of the paper that was presented to me by Georg Henrik von Wright, my predecessor as Swedish Professor of Philosophy at the University of Helsinki. He had received the offprint, together with the one on completeness, from his professor Eino Kaila, who was associated with the Vienna Circle and came to know Gödel in the 1930s.

Concerning the enigmatic Jacques Herbrand, whose life came to such an abrupt end at twenty-three, my warm thanks go to Giuseppe Longo for his help with my visit to the archives of the Ecole normale, and to Warren Goldfarb and Peter Roquette for their generous sharing of copies of Herbrand's letters with me, Peter in particular for the amazing letter of 3 December 1930.

9

THE PERFECTION OF PURE LOGIC

Gerhard Gentzen, a student of Paul Bernays with whom Hilbert was working, set as his objective in the early 1930s "to study the structure of mathematical proofs as they appear in practice." He presented the general logical structure of mathematical proofs as a system of rules of proof by which a path is built from the assumptions of a theorem to its conclusion. Earlier formalizations of logic had given a set of axioms and just two rules of inference. Another essential methodological novelty in Gentzen's work was that he presented proofs in the form of a tree instead of a linear succession from the given assumptions to the claim of a proof. Each step in a proof determined a subtree from the assumptions that had been made down to that point, and these parts could be studied in isolation. Most importantly, such parts of the overall proof could be combined in new ways, contrary to the earlier linear style of proof. Gentzen was able to give for proofs in pure logic, that is, without any mathematical axioms, combinatorial transformations that brought these proofs into a certain direct form. Questions such as the consistency and decidability of a system of rules of proof could then be answered.

Gentzen's first attempt at a logic of mathematical proofs led to the system of *natural deduction* and a fundamental result by which all formal proofs in the system can be brought into a certain "normal form," with the property that all formulas in a proof in normal form are parts of the assumptions or conclusion of the proof. Consistency is an immediate consequence. To extend the result to cover classical logic, Gentzen formulated the logical rules in a *sequent calculus*. Later refinements of the latter have led to the most powerful methods for analyzing the structure of proofs known today. In addition to the analysis of a given proof, these systems support *proof search*, often in a way that makes proof search an algorithmic task.

9.1. NATURAL DEDUCTION

(A) THE SITUATION IN 1932. A perplexing situation regarding the consistency problem of arithmetic had arisen with the arrival of Gödel's incompleteness theorem, a result that had become known during the fall of 1930. As we just saw in Chapter 8, it was at once well received, especially through the forceful endorsement of Johann von Neumann. Bernays had been in contact with Gödel to clarify the consequences of the result for Hilbert's enterprise of "securing the foundations of mathematics" through a consistency proof. In fact, the preface Bernays wrote to the first volume of the *Grundlagen der Mathematik*, dated March 1934 and published later that year, tells that the manuscript had in practice been finished in 1930, but the whole project had to be thought through again when Gödel's result became known. A finitary consistency proof of the kind envisaged by Hilbert turned out to be an impossibility. Some, von Neumann as foremost, declared the foundational enterprise dead.

Gentzen, who was just twenty-two years old in 1932, would have nothing of the defeatism of von Neumann. Where his confidence came from is not known, but it was confirmed in less than a year by the interpretation of classical Peano arithmetic in intuitionistic Heyting arithmetic. This result proved Bernays' admission that the help of Brouwer's intuitionistic mathematics would be needed to overcome the dead end of Hilbert's *Beweistheorie*. The consistency proof itself was finished late in 1934.

The perfection of pure logic through Gentzen's proof systems of natural deduction and sequent calculus was a by-product of his consistency program. Sometime in early 1933, he realized that his grand plan of proving the consistency of formal arithmetic through an analysis of derivations in natural deduction would not go through as such. He was able to bring natural derivations into a normal form from which consistency would follow, but only in an intuitionistic formulation, and with the scheme of induction left out. These failures led to revisions in Gentzen's plan for a doctoral thesis on the consistency of arithmetic, including a submission of his discovery of the translation from classical to intuitionistic arithmetic as a separate article, and his concentration on the part that dealt with pure logic. The latter soon led to the sequent calculi for intuitionistic and classical logic that became the main topic of the finished thesis. Gentzen developed the calculi and proved the main result, Gentzen's *Hauptsatz*, between March and late May, when the finished thesis was handed in; a sort of emergency solution in a situation in which his teacher Bernays had already been expelled from the university on racial grounds.

Next to Gentzen's published thesis, the discovery of its early handwritten version and related documents have brought to light the discovery of proof systems. All sources are found in English translation in the collection *Saved from the Cellar: Gerhard Gentzen's Shorthand Notes on Logic and Foundations of Mathematics.*

(B) ACTUAL PROOFS IN MATHEMATICS. Gentzen began by studying how "one actually carries through proofs in mathematics." He observed that the prevailing method of formally presenting proofs did not match the practice: Mathematical statements were formalized in the language of logic, and especially the starting points of proofs, the mathematical axioms. Logic itself was also axiomatized, with axioms exemplified by $(A \supset (B \supset C)) \supset (B \supset (A \supset C))$. The "horseshoe" implication symbol, invented by Giuseppe Peano, was just an inverted capital letter C that was later stylized into $\supset$. It reveals what the preceding axiom does: If you read $\supset$ as "consequence" (for the C inverted), "follows," or "if..., then ...," whatever is handiest, you get:

If from A it follows that C follows from B, then from B it follows that C follows from A.

Think of A and B as assumptions, and the axiom prescribes that C follows, in whichever order you take the assumptions A and B.

The rest of the logical axioms have similar intuitive meanings. They were clear to Frege, who was the main inventor of the axioms. Later, his identification of the principles of proof turned into "symbolic logic," interpreted as a formal game, and the meaning of the axioms was by and large forgotten.

There were just two rules of inference in axiomatic logic: From $A \supset B$ and A to infer B was the propositional one, and universal generalization was the other. In the latter, a universal quantifier could be introduced if a statement was proved for an arbitrary object, as denoted by an *eigenvariable*. The precise statement of conditions for universal generalization was a great achievement of Frege's.

The application of Frege's logic to mathematical proofs, as in the work of Peano and Russell, proceeds through expressing the mathematical axioms with the language of logic, and in the application of the two principles of proof. Here is a simple example, the axiomatic theory of equality. The axioms are reflexivity, symmetry, and transitivity:

$$a = a, \quad a = b \supset b = a, \quad a = b \,\&\, b = c \supset a = c.$$

It is next to impossible to put the logical codification of mathematical proofs in terms of axiomatic logic into actual use. Say the expression of transitivity of equality in the Euclidean style, $a = c \,\&\, b = c \supset a = b$, already has an axiomatic proof that cannot be shown here because it is too broad to be printed. The proofs are often so wicked that the only feasible way to construct them would be to do them first in a calculus of natural deduction and then apply a translation algorithm into proofs in axiomatic logic.

When Gentzen started his program in early 1932, he had no difficulty in putting the ruling axiomatic logical tradition aside. The aim of axiomatic logic had been dictated by Frege's and Russell's doctrine of *logicism*, by which logical axioms express the most basic logical truths and logical proofs just add more truths to the basic stock. Logical principles, Gentzen's rules of proof, show how to move from given assumptions to a conclusion. Gentzen would grant, at most, that if the assumptions are *correct* (richtig), also the conclusion should be.

The conceptual order in Gentzen is different from that of logicism. In the latter there cannot be any doubt that logical proofs preserve correctness, because, if we take the doctrine seriously, such proofs are based on the ultimate notion of logical truth in the simplest possible manner. The axioms are such truths, and if $A \supset B$ and A are, also B is. There are no hypotheses, so this inductive argument is strictly local in character.

In logicism, mathematical truth is subordinate, and perhaps even reducible, to logical truth. If the reduction succeeds, the foundational problems of mathematics are solved for good. In Gentzen, instead, the very problem is to find a notion of correctness, in the first place for arithmetic, that is supported by logical inferences.

By September 1932, Gentzen had finalized his set of logical principles of proof, what is known as natural deduction (*natürliches Schliessen*, perhaps more properly rendered as natural inference, or even natural reasoning). His analysis of "actual proofs" in mathematics led to intuitionistic logic, a topic well defined after Arend Heyting's 1930 axiomatization that had the axiomatizations of *Principia Mathematica* and Hilbert-Bernays (table 7.1) as a basis.

A year later, Heyting explained the logical connectives in terms of proof, or perhaps better, sufficient conditions for proof: $A \& B$ is proved whenever A and B have been proved separately, $A \vee B$ is proved whenever one of A and B has been proved, and $A \supset B$ is proved whenever any proof of A turns into some proof of B. For the quantifiers, $\forall x\, A(x)$ is proved whenever $A(y)$ is proved for an arbitrary y, and $\exists x\, A(x)$ is proved whenever $A(a)$ is proved for some object a. It was soon realized that the explanation of implication need not reduce a proof of $A \supset B$ into something simpler, for A could have been obtained by any proof.

In the collection of stenographic notes that Gentzen wrote, there is a set from the fall of 1932 consisting of some 25 big stenographic pages, with a few pages added the following spring and ten more in October 1934. The title is "Formal conception of the notion of contentful correctness in pure number theory, relation to proof of consistency" (*Die formale Erfassung des Begriffs der inhaltlichen Richtigkeit in der reinen Zahlentheorie, Verhältnis zum Widerspuchsfreiheitsbeweis*). — I translate *inhaltlich* as contentful. Gödel suggested in the 1960s "contentual," but my translation is at least an English word. Georg Kreisel disliked it: He told me in July 2010 that one should just use the word *meaning*. *Inhaltlich*, then, would be *meaningfully*, or perhaps *in terms of meaning*. I regret not having asked what he thought of Gödel's invented word.

Most of Gentzen's manuscript was written within a month in October and November, and it was meant to be a groundwork for systematic formal studies, after the basic structure of mathematical reasoning had been clarified in September. I abbreviate the manuscript in the same way he did, as **INH**. The first task in it is to explain the notion of *correctness* for intuitionistic logic, quite similarly to Heyting's explanations. In the case of $A \& B$ and $A \lor B$, a reduction is achieved, but $A \supset B$ remains problematic.

Bernays was well aware of the problem that in a case of iterated implications such as $(A \supset B) \supset C$, the correctness of C depends on the correctness of another conditional statement, $A \supset B$. This is a problem of well-foundedness. A related problem is circularity: If, as in Heyting's explanation, a proof of $A \supset B$ takes any proof of A and gives as a result some proof of B, the notion to be explained, proof, is already assumed.

Once correctness for statements has been explained, it can be applied to statements in proofs. Here is the lesson from Gentzen's analysis.

Reduction to components. *If $A \supset B$ is provable, it should have a proof that is somehow made up from the components of $A \supset B$.*

The correctness of a notion of proof with this property would not be circular.

What is the notion Gentzen was searching after? Looking at his rules of natural deduction, a specific feature of most of the rules strikes the eye:

$$\frac{A \quad B}{A \& B} \, \&I \quad \frac{A \& B}{A} \, \&E \quad \frac{A \& B}{B} \, \&E \quad \frac{A}{A \lor B} \, \lor I \quad \frac{B}{A \lor B} \, \lor I \quad \frac{A \supset B \quad A}{B} \, \supset E$$

In the introduction rules, the premisses are *subformulas* of the conclusion, whereas in the elimination rules, it is the other way around. There remain the introduction rule for implication and the elimination rule for disjunction that have a schematic character different from the preceding:

$$\frac{\begin{array}{c} [A] \\ \vdots \\ B \end{array}}{A \supset B} \, \supset I \qquad \frac{A \lor B \quad \begin{array}{c} [A] \\ \vdots \\ C \end{array} \quad \begin{array}{c} [B] \\ \vdots \\ C \end{array}}{C} \, \lor E$$

Intuitionistic propositional logic results when the rule of *falsity elimination* is added to these rules: There is a constant proposition called falsity and denoted $\bot$, with negation defined by $\neg A \equiv A \supset \bot$, and with the rule $\bot E$ by which any formula can be concluded from $\bot$. Intuitionistic predicate logic is obtained by adding the quantifier rules:

$$\frac{A(y)}{\forall x\, A(x)}\;\forall I \qquad \frac{\forall x\, A(x)}{A(t)}\;\forall E \qquad \frac{A(t)}{\exists x\, A(x)}\;\exists I \qquad \frac{\exists x\, A(x) \quad \overset{[A(y)]}{\overset{\vdots}{C}}}{C}\;\exists E$$

In rules $\forall I, \exists E, y$ is an eigenvariable.

The introduction rules of Gentzen's natural deduction are formal versions of Heyting's explanations. For the elimination rules, different motivations and criteria have been presented, as discussed in von Plato (2012).

(C) THE DISCOVERY OF NATURAL DEDUCTION. The setup of natural deduction in Gentzen's printed thesis begins with the following one-sentence paragraph:

> *Wir wollen einen Formalismus aufstellen, der möglichst genau das wirkliche logische Schliessen bei mathematischen Beweisen wiedergibt.*

in English, it reads: *We want to set up a formalism that reproduces as precisely as possible the real logical reasoning in mathematical proofs.* Prior to 2005, knowledge of the discovery of natural deduction was limited to such remarks in Gentzen's published thesis. There he goes through three examples of theorems of logic, taken from Hilbert and Ackermann's axiomatic formulation, trying to "see their correctness in the most natural way possible." Then he states that the characteristic property of such logical reasoning is that it starts from assumptions, contrary to the "logistic" tradition. How he actually found the calculus, presented as a finished set of rules, is not detailed; it is only verified that the intuitive reasoning with the three examples can be reproduced in the calculus. A similar path is taken in his proof of the consistency of arithmetic (Gentzen 1936). It uses natural deduction with a notational variant taken from sequent calculus. A lengthy example is used for determining the modes of actual inference in mathematical proofs.

One assumption about the origins of natural deduction is that Gentzen converted the handsome axioms of Hilbert and Bernays into rules. Yet another possibility is the explanation of the logical connectives by Heyting, which came to be known as the "BHK explanation," but it is unknown if Gentzen knew about it. If not, the letter G could be added to BHK. It is instead clear that Gentzen had carefully studied Heyting's (1930a) formalization of intuitionistic logic, the notation of which he mostly followed. Heyting produced an axiomatization with 11 axioms for the propositional part, not as much along the model of *Principia Mathematica*, as has often been said, but along the lines of the Hilbert-Bernays axioms, in which all connectives and quantifiers are present. Contrary to earlier practice, implication elimination was not the only propositional rule, but there was also an explicit rule of conjunction introduction. Heyting writes that it could be dispensed with but that the proofs would then be "even more intricate" (*verwickelt*, a word with a negative connotation). Indeed, a little reconstruction shows that this is the case. Heyting's first theorem is $A \& B \supset A$, and three axiom instances and a step of implication elimination are given as a proof outline. A complete proof, as in section 6.4(A), has 11 lines, 3 of them conjunction introductions. Such steps can be replaced by a combination of axioms and just implication eliminations.

As noted in Section 6.4(A), the turning of Heyting's proof outlines into formal derivations involves so much work that one lets it be, but the way to proceed becomes clear: one starts reading Heyting's axioms and theorems in terms of their intuitive content. For example, the axiom $(A \supset B) \supset ((B \supset C) \supset (A \supset C))$ is seen as a rule by which two previous axioms or theorems $A \supset B$ and $B \supset C$ combine into a new one, exactly as in the work of Bernays (1926) that Heyting refers to. His first theorem, $A \& B \supset A$, becomes: from $A \& B$ follows A. From this perspective, Gentzen's discovery of natural deduction was nothing more than a systematic undoing of the intricate axiomatic formal proofs *à la Heyting* into the intuitive terms of what follows from what.

In February 2005, during a visit to Erlangen, I found among the articles and documents Christian Thiel had collected about Gentzen a photocopy of a handwritten early version of Gentzen's thesis that he had obtained through the good offices of Eckart Menzler-Trott in

1988. Remarkably, part of page two of the manuscript is reproduced photographically among the frontmatter of Gentzen's *Collected Papers* of 1969, with the indication that the manuscript comes from Bernays. Most of the manuscript is written in a formal hand ("Sütterlin-Schrift") that is not so easy to read, as one can gather from the photograph there. In January 2008, I recovered the missing last page from the original manuscript, checked some symbols that were "indicated in red" as Gentzen wrote but not clearly visible in a black and white photocopy, and so on. The piece is kept in the Bernays collection of the ETH.

Another piece I found in Erlangen back in 2005 were a few pages of Gentzen's in shorthand notation that I tried to decipher. A complete translation to German was soon prepared by Thiel, who reads fluently Gentzen's "unified shorthand" (*Einheitskurzschrift*) that had replaced Gabelsberger and some other earlier systems. The piece is titled *Five forms of natural calculi*, dated 23 September 1932, and I saw that it made understandable the way in which Gentzen arrived at the specific form of natural deduction. The easy answer is that he tried out all possible ways consonant with his idea that actual mathematical reasoning proceeds by hypotheses or assumptions, rather than instances of axioms, and that he was a logical genius who got things right.

The handwritten thesis indicates a proof of the equivalence of natural deduction and axiomatic logic, but the axioms of the latter are those of Hilbert and Ackermann, not the elegant axioms of Hilbert and Bernays that Heyting also used in part. Direct translations are given from the calculus of Hilbert and Ackermann to natural deduction and back. In the printed version of the thesis, the logical axioms come from the Hilbert-Bernays formulation, and there is no direct translation from natural deduction to axiomatic logic. First, a straightforward translation from derivations in axiomatic logic to natural deduction is given: It is sufficient to substitute the axioms by their derivations in natural deduction and to leave implication elimination intact. Then follows, instead of a translation back, a translation from natural deduction to sequent calculus, and finally the hard task that closes the circle, the translation from sequent calculus to axiomatic logic. The part of the last one that shows how to convert to axiomatic form a step of derivation that moves from B concluded from the assumptions Γ and

A into $A \supset B$ concluded from Γ, so in terms of sequent calculus, the step from $\Gamma, A \to B$ to $\Gamma \to A \supset B$ contains in fact a proof of the deduction theorem.

As said, what is left of the road to natural deduction are the few pages of *Five forms of natural calculi*. The rules appear in the beginning of that document, with a "boxed" date typical of the stenographic notes. The letters in the rules are direct adaptations of Gentzen's German nomenclature.

Table 9.1. The *"Five forms"* rules of natural deduction

$\boxed{23.\text{IX}.32}$ *Inference schemes:*

$$
\begin{array}{cccccc}
AI & AE & OI & OE & FI & FE
\end{array}
$$

$$
\cfrac{A \quad B}{A \& B} \quad \cfrac{A \& B}{A} \quad \cfrac{A \& B}{B} \quad \cfrac{A}{A \vee B} \quad \cfrac{B}{A \vee B} \quad \cfrac{A \vee B \quad \begin{array}{c} A \\ \vdots \\ C \end{array} \quad \begin{array}{c} B \\ \vdots \\ C \end{array}}{C} \quad \cfrac{\begin{array}{c} A \\ \vdots \\ B \end{array}}{A \to B} \quad \cfrac{A \quad A \to B}{B}
$$

$$
\begin{array}{ccccccc}
UI & UE & EI & EE & RA & REND & CI
\end{array}
$$

$$
\cfrac{Pa}{x\,Px} \quad \cfrac{x\,Px}{Pa} \quad \cfrac{Pa}{Ex\,Px} \quad \cfrac{Ex\,Px \quad Pa}{Pa} \quad \cfrac{\begin{array}{c} A \\ \vdots \\ B \quad \neg B \end{array}}{\neg A} \quad \cfrac{\neg\neg A}{A} \quad \cfrac{\begin{array}{c} Pa \\ \vdots \\ P1 \quad Pa' \end{array}}{Ph}
$$

$$
\begin{array}{ccccc}
NEE & W & D & RA2 & H
\end{array}
$$

$$
\cfrac{\begin{array}{c} A \\ \vdots \\ B \quad \neg B \end{array}}{\neg A} \qquad \neg.A \& \neg A \qquad A \vee \neg A \qquad \cfrac{\begin{array}{c} \neg A \\ \vdots \\ B \quad \neg B \end{array}}{A} \qquad \cfrac{A \quad \neg A}{B}
$$

The basic rules are from *AI* (and introduction) to *EE* (existence elimination). Then follow rules for negation, the rule of complete induction, and so on. *REND* should be something like "reduction of negation doubled" (almost the same in German). Rules *W* (for *Widerspruch*, contradiction) and *D* (for *Dilemma*) have zero premises. *H* stands for "Heyting's rule." The inference schemes for the connectives and quantifiers are the ones we are used to, with the letters that indicate the connective and whether it is an introduction or an elimination.

The presence of a rule of induction stems from the main aim of the exercise: to find a proof of the consistency of arithmetic. It was probably by January or February 1933 that Gentzen, for the time being, dropped arithmetic from his scheme and began to consider pure logic in isolation.

There follows a division into three calculi, beginning with the weakest calculus, designated *C*, possibly for *consequentia*, the pure logic of consequence, and called minimal logic after Johansson (1936):

> *Division of the inference schemes*:
> (*C*) *AI* to *EE*.
> Intuitionistic (*I*): *AI* to *RA, H*.
> Classical (*K*): *AI* to *REND*.

It goes exactly as it should, especially that falsity elimination, or rule *H*, is not needed in *K*. Gentzen writes *J* for *I*, but this is pure typography. The same goes for the names of his calculi *NI* and *LI*, as well as rule *CI*, which is *VJ* in the original German.

There are two basic forms of proofs: tree proofs and net proofs. Gentzen uses for the former the word *Stammbaum*, family tree:

> *Formal shape of proof: each proposition conclusion of at most one inference, no circle in inference.*
>
> **T** *Tree form: each proposition except for the endproposition a premiss in exactly one inference.*
>
> **N** *Net form: each proposition except for the endproposition premiss in at least one inference, eventually several.*

I have tried to find out who was the first to put logical inferences in the form of a tree. It is the kind of question almost no one asks, but I found the paper of the Göttinger Paul Hertz (1923) as the best candidate, suggested by Schroeder-Heister (2002, p. 251). The notation in Hertz is somewhat awkward, with premisses partially stacked. Gentzen's (1932) paper on Hertz systems, as well as all of his later papers, take it as a matter of course that one does derivation trees. Since the papers of Hertz remained unread, our habit of drawing derivation trees stems from the work of Gentzen, and it would be of some interest to chart how that habit spread out. In "Five forms" we see that he considered also what are today called directed acyclic graphs ("dags," from the first letters) as an alternative.

Derivation trees have several extremely useful properties that the earlier linear derivations do not have. Assumptions are top formulas so that it can be seen what depends on what, and it is possible to prove that two derivations can be composed into one; that is, that composition is an admissible rule. Further, the order of application of rules can be permuted, a thing noted and used already by Hertz.

Permutation of rules became the most central method in Gentzen's work on proof theory. The aim is to get a hold of the combinatorial possibilities offered by an axiom system, pure logic in the first place. One shows through permutability arguments that formal proofs by rules that correspond to an axiomatic system can be brought into a standard form, such as normal derivations in natural deduction or cut-free derivations in sequent calculus. Thus, we do not say that we can master the structure of an arbitrary proof but instead that any proof transforms into one that has a transparent structure.

Logicians are struck by the purity of Gentzen's method and understand and learn to use it. A comparison to the work of Skolem is instructive: he gave a profound analysis of proofs by the axioms of lattice theory and projective geometry in the 1920 paper that began with the Löwenheim-Skolem theorem, and as much as any paper in logic, must have been studied by many. No one understood his arguments in the latter parts of the paper, and his results, such as a polynomial-time algorithm for what is known as the word problem for free lattices, had to be reinvented in 1988! His geometrical results were used once between 1920 and 1995.

As to the tree form, Gentzen took it to be a deviation from actual reasoning that is linear. The thought seems to have been that proofs are texts that consist of sentences written on lines after lines.—Modern theories of sentence structure analyze them in a two-dimensional syntax tree form and consider the linear writing a codification of a deeper tree structure. Such is the case even with linear derivations: The lines are numbered and the justifications of the steps, through reference to the line numbers of the premisses, codify a tree derivation.

Gentzen's *Five forms* continues with various ways of treating free and bound variables, of closing open assumptions, and so on. Each alternative has a number, and he studies the combinations of the alternatives, as shown by the following passage:

Variable conditions...

Condition 10: There is a linear order of propositions in a proof, in which each conclusion comes after its premises, and in which no common free variable appears before its EE.

Condition 30: In a *UI*, there must not occur any common variable that depends on the all-variables.

Condition 34: There is to each free variable an eigenvariable equal to it.

Pages that explain these conditions have not been saved; there are a minimum of 34 different conditions on formal proofs!

Next, Gentzen starts combining the various possibilities, rules for variables, and so on, to pin down his *five forms of natural calculi*:

Important forms of proof:

NNA: natural net proof general form, i.e., net form with conditions 6.2; 11, 10, 30, 33. (sheet 78.3–79)

N1: Tree form, with conditions 6.2; 27, Bb4. (Bb 1, Bb3, Bb 4.) (sheet 66.3)

N2: Tree form, inference forms *AI,AE,UI,UE,RA* and *REND*, with condition 6.2; Bb 4 (only the special case of condition 32) (sheet 80.2)

NN2: Net form, inference forms *AI,AE,UI,UE,RA* and *REND*, with condition 6.2; 11. (sheet 84.1)

NO: Tree form, with condition 6.2; 10, 30, 33. Corresponds to *NNA*, condition 11 and superfluous because of the tree form. For complete "generality" there is still missing the admission of several *EE*'s with the same variable.

Fragments of the full language are considered, as in *N2* and *NN2*, and there appear mysterious codes such as "Bb" (most probably for *Beweisbedingung*, proof condition). Numbers such as "sheet 84.1" refer to Gentzen's notes from a series he called **D**, possibly for "dissertation." Each sheet was a four-page legal size paper (*Kanzleiformat* in German), written in stenographic notation, with the sheet number followed by a dot followed by a page number from 1 to 4. Thus, 84 sheets, say, could amount to anything up to 500 or more printed pages! Moreover, **D** was not the only series. There is even a reference to "my book on propositional logic."

A certain attitude of starting from scratch, as well as intelligence and concentrated work, produces results. This is the essential secret behind the discovery of natural deduction, if one has to guess from the few pages that have remained of the series **D**.

The first, introductory chapter of Gentzen's thesis manuscript ends with a summary of his five central results. The fourth of these is:

IV. Theorem:

If a logical proposition is provable in the calculus *NII*, there is a proof for it in which only subpropositions of it appear, eventually with different variables.

The word "conjectured" has been cancelled from the theorem. The calculus *NII* of intuitionistic natural deduction is given in the first chapter. The rules are now, in comparison to those in the *Five forms*, stable and fixed, and notational improvements have been made. Even if there is no date in the manuscript, the changes indicate that it is clearly a later product than the *Five forms* of 23 September.

Table 9.2. Rules of natural deduction in Gentzen's thesis

AI	*AE*		*OI*		*OE*		
$\dfrac{A\ B}{A\&B}$	$\dfrac{A\&B}{A}$	$\dfrac{A\&B}{B}$	$\dfrac{A}{A\vee B}$	$\dfrac{B}{A\vee B}$	$A\vee B$	$\dfrac{\overset{[A]}{C}}{}$	$\dfrac{\overset{[B]}{C}}{C}$

UI	*UE*	*EI*	*EE*
$\dfrac{Dy}{(x)Dx}$	$\dfrac{(x)Dx}{Da}$	$\dfrac{Da}{(Ex)Dx}$	$\dfrac{(Ex)Dx \quad \overset{[Dy]}{C}}{C}$

FI	*FE*	*V*
$\dfrac{\overset{[A]}{B}}{A\supset B}$	$\dfrac{A \quad A\supset B}{B}$	$\dfrac{\curlywedge}{A}$

Rules *UI* and *EE* have the standard variable restrictions. Square brackets indicate any number $\geqslant 0$ of discharged assumption formulas. In actual derivations, of which Gentzen gives some examples, instances of rules are indicated by writing the name of a rule next to the inference line, and assumptions are indicated as closed by numerical markers next to the rule symbol and above the closed assumptions.

Three rules in the table are canceled: a rule *T* that concludes *A* from *A*, and rules *R* and *V* that are like rules *RA* and *H* of table 9.1. Rule *T* is perhaps there to make the derivation of $A \supset A$ look less awkward. The derivations with and without the rule would be

$$\cfrac{\cfrac{\overset{1}{A}}{A}\ T}{A \supset A}\ FI\ 1 \qquad\qquad \cfrac{\overset{1}{A}}{A \supset A}\ FI\ 1$$

It takes some mental effort to see the premiss *A* in the right derivation simultaneously as an assumption and a conclusion. As to the letter *T*, Gentzen notes later in the manuscript that the assumption of *A* in natural deduction is written as $A \to A$ in sequent calculus and calls it "the tautological sentence."

The new rules *R* and *V* are two primitive rules for negation, and a transformation is given that reduces two successive instances of *R* into one such instance (page 9 of the manuscript):

$$\cfrac{\begin{bmatrix} 2 \\ C \end{bmatrix} \quad \cfrac{\begin{bmatrix} 1 \\ A \end{bmatrix},\begin{bmatrix} 2 \\ C \end{bmatrix} \quad \begin{bmatrix} 1 \\ A \end{bmatrix},\begin{bmatrix} 2 \\ C \end{bmatrix}}{\cfrac{\overset{\gamma}{B} \qquad \overset{\delta}{\neg B}}{\neg A}\ R1}}{\cfrac{\zeta}{A} \qquad \neg A}{R2}}{\cfrac{\neg C}{\varepsilon}}$$

becomes:

$$\cfrac{\cfrac{\begin{bmatrix} \begin{bmatrix} 2 \\ C \end{bmatrix} \\ \zeta \\ A \end{bmatrix},\begin{bmatrix} 3 \\ C \end{bmatrix}}{\overset{\gamma}{B}} \qquad \cfrac{\begin{bmatrix} \begin{bmatrix} 2 \\ C \end{bmatrix} \\ \zeta \\ A \end{bmatrix},\begin{bmatrix} 3 \\ C \end{bmatrix}}{\overset{\delta}{\neg B}}}{\cfrac{\neg C}{\varepsilon}}\ R3$$

An instance of *R* followed by *V* reduces similarly into just an instance of *V*:

$$\cfrac{\overset{\alpha}{A} \qquad \cfrac{\overset{\gamma}{\underset{B}{[A]}} \qquad \overset{\delta}{\underset{\neg B}{[A]}}}{\neg A}\ R}{\cfrac{C}{\varepsilon}}\ V$$

becomes:

$$\cfrac{\cfrac{\begin{bmatrix} \alpha \\ A \end{bmatrix}}{\overset{\gamma}{B}} \qquad \cfrac{\begin{bmatrix} \alpha \\ A \end{bmatrix}}{\overset{\delta}{\neg B}}}{\cfrac{C}{\varepsilon}}\ V$$

Gentzen now notes that one can define negation through the use of a "false proposition F": $\neg A$ becomes $A \rightarrow F$, and "the inference R can be expressed as follows with the help of FE and FI":

$$
\text{In place of} \quad
\begin{array}{cc}
\overset{1}{[A]} & \overset{1}{[A]} \\
\gamma & \delta \\
B & \neg B \\
\hline
\multicolumn{2}{c}{\neg A}
\end{array} R1
\qquad \text{enters:} \qquad
\begin{array}{c}
\begin{array}{cc}
\overset{1}{[A]} & \overset{1}{[A]} \\
\gamma & \delta \\
B & B \rightarrow F \\
\hline
\multicolumn{2}{c}{F}
\end{array} FE \\
\hline
A \rightarrow F
\end{array} FI\ 1
$$

Thus, the rules with primitive negation become derivable by the pairs $\supset E, \supset I$ and $\supset E, \perp E$, respectively. It is now seen clearly that the two successive instances R, R and R, V pinpoint precisely the possible convertibilities and the transformations Gentzen gives their normalization steps. Writing out the preceding derivations with the defined notion of negation with R followed by another R and R followed by V, respectively, we find a detour convertibility on implication in both. Writing out Gentzen's reduced derivations, we find that the detour conversions have been done. Thus, the defined notion of negation perfectly fits Gentzen's scheme of introduction and elimination rules and their conversion into normal form, whereas the conversions with the primitive rules for negation do not follow any general pattern.

In a chapter on normalization that was added later, negation is defined through *falsum* and a modified rule V with conclusion λ added next to the cancelled rules. The printed thesis is halfway between primitive and defined negation, with I- and E-rules for negation and rule V, all written with λ.

The implication symbol in the first chapter was an arrow but got changed into the horseshoe in the new chapter on normalization. The universal and existential quantifiers were still written as $(x), (Ex)$. One earlier chapter became the article on translation from classical to intuitionistic arithmetic and has the older notation, whereas the printed thesis instead takes one step further from the manuscript and uses the now standard quantifier symbols $\forall, \exists$, the former of which is originated with to Gentzen. It can be used in places to date his manuscripts. The translation part was submitted as an article in mid-March, but one cannot simply conclude that the normalization proof would have been produced that late. The translation uses axiomatic logic, and the notation is that of Heyting's 1930 article.

In Gentzen's original presentation of natural deduction, finished formal derivations ended with the closing of all open assumptions. The same is in fact true of his printed thesis, although this aspect is always ignored. Thus, implication introductions build up the theorem to be proved. At some point, Gentzen must have realized that the rules of natural deduction move, by and large, in the *I*-rules to conclusions that contain the premises as components, and the other way around for the *E*-rules. With the closing of all open assumptions in the end, the conclusion became the longest formula in a derivation that also contained all the formulas in the derivation as subformulas, if no detours had been taken. It is seen from Gentzen's summary that he first conjectured the normalization theorem for derivations in intuitionistic natural deduction, then found a proof and cancelled the word "conjecture." So much is even stated in his printed thesis, even if not directly (introduction, paragraph 2):

> A closer inspection of the natural calculus led me in the end to a general theorem that I want to call the "Hauptsatz" in what follows.
>
> ⋮
>
> To be able to express and prove the Hauptsatz in a convenient form, I had to lay as a basis a logical calculus especially adapted to the purpose. The natural calculus proved not to be suitable. It is true that it has the essential properties needed for the validity of the Hauptsatz, but only in its intuitionistic form.

Gentzen is not exactly stating that he has a detailed proof. Today, knowing better, we can draw conclusions from his various remarks. Subsequently, the emphasis in research was on Gentzen's systems of sequent calculi until 1965, when Dag Prawitz gave a direct proof of normalization for a system of classical predicate logic without disjunction and existence. Then he extended it to the full language of intuitionistic predicate logic. Another, less known proof was given by Andres Raggio, also in 1965.

Gentzen's handwritten thesis manuscript contained as its greatest surprise a detailed proof of normalization for intuitionistic logic. It is clearly a later addition, because it contains the result conjectured in the summary. The proof forms a separate chapter, with Gentzen's name on it but with no title other than "Dissertation. Part III." The writing is different from the formal hand of the rest of the thesis and could be described as a schoolboy's hand, rather clumsy, but eminently legible, and perhaps intended to be read by a printer. The text is perfect and finished, "ready for publication" as Gentzen might have said.[1] What happened to this proof, and why Bernays ignored and forgot about it, is not known. As Gentzen at least intimated in the printed thesis, he was unable to extend the result to classical natural deduction and therefore invented classical sequent calculus and proved the *Hauptsatz* that applied to both a classical and an intuitionistic calculus. There was no need for the normalization proof, and he could afford to never mention it again.

Szabo's translation of Gentzen's thesis was published first in the *American Philosophical Quarterly* in 1964–65. It has a brief introductory comment by Bernays, but nothing new is revealed. There is also a letter in the ETH collections from Bernays to Szabo, dated 14 May 1968 (manuscript Hs. 975:4536, ETH Zurich), that gives a lengthy commentary on various issues, but the normalization proof is not mentioned. Commenting on the latter's introduction to Gentzen's *Collected Papers*, Bernays asks:

> How do you mean that Gentzen was led to his Hauptsatz through an investigation of the "natural calculus"? Consider that modus ponens is an essential elimination rule in the natural calculus.

Further, there is the somewhat disturbing set of lectures Bernays gave in Princeton in 1935–36, titled *Logical Calculus*. These lectures explain natural deduction, but in a messy and awkward style of notation for derivation trees. Only classical logic is covered, and none of the aims that Gentzen had—normalization, subformula property, consistency—is mentioned. In the end of the part on natural deduction (p. 34), Bernays suggests that we might as well stick to truth tables.

---

[1] In a couple of places in his stenographic notes, Gentzen describes a piece in such terms (*publikationsreif verfasst*).

Reflecting on Bernays, I see a clear lack of appreciation of Gentzen's discovery of natural deduction in his writings. One may ask what effect the failure to prove normalization for the classical calculus could have had on his judgment and later recollection, as opposed to the clear-cut situation in sequent calculus, with the *Hauptsatz* covering in one result both intuitionistic and classical logic. So, natural deduction seems to have been for Bernays just a part of the way to sequent calculus, and his question and comment to Szabo seem to indicate further that he thought that the subformula property fails in natural deduction because of the indispensability of rule $\supset E$. He must have been led to this thought by Herbrand's result in which *modus ponens* plays the role of cut.[2] Thus he did not realize that Gentzen had proved a normalization theorem for intuitionistic natural deduction, even though he had been in possession of the detailed proof for decades.[3] He was instead careful to mention in connection with natural deduction what he considered to be a simultaneous discovery and development of natural deduction by Jaśkowski (1934), and suggested the same in the mentioned letter to Szabo. If one reads Jaśkowski's paper and compares it to Gentzen's work, it will be very difficult to see anything at all developed in a way that could be set on a par with Gentzen in the way Bernays did. The calculus is classical and the derivations fashioned in a linear form. The latter feature especially makes it practically impossible to obtain any profound insights into the structure of derivations. Consequently, there are no results in this work beyond what, more or less, Aristotle could have said about the hypothetical nature of proof. Jaśkowski's idea was remarkable enough, but a comparison to Gentzen's perfect formulation is not the fairest way to look at his article.

From today's perspective, Gentzen's proof of normalization is exactly what it should be. First, permutative convertibilities are eliminated, so that no *I-E*-pairs separated by an *E*-rule remain, and then detour convertibilities are eliminated. This is the standard

---

[2] Let the premises be A and $A \supset B$, so the conclusion of *modus ponens* is B. The premises in sequent notation are $\rightarrow A$ and $\rightarrow A \supset B$. Rule $R\supset$ is invertible, so $A \rightarrow B$ is derivable. Then, a cut with cut formula A gives the conclusion $\rightarrow B$ and trace of A may get lost.

[3] The package of Gentzen's handwritten manuscript pages is folded, and at the back of the normalization proof there is written obliquely across the blank page in Bernays' hand: "Konzepte von Herrn Gentzen."

proof procedure of normalization. Gentzen's almost manic attention to details such as the avoidance of a clash of variables through the "purification of derivations" strikes the eye, for this was in early 1933, the very beginnings of proof theory as we know it today.

Finally, a few words about the reception of Gentzen's logical calculi by the intuitionists. Gentzen had sent his double-negation translation work to Heyting early in 1933, and mentioned some results from his thesis in later correspondence. A January 1934 letter to Heyting mentions that Gentzen can prove the intuitionistic underivability of what today is called "double-negation shift" in predicate logic: $\forall x \neg\neg A \supset \neg\neg\forall x A(x)$. This is just another indication of his careful reading of Heyting's 1930 paper: Its second part mentioned two classical theorems of predicate logic, the double negation of the universal form of excluded middle discussed earlier and the double-negation shift, both of which fail intuitionistically, as shown by Brouwerian counterexamples. Gentzen was writing the letter in response to a card of Heyting's in which Heyting asked about the decidability of intuitionistic propositional logic, mentioned in a letter of Gentzen's from November 1933. Heyting was obviously struck by Gentzen's proof of the said decidability and by natural deduction in general. In the first of a series of Dutch papers that began in 1935, titled "Intuitionistische wiskunde" (Intuitionistic mathematics), he explains this new formulation of the logical principles. The naturalness of formalization after the model of Gentzen's calculus is one conclusion (Heyting 1935, p. 81):

> If one then introduces in the logical calculus the usual mathematical signs and writes the axioms of the mathematical theory that one wishes to develop, with the aid of the logical signs, it would be possible to write a mathematics book with only formulas, without a connecting text, because the reasonings that are usually included in the text, we can now also write with formulas.
>
> ⋮
>
> It is in principle remarkable that a proof can be assessed independently of its content in thought, exclusively on the basis of the intuitive properties of the collection of symbols that make up the proof.

It was a real pity that Gentzen did not present his normalization theorem for natural deduction in the published thesis or explain it to Heyting in correspondence. The result would have cut short the talk about the possible circularity of Heyting's explanation of implication, at least for first-order logic. Heyting's initial enthusiasm for Gentzen's natural deduction is not reflected in his later work; Heyting's 1956 book looks as if it had gone through the hands of a censor who deleted from the ample references all those to Gentzen.

One can also ask why Brouwer remained silent in front of one who came from Hilbert's Göttingen but agreed with the intuitionists on all essential points, as in the "four intuitionistic insights on formalism" listed in Brouwer (1928) and referred to and endorsed in Gentzen (1936a).

(D) THE NORMALIZATION PROCEDURE AND SOME CONSEQUENCES. At this point, in October 1932, the task was to establish a subformula property for formal proofs, or *derivations* (Herleitungen), by the new rules of natural deduction. Going through the combinatorial possibilities, one notices cases such as

$$
\cfrac{\cfrac{A \quad B}{A \,\&\, B}\ {}_{\&I}}{A}\ {}_{\&E}
\qquad\qquad
\cfrac{\cfrac{[A]}{\vdots}{B}\ {}_{\supset I}\quad \vdots}{B}\ {}_{\supset E}
$$

There is a local "peak" (*Gipfel*) in a derivation, $A \,\&\, B$ or $A \supset B$, that need not belong as a part to the conclusion of the whole derivation or some open assumption the conclusion depends on. These peaks can be eliminated:

$$
\cfrac{\cfrac{A \quad B}{A \,\&\, B}\ {}_{\&I}}{A}\ {}_{\&E}
\qquad \text{becomes} \qquad
\vdots\; A
\qquad\qquad
\cfrac{\cfrac{[A]}{B}\ {}_{\supset I}\quad A}{B}\ {}_{\supset E}
\qquad \text{becomes} \qquad
\begin{matrix} A \\ \vdots \\ B \end{matrix}
$$

There is a subtlety with the second proof transformation: Rule $\supset I$ is displayed schematically, with an arbitrary number of copies of the open assumption A closed by the introduction of $A \supset B$. If A was used

n times in the derivation, the transformed derivation can be presented by the scheme

The derivation of A and what it depends on get multiplied any number of times.

Things are not so obvious with disjunction (nor with existence; universality is easy). There are the transformations for $\vee I$ followed by $\vee E$, as in the first of the I-rules:

$$\frac{\dfrac{A}{A \vee B}\,{}^{\vee I} \quad \overset{[A]}{\underset{C}{\vdots}} \quad \overset{[B]}{\underset{C}{\vdots}}}{\underset{\vdots}{C}}\,{}_{\vee E} \qquad \text{becomes} \qquad \overset{\vdots}{\underset{\vdots}{\underset{C}{A}}}$$

There is in addition the possibility that a disjunction or existence elimination separates an introduction from an elimination; say, if C is of the form $D \& E$ and has been derived in a minor premiss by rule $\& I$, then to be eliminated by $\& E$ applied to the conclusion. The hidden nonnormality is made explicit by a *permutative* conversion:

$$\frac{A \vee B \quad \dfrac{\dfrac{D \quad E}{D \& E}\,{}^{\& I}}{D \& E}\,{}_{\vee E, 1} \quad \overset{\overset{1}{B}}{\underset{D \& E}{\vdots}}}{\underset{\vdots}{\dfrac{D \& E}{D}\,{}_{\& E_1}}} \qquad \text{becomes} \qquad \frac{A \vee B \quad \dfrac{\dfrac{D \quad E}{D \& E}\,{}^{\& I}}{D}\,{}_{\& E_1} \quad \dfrac{D \& E}{D}\,{}_{\& E_1}}{\underset{\vdots}{D}}\,{}_{\vee E, 1}$$

Now the I-E pair in the derivation of the first minor premiss can be eliminated.

Gentzen first left out $\vee$ and $\exists$ by translating $A \vee B$ into $\neg(\neg A \& \neg B)$ and $\exists x\, A(x)$ into $\neg \forall x \neg A(x)$. That gave him the *normalization theorem* for the $\vee, \exists$-free fragment of predicate logic.

Normalization theorem. *All derivations can be so transformed that no I-rule is followed by the corresponding E-rule.*

The main difficulty in the proof is to give a measure or weight to derivations such that the elimination of a local peak such as $A \supset B$ (a nonnormality) reduces the weight more than the multiplication of the derivation of A by any number n. Gentzen solved the problem sometime in late 1932 and at some stage even included a treatment of the rules for disjunction and existence with the permutative conversions as in the preceding example. Namely, as already indicated, the handwritten version of a plan and partial execution of his thesis contained as its greatest surprise a detailed proof of normalization for intuitionistic natural deduction. An English translation of Gentzen's proof, requiring 13 journal pages, together with my introduction, was published in von Plato (2008).

The thesis manuscript contains a stenographic addition by which the *subformula property* of normal derivations is an immediate corollary to normalization.

Subformula property. *All formulas in a normal derivation are subformulas of the conclusion or some open assumption.*

Consistency is an immediate consequence of these results. If $A \& \neg A$ were derivable, $\bot$ also would be derivable, and therefore so would any formula, but $\bot$ has no normal derivation by the subformula property and therefore no derivation at all.

How to extend all of this to arithmetic, was the new formulation of the consistency problem.

Elimination of indirect proofs. The proof of the normalization theorem in two stages, first without $\vee, \exists$, then for the full language, bore unexpected fruit when Gentzen noticed that the *principle of indirect proof* could be dispensed with if $\vee$ and $\exists$ were absent, subject to a little adjustment.

Two treatments of negation were given in the thesis manuscript, either as a primitive notion with separate rules or as defined by $\neg A \equiv A \supset \bot$. Even the printed thesis lists both notions and their respective rules. They are, for the defined notion, special cases of the implication rules:

$$\frac{\begin{array}{c}[A]\\ \vdots\\ \bot \end{array}}{\neg A}\supset I \qquad\qquad \frac{\neg A \quad A}{\bot}\supset E$$

The rules of primitive negation are:

$$\frac{\begin{array}{cc}[A] & [A]\\ \vdots & \vdots\\ B & \neg B\end{array}}{\neg A}\,\neg I \qquad\qquad \frac{\neg A \quad A}{C}\,\neg E$$

Both rules are derivable if the defined notion of negation is used. The introduction rule is not pure, in the sense that it already contains a negation in a premiss. All other rules are such that no connective other than the one introduced or eliminated appears in the rule schemes.

As noted above Gentzen's thesis manuscript gives transformations to repeated applications of the primitive rules of negation, but these transformations do not follow any general pattern for the simplification of derivations. If the transformations are reproduced with the use of the rules for the defined notion of negation, they turn out to be instances of standard conversion patterns of natural deduction (see von Plato 2012 for a detailed presentation). In conclusion, the defined notion of negation is the well-behaving one.

Classical natural deduction results if the rule of *indirect proof* is added to intuitionistic logic:

$$\frac{\begin{array}{c}[\neg A]\\ \vdots\\ \bot \end{array}}{A}\,DN$$

The nomenclature *DN* stands for double negation, which is explained as follows: If instead of *DN* rule $\supset I$ is applied, the conclusion is $\neg\neg A$, and double-negation elimination gives the conclusion A.

If the conclusion of rule *DN* is a premiss in an elimination rule, there is no direct guarantee for the subformula property. This problem is clear from a text fragment Gentzen later dated as being from "about January 1933." It is titled "Decision in classical predicate calculus reducible to decision in intuitionistic calculus with only $\supset$ and ( )?" ($\triangleright\forall\triangleleft$ is found above the notation for the universal quantifier ( ), where the triangles indicate a later addition.) The object of the piece is to translate derivations in classical natural deduction to derivations by the rules for implication and universal quantification and with an added constant proposition $\mathcal{F}$ that stands for the false formula (i.e., a fragment of what today is called minimal logic). To this purpose, Gentzen first transforms the formulas of classical predicate logic into equivalent ones that contain only implication, universal quantification, and $\mathcal{F}$. The rules for negation are

$$
RA: \quad \cfrac{\cfrac{\overset{1}{[\mathfrak{A}]}}{\mathfrak{B}} \quad \cfrac{\overset{1}{[\mathfrak{A}]}}{\mathfrak{B} \supset \mathcal{F}}}{\cfrac{\mathcal{F}}{\mathfrak{A} \supset \mathcal{F}} \, FI1} FE
\qquad
REND: \quad \cfrac{\mathfrak{A} \supset \mathcal{F} . \supset \mathcal{F}}{\mathfrak{A}} \text{ "}DN\text{" (law of double negation)}
$$

This is directly from the manuscript. Numerical labels identify occurrences of closed assumptions, *RA* stands for *reductio* and *REND* signifies something like "reduction of negation doubled." The order of premisses in rule *FE* ("follows elimination") was changed later in the winter of 1932–33.

The last point is to change every atomic formula into its double negation. Now derivations can be transformed so that rule *DN* is applied to the components of its conclusion. If *DN* has been applied to conclude an implication $\mathfrak{B} \supset \mathfrak{C}$, the transformation, again from the manuscript, is

$$
\cfrac{\cfrac{\cfrac{\overset{1}{\mathfrak{B}} \quad \overset{3}{\mathfrak{B} \supset \mathfrak{C}}}{\mathfrak{C}} FE \quad \overset{2}{\mathfrak{C} \supset \mathcal{F}}}{\cfrac{\cfrac{\mathfrak{F}}{\mathfrak{B} \supset \mathfrak{C}. \supset \mathcal{F}} FI\,3 \quad \mathfrak{B} \supset \mathfrak{C}. \supset \mathcal{F} : \supset \mathcal{F}}{\cfrac{\mathcal{F}}{\cfrac{\mathfrak{C} \supset \mathcal{F}. \supset \mathcal{F}}{\cfrac{\mathfrak{C}}{\mathfrak{B} \supset \mathfrak{C}} FI\,1} DN \text{ for } \mathfrak{C}} FI\,2} FE}}{}
$$

A similar transformation is made if *DN* is applied to a universally quantified formula. In the end, *DN* is applied to double negations of what were atomic formulas before the transformation added two negations. With four negations at the head of each atomic formula, rule *DN* just eliminates two of them, but this can be done without the classical rule. Therefore, as Gentzen concludes: "It is obvious that the inference *DN* can be completely eliminated by these steps." The procedure and result are explained with great clarity in Bernays (1935a), so we can conclude that Gentzen had explained the matter to him in detail.

The atomic formulas of arithmetic are equations. If they don't contain free variables, they are decidable, as Gentzen well understood. Rule *DN* applied to the double negation of a numerical equation $n = m$ has the same force as the law of excluded middle, $n = m \lor \neg n = m$, and which of the disjuncts is the case can be decided. Therefore, *DN* need not be applied to atomic formulas without free variables. In particular, in January 1933, Gentzen could conclude the following.

Relative consistency. *If a contradiction is derivable in classical Peano arithmetic, it is already derivable in a fragment of intuitionistic Heyting arithmetic.*

This was, of course, not Gentzen's terminology, but the result was clear. As mentioned, one of the central aims of the Hilbert school had been to "secure the transfinite arguments of arithmetic." These contain in particular the indirect existence proofs, with $\exists x\, A(x)$ concluded if $\forall x \neg A(x)$ led to a contradiction. Gentzen's result showed that such steps were not a "dubious" component in arithmetic proofs.

The general conclusion from Gentzen's result, obtained at the same time by Gödel, was that the consistency problem of arithmetic has an intuitionistic sense and therefore possibly an intuitionistic solution. A further conclusion was that intuitionism does indeed go, as described by Bernays in general terms, beyond Hilbert's "strictly finitistic methods."

The surfacing of transfinite ordinals. Gentzen found out, probably in early 1933, that his proof idea for the consistency of intuitionistic arithmetic, and therefore also for Peano arithmetic, would not be realizable. He had added, right at the start when he developed intuitionistic natural deduction, a *rule of induction:*

$$\frac{A(1) \quad \begin{array}{c} A(y) \\ \vdots \\ A(y+1) \end{array}}{A(t)} \, CI$$

The conclusion of CI (for complete induction, *vollständige Induktion*) gives by rule $\forall I$, when a fresh variable x is chosen for the arbitrary term t, $\forall x \, A(x)$. As with indirect proof, there need not remain any trace of the conclusion of CI in any of the open assumptions or in the endformula of a finished derivation in arithmetic, so the subformula property is not guaranteed to hold.[4] Neither can one restrict the induction formula to some specific class of formulas to get sufficient control over the structure of derivations.

The first occurrence of transfinite induction in Gentzen's work had appeared is already in 1932 in **INH** (date 9.X.):

> A new idea: Is it possible to perform appropriate reductions so that one takes the longest proposition, or a proposition that is of the highest value according to some other assignment that is invariant under reductions, and eliminates it in all its places of occurrence, without multiplying propositions of the same value?

> The assignment of values will go into the transfinite with CI's.

It is not difficult to see where the last idea comes from. If instead of the universal generalization of the conclusion of rule CI a numerical instance $A(n)$ is concluded by CI, the derivation should have a lower value than the derivation of $\forall x \, A(x)$. (This is mentioned explicitly in the popular article Gentzen 1936a.) The only way out is for an uppermost CI with a fresh variable in the conclusion to have the value ω. The next thing to determine is what happens when there are several nested CI's. There are some remarks about the possible ordinal assignments made during the spring of 1933, but nothing definitive. It seems to be a line of study abandoned for the time being.

---

[4] A notion of normal derivability can be applied even in the absence of the subformula property, with easy proofs of the disjunction and existence properties (von Plato 2006).

With the original aim of Gentzen's study temporarily lost, he concentrated on pure classical logic, found his sequent calculus, and proved the famous *Hauptsatz*, the cut elimination theorem, during the rest of the spring of 1933.

(E) FROM NATURAL DEDUCTION TO AXIOMATIC LOGIC. A narrow path led Gentzen to the discovery of sequent calculus. The first component was a translation from a classical natural calculus to the axiomatic system of logic of Hilbert and Ackermann. Secondly, there was Gentzen's experience with the sequents of Hertz systems, which made him see a similarity of structure in the translation and in sequent notation.

A four-page text titled "The N2-formalism" ends the handwritten thesis manuscript. It has been shifted to the end from before other parts, the reason clearly being a decision to present the translations between the different logical calculi at the end of the thesis. The numbering of the calculi indicates that it was written before the part that has the title "Calculus NL3, 'natural-logistic' parallel calculus." *N2* is the classical ¬-&-∀-fragment of natural deduction. The rules are:

Table 9.3. The classical natural calculus *N2*

$$
\begin{array}{cccc}
AI & AE & UI & UE \\[4pt]
\dfrac{A \; B}{A\&B} & \dfrac{A\&B}{A} \quad \dfrac{A\&B}{B} & \dfrac{Pa}{(x)Px} & \dfrac{(x)Px}{Ph}
\end{array}
$$

$$
\begin{array}{ccc}
DN & TND & CI \\[4pt]
\begin{array}{cc} A & A \\ \vdots & \vdots \\ B & \neg B \end{array} & & \begin{array}{c} Pa \\ \vdots \end{array} \\[4pt]
\dfrac{}{\neg A} & \dfrac{\neg\neg A}{A} & \dfrac{P1 \quad Pa'}{Ph}
\end{array}
$$

Schematic letters $A, B, C, \ldots$ are used in the rules, except when free and bound variables appear and the writing becomes Pa, Ph, and so on. The variable condition is that a is an eigenvariable in rules *UI* and *CI*. Both *CI* and *UE* have an arbitrary term h in the conclusion. Furthermore, an eigenvariable must not occur in formula A in rule *DN*.

Gentzen now uses an axiomatic formulation of logic that he calls the "*H-A*-formalism," from Hilbert and Ackermann. The axioms are (Hilbert and Ackermann 1928, pp. 22 and 53):

Table 9.4. The H-A axioms of Gentzen's thesis manuscript

a) $A \vee A \rightarrow A$

b) $A \rightarrow A \vee B$

c) $A \vee B \rightarrow B \vee A$

d) $(A \rightarrow B) \rightarrow (C \vee A \rightarrow C \vee B)$

e) $(x)Px \rightarrow Ph$

f) $Ph \rightarrow (Ex)Px$

The rules of inference are *modus ponens*, or *FE* in Gentzen's notation, and for the quantifiers the following two (ibid., p. 54):

Let $B(x)$ be an arbitrary logical expression that depends on x, and A one that doesn't. If now $A \rightarrow B(x)$ is a correct formula, also $A \rightarrow (x)B(x)$ is. One obtains similarly from a correct formula $B(x) \rightarrow A$ the new $(Ex)B(x) \rightarrow A$.

To prove that derivations in the axiomatic calculus of *H-A* can be reproduced in *N2*, Gentzen shows the following:

I define:
$A \vee B$ is $\neg.\neg A \& \neg B$, $(Ex)Px$ is $\neg(x)\neg Px$, $A \rightarrow B$ is $\neg.A \& \neg B$.
The axioms of *H-A* become through the substitution of $\vee$, $\rightarrow$, Ex by the defining expressions:

a) $\neg ... \neg.\neg A \& \neg A .. \& . \neg A$
b) $\neg ..A \& \neg\neg. \neg A \& \neg B$
c) $\neg ... \neg.\neg A \& \neg B..\& ..\neg\neg.\neg B \& \neg A$
d) $\neg\{\neg.A \& \neg B..\& \neg\neg...\neg.\neg C \& \neg A..\& ..\neg\neg.\neg C \& \neg B\}$
e) $\neg.(x)Px \& \neg Ph$
f) $\neg.Ph \& \neg\neg(x)\neg Px$

All of these are easily proved in *N2*; as an example I prove *b*:

$$\cfrac{\cfrac{A\&\neg\neg.\neg A\&\neg B}{A}\ {\scriptstyle AE} \qquad \cfrac{\cfrac{\cfrac{\overset{1}{A\&\neg\neg.\neg A\&\neg B}}{\neg\neg.\neg A\&\neg B}\ {\scriptstyle AE}}{\cfrac{\neg A\&\neg B}{\neg A}\ {\scriptstyle AE}}\ {\scriptstyle TND}}{\neg A}}{\text{Axiom } b}\ {\scriptstyle DN\ 1}$$

(I have carried through even the remaining proofs.)

The inference schemes of *H-A* are reproducible in *N2*:

"Inference schema": We have $\dfrac{A \quad \neg.A\&\neg B}{B}$

Reproduction in *N2*:

$$\cfrac{\cfrac{\cfrac{A \quad \overset{1}{\neg B}}{A\&\neg B} \qquad \neg.A\&\neg B}{\neg\neg B}\ {\scriptstyle DN\ 1}}{B}\ {\scriptstyle TND}$$

The generalization scheme $\dfrac{\neg.A\&\neg Pa}{\neg.A\&\neg(x)Px}$

Reproduction in *N2*:

$$\cfrac{\cfrac{\overset{1}{A\&\neg(x)Px}}{\neg(x)Px}\ {\scriptstyle AE} \qquad \cfrac{\cfrac{\cfrac{\cfrac{A\&\neg(x)Px \atop A}{A\&\neg Pa}\ {\scriptstyle AI} \qquad \neg.A\&\neg Pa}{\neg\neg Pa}\ {\scriptstyle DN\ 2}}{\cfrac{Pa}{(x)Px}\ {\scriptstyle UI}}\ {\scriptstyle TND}}{}}{\neg.A\&\neg(x)Px}\ {\scriptstyle DN\ 1}$$

UI fulfills the variable condition because *a* does not occur in $\neg.A\&\neg(x)Px$.

The scheme of existential instantiation is treated analogously. The induction axiom is derived by

$$\cfrac{\cfrac{P1\&.(x)\neg.Px\&\neg Px'..\&\neg Ph}{P1\&.(x)\neg.Px\&\neg Px'}\ {\scriptstyle AE}}{P1}\ {\scriptstyle AE} \qquad \cfrac{\overset{3}{Pa}\quad\overset{2}{\neg Pa'}}{Pa\&\neg Pa'}\ {\scriptstyle AE} \qquad \cfrac{\cfrac{\cfrac{\cfrac{\overset{1}{P1\&.(x)\neg.Px\&\neg Px'..\&\neg Ph}}{P1\&(x)\neg.Px\&\neg Px'}\ {\scriptstyle AE}}{\cfrac{(x)\neg.Px\&\neg Px'}{\neg.Pa\&\neg Pa'}\ {\scriptstyle UE}}\ {\scriptstyle AE}}{\neg\neg Pa'}\ {\scriptstyle DN2}}{Pa'}\ {\scriptstyle TND}$$

$$\cfrac{\cfrac{Ph}{\neg...P1\&.(x)\neg.Px\&\neg Px'..\&\neg Ph}}{}\ {\scriptstyle CI\ 3} \qquad \cfrac{\overset{1}{P1\&.(x)\neg.Px\&\neg Px'..\&\neg Ph}}{\neg Ph}\ {\scriptstyle AE}\ {\scriptstyle DN\ 1}$$

By this, the calculus *N2* is at least as strong as the standard axiomatic calculus of Hilbert and Ackermann. In the other direction, derivations in *N2* are reproduced in axiomatic logic as follows:

One finds for each proposition A of the *N2*-proof all those assumptions above it the respective *DN*'s (*CI*'s) of which still stand under A. These are $B_1, \ldots B_\varphi$. Then one substitutes A by $B_1 \& \ldots \& B_\varphi \to A$.

If A is an assumption, $A \to A$ takes its place. The steps of inference of natural deduction become as shown in the table.

Table 9.5. Gentzen's translation from *N2* to axiomatic logic

AI	AE	UE	TND

$$\frac{D \to A \qquad E \to B}{D\&E \to A\&B} \qquad \frac{C \to A\&B}{\begin{array}{c} C \to A \\ B \end{array}} \qquad \frac{C \to (x)Px}{C \to Ph} \qquad \frac{C \to \neg\neg A}{C \to A}$$

UI	DN	CI

$$\frac{C \to Pa}{C \to (x)Px} \qquad \frac{D\&A \to B \quad E\&A \to \neg B}{D\&E \to \neg A} \qquad \frac{D \to P1 \quad E\&Pa \to Pa'}{D\&E \to Ph}$$

Now it remains to show that these steps can be reproduced in axiomatic logic. With *AI*, first one proves
$$(D \to A) \to ((E \to B) \to (D\&E \to A\&B))$$
to be a theorem of axiomatic logic. Then the translation of *AI* in table 9.5 is replaced by two implication eliminations:

$$\frac{\dfrac{(D \to A) \to ((E \to B) \to (D\&E \to A\&B)) \quad D \to A}{(E \to B) \to (D\&E \to A\&B)} \qquad E \to B}{D\&E \to A\&B}$$

The pattern is the same for the rest of the translations.

The quoted passage shows how the derivability of a formula A from the open assumptions $B_1, \ldots, B_\varphi$ in natural deduction becomes expressed as an implication $B_1 \& \ldots \& B_\varphi \to A$ in axiomatic logic. The next step is to read the implication arrow as the derivability relation instead of as a connective. In this step, Gentzen was greatly helped by his work from the summer of 1931 on what are known as Hertz systems (see von Plato 2012 for these).

The calculus of Hertz was one of the two main sources of Gentzen's sequent calculus. It suggested the transition from an implication to something that lies one level higher, so to speak, namely a formal derivability relation. Similarly, in place of a connective, the commas of this higher level represent a collection of assumptions that can be used conjunctively. Further discussion of the background of sequent calculus in the work of Hertz and in Gentzen's first paper can be found in Bernays (1965).

The conceptual ground was now sufficient to support the idea of a formal calculus for the derivability relation. Before I go to that, a final observation on *N2*: It appears from the summary of results of the handwritten thesis that Gentzen had first obtained the normalization theorem for what he called calculus *N21*. There are no permutative conversions in the absence of disjunction and existence, and assumptions are closed only in a rule for negation. The addition of a rule of double-negation elimination gives *N2* and makes the fragment complete for classical logic. The definitions of disjunction, implication, and existential quantification and the realization that double-negation elimination can be restricted to negative formulas, an intuitionistically justified step, led to the reduction of classical arithmetic to intuitionistic arithmetic.

9.2. SEQUENT CALCULUS

(A) FROM NATURAL DEDUCTION TO SEQUENT CALCULUS. In the order in which it was left, the proof of normalization in the thesis manuscript is followed by the presentation of a "natural-logistic parallel" calculus, denoted *NL3*. The parallelism comes from the way Gentzen arrived at the sequent rules. He writes:

> The construction of a proof is not, as so far, undertaken from top to downwards, i.e., from the conditions to the final result, but from *inside* out. New propositions are added partly to below, partly to above, and such parts of proof are also "hung one beside the other."

This vivid language may leave something to be desired; Gentzen continues with a detailed exposition:

I explain the concept: "Proof of a proposition B from the propositions $A_1, \ldots A_r$."[5]

I shall use the following symbolic writing when one such is at hand: $A_1, \ldots A_r \supset B$ to be read as: B is provable from $A_1, \ldots A_r$. This is a "sentence" in the sense of P. Hertz. I shall use the word "sentence" only in this sense.[6]

I designate by $\Gamma, \Delta, \Theta, \Lambda$ complexes of propositions; they can be empty.

One sees that the arrow still stands for implication, so the inverted C of Peano is used for derivability. The latter symbol is soon used for implication, in the chapter on normalization. Formal derivations in the two systems run parallel and are listed by Gentzen in two columns, the left for a "natural writing" of a step of proof, the right for a "symbolic writing for its formation." The parallel begins with "A is a proof of A from A." It is rendered as A in the natural writing and as $A \supset A$ in the symbolic writing. Rule UE of natural deduction becomes the sentence $(x)Px \supset Ph$. Next, the rest of the rules of natural deduction are turned into the symbolic writing:

$$EI) \quad \frac{Ph}{(Ex)Px} \qquad\qquad\qquad Ph \supset (Ex)Px$$

$$AE) \quad \frac{A \& B}{A} \quad \frac{A \& B}{B} \qquad\qquad A \& B \supset A \qquad A \& B \supset B$$

$$OI) \quad \frac{A}{A \vee B} \quad \frac{B}{A \vee B} \qquad\qquad A \supset A \vee B \qquad B \supset A \vee B$$

$$FE) \quad \frac{A \quad A \to B}{B} \qquad\qquad A, A \to B \supset B$$

These were the single-step natural rules and their corresponding symbolic writings. The latter received the name *groundsequents* sometime later. For the rest of the rules, Gentzen first explains rule UI:

A proof of Pa from $A_1 \ldots A_v$ gives (through the joining of $(x)Px$, to below) rise to a proof of $(x)Px$ from $A_1 \ldots A_v$. Here a must not occur in $A_1 \ldots A_v, (x)Px$.

---

[5] A marginal addition reads: The order and [changed into: though not the] multiplicity of the A is irrelevant. The A could be empty; B not.

[6] The German is *Satz*; that for proposition is *Aussage*.

The two parallel writings are, with $\Gamma = A_1 \ldots A_\nu$:

$$\frac{A_1 \ldots A_\nu}{A_1 \ldots A_\nu}$$

$$\vdots$$

$$\frac{Pa}{(x)Px} \qquad\qquad \frac{\Gamma \supset Pa}{\Gamma \supset (x)Px}$$

The variable condition is that a must not occur in $A_1 \ldots A_\nu, (x)Px$.[7]
Rule *EE* has the parallel writing

$$\frac{\Gamma\,(Ex)Px}{\Gamma\,[Pa]}$$

$$\vdots$$

$$\frac{B}{B} \qquad\qquad \frac{\Gamma\,[Pa] \supset B}{\Gamma\,(Ex)Px \supset B}$$

The condition is that a must not occur free in Γ, B. This is the first rule in which the principal formula appears in the antecedent part of a sequent.

The rest of the parallel writings produce the following:

$$\frac{\Gamma \supset A \quad \Delta \supset B}{\Gamma\,\Delta \supset A\&B} \quad \frac{[A]\Gamma \supset C \quad [B]\Delta \supset C}{(A \vee B)\Gamma\,\Delta \supset C} \quad \frac{[A]\Gamma \supset C}{\Gamma \supset A \to B} \quad \frac{\Gamma\,[Pa] \supset Pa'}{\Gamma\,P1 \supset Ph}$$

The structural rules come last. They consist in the addition of a condition and in the "hanging together" of two proofs. The parallel writing for the rule of weakening is

$$\Gamma \qquad\qquad \Gamma\,A$$
$$\vdots \qquad\qquad \vdots \qquad\qquad \frac{\Gamma \supset B}{\Gamma\,A \supset B}$$
$$\text{from} \quad B \quad \text{becomes} \quad B$$

It is not told what (if any) the sense of the left part is. The word *Verdünnung* (thinning) is added to the right part. The second rule is

---

[7] A circle is drawn in the left column around the subderivation from $A_1 \ldots A_\nu$ to Pa. It is further explained that "$[A]$ means $A \ldots A$, i.e., A occurs in an arbitrary number, even 'not at all' is allowed."

cut (*Schnitt*):

$$
\text{from } \begin{matrix} \Gamma \\ \vdots \\ A \end{matrix} \quad \text{and} \quad \begin{matrix} \Delta\ A \\ \vdots \\ B \end{matrix} \quad \text{becomes}^8 \quad \begin{matrix} \Gamma \\ \vdots \\ \Delta\ A \\ \vdots \\ B \end{matrix} \qquad \frac{\Gamma \supset A \quad \Delta\ A \supset B}{\Gamma\ \Delta \supset B}
$$

(B) THE GROUNDSEQUENT CALCULUS. The upshot of Gentzen's parallel, really a translation from natural deduction, is a specific sequent calculus that he used actively in his work for the next ten years or so but on which he never published anything separate. He called it *LIG*, an intuitionistic *sequent calculus with groundsequents* (*LJG* in the German original, a calculus with "Grundsequenzen"). There was, as one can expect, a corresponding classical calculus, denoted *LKG*.

A careful reading of Gentzen's thesis is again revealing, when one knows what to look for. The printed thesis has the two sequent calculi, *LI* and *LK*, that have become standard. Looking at the translation from *LI* to axiomatic logic (chapter 5, §5), you will find there the list of the groundsequents. The translation is in two stages. First, derivations in *LI* are translated into ones in *LIG*, and then in the next stage they are translated from *LIG* to axiomatic logic. The first stage of the translation begins with rules *L&* and *L¬* of *LI*:

Table 9.6. Gentzen's translation from *LI* to *LIG*

$$
\frac{A,\Gamma \to \Theta}{A\&B,\Gamma \to \Theta}\ L\& \qquad \rightsquigarrow \qquad \frac{\overset{Gs1}{A\&B \to A} \quad A,\Gamma \to \Theta}{A\&B,\Gamma \to \Theta}\ Cut
$$

$$
\frac{\Gamma \to A}{\neg A,\Gamma \to}\ L\neg \qquad \rightsquigarrow \qquad \frac{\Gamma \to A \quad \dfrac{\overset{Gs7}{\dfrac{\neg A,A \to}{A,\neg A \to}}\ Ex}{\dfrac{\Gamma,\neg A \to}{\neg A,\Gamma \to}\ Ex^*}}{}\ Cut
$$

Here Θ is either one formula or empty, and *Gs*1 and *Gs*7 stand for the first and seventh groundsequent in Gentzen's list, respectively.

---

[8] The derivation of *A* from Γ after composition has an extra smaller *A* between Γ and *A*.

*Ex** (my notation) stands for possibly repeated instances of the rule of exchange of order in a list. The rest of the translations are similar.

LIG with its equal share of rules and groundsequents is a calculus halfway between sequent calculus and an axiomatic calculus.

The discussion of sequent calculi in §2 of chapter II of the printed thesis, right after the rules have been given, makes two remarks. The first one states that the rules are not independent, but if some were left out, the *Hauptsatz* would be lost.[9] The second remark is that "if one put no value on the *Hauptsatz*," one could simplify the calculus by replacing some rules by groundsequents. Had Gentzen just given a name to the ensuing calculus defined on top of p. 194 of his thesis, it would have become common property long ago.

The first sequent calculus LIG. The rules of sequent calculus in Gentzen's first sketch have what are today called *independent contexts.* The side formulas in the "complexes of propositions" Γ and Δ in the rules are added up in the conclusion, exactly as in natural deduction. In the rules of the printed thesis, the contexts are *shared*, i.e., the same in all two-premiss rules except implication at the left. The printed thesis notes that the proof of cut elimination is simplified somewhat if the side formulas are the same. I shall now change the system with groundsequents a bit so that it is in harmony with the structural features of the rules in the printed thesis. The aim is to have a better understanding of Gentzen's path to the cut elimination theorem. I also start using the notation of the printed thesis that has become the standard one in proof theory. Gentzen himself indicated a change of the roles of the arrow and the horseshoe in a marginal note and carried it through in the third chapter of the handwritten manuscript, the one on the normalization proof.

Sequents have the general form $A_1, \ldots, A_n \to C$, with the possibility that the antecedent is empty, and the same for the succedent. The antecedent formulas form a *list* or *sequence*, the latter of which is the origin of the name sequent calculus. There was some variation in how the formulas in the antecedent were treated. At first, Gentzen stated that the order and multiplicity of the A_i do not matter, but then he changed it so that multiplicity matters but not order. In the printed

---

[9] Difficult to understand; one suggestion is in my 2009a, p. 688.

thesis, both order and multiplicity were handled formally by rules of *exchange* and *contraction*. Today, one tends to work with multisets in which multiplicity but not order is counted, the middle one of the three possibilities that Gentzen considered.

By 1934, Gentzen had become aware of the crucial role of the rule of contraction when attempting to prove the consistency of arithmetic. It can be shown, in fact, that if multiplicities of formulas could be dispensed with or, equivalently, a rule of contraction left out, then the consistency of arithmetic would have a proof by ordinary arithmetic induction.

In the rules of the calculus *LIG*, *L* and *R* stand for left and right (*A* and *S* in Gentzen).

Table 9.7. Logical rules and groundsequents of the calculus *LIG*

$$A\&B \to A, \quad A\&B \to B, \quad \frac{\Gamma \to A \quad \Gamma \to B}{\Gamma \to A\&B} \; R\&$$

$$\frac{A,\Gamma \to C \quad B,\Gamma \to C}{A \vee B, \Gamma \to C} \; L\vee \quad A \to A \vee B, \quad B \to A \vee B$$

$$\frac{A,\Gamma \to B}{\Gamma \to A \supset B} \; R\supset \quad A \supset B, A \to B$$

$$A, \neg A \to$$

$$\forall x\, A(x) \to A(t), \quad \frac{\Gamma \to A(y)}{\Gamma \to \forall x\, A(x)} \; R\forall$$

$$\frac{A(y),\Gamma \to C}{\exists x\, A(x), \Gamma \to C} \; L\exists \quad A(t) \to \exists x\, A(x)$$

The sequents have at most one formula at the right, in the succedent part. Thus, falsity elimination of natural deduction is replaced by the groundsequent $A, \neg A \to$. The quantifier rules have the usual variable restriction. The quantifier groundsequents have an arbitrary term t.

Next we come to the *structural rules* of *LIG*. There, *W* stands for weakening, *C* for contraction, and *E* for exchange. Rule *Cut* makes possible the *composition* of two derivations, one with the conclusion *A*, the other with the assumption *A*.

Table 9.8. The structural rules of the sequent calculus *LIG*

$$\frac{\Gamma \to C}{A, \Gamma \to C} \; LW \qquad \frac{\Gamma \to}{\Gamma \to C} \; RW$$

$$\frac{A, A, \Gamma \to C}{A, \Gamma \to C} \; LC \qquad \frac{\Gamma, A, B, \Delta \to C}{\Gamma, B, A, \Delta \to C} \; LE$$

$$\frac{\Gamma \to A \quad A, \Delta \to C}{\Gamma, \Delta \to C} \; Cut$$

To the structural rules can be added what Gentzen sometimes called axioms, sometimes initial sequents: $A \to A$. These and groundsequents start derivation branches.

Looking at the translation from *LI* to *LIG* in table 9.6, one might think that *LIG* is just a trivial variant of *LI*. To recover a derivation in the latter, it is sufficient to substitute each groundsequent by its derivation in *LI*. Somewhat of a mess is produced instead, for the calculus with groundsequents is not complete if the rule of cut is left out. Cuts, on the other hand, do not in general maintain the subformula property. The indispensability of cuts in *LIG* is seen from the derivation

$$\frac{A \to A \vee B \quad A \vee B \to (A \vee B) \vee C}{A \to (A \vee B) \vee C} \; Cut$$

It is clear that there is no cut-free derivation of the conclusion of this cut. However, cuts can be so permuted that the cut formula is a subformula of the conclusion.

Cut elimination for *LIG*. *A derivation in LIG can be so transformed that it contains only cuts in which the cut formula is a subformula of the conclusion of the cut.*

The most interesting aspect of *LIG* and cut elimination concerns the rule of contraction. In the printed thesis, the calculus is *LI*, and *LK* for classical logic, and a *mix rule* is introduced in place of cut. It allows cutting more than one occurrence of the cut formula. The problem arose as follows. Assume that the right premiss of cut has been derived by contraction. If one tries to permute cut upward by cutting twice the contracted formula, there is no reduction in the cut elimination parameters. With *LIG*, instead, a cut formula is never principal in both premisses of cut. Therefore, if the right premiss has been derived by contraction, either cut can be permuted up at left until the left premiss is a groundsequent and the cut is harmlessly repeated twice, or else the duplication in the right premiss can be deleted because it comes from weakening.

Gentzen did not leave *LIG* aside but used it, for example, in 1938 during attempts at doing consistency proofs for an intuitionistic sequent calculus for arithmetic, with the rules written out very prominently exactly as in table 9.7. The calculus keeps appearing in his notes until 1944.

As mentioned, in 1965 Bernays published a rather long paper on sequent calculus that, in my opinion, shows that Gentzen had all the properties of *LIG* under control. Bernays first discusses at length the work of Hertz, how Gentzen contributed to it, and how all this was one of the two components in the discovery of sequent calculus. The presentation is careful; Bernays shows, for example, the following result (p. 8): In a single-succedent calculus, weakening and context-sharing cut render the rest of the structural rules derivable, including the rule of exchange.

The second component on the path to sequent calculus is "the translation of Gentzen's rules of natural deduction into sequent calculus" (p. 9). The result, as given by Bernays:

Table 9.9. Variant of *LIG* from Bernays (1965)

$$A, B \to A \& B \qquad A \& B \to A \qquad A \& B \to B$$

$$A \to B \vee C \qquad B \to A \vee B \qquad \frac{A, B \to D \quad A, C \to D}{A, B \vee C \to D}$$

$$\frac{A, B \to C}{A \to B \supset C} \qquad A, A \supset B \to B$$

Only the propositional rules are discussed in Bernays. The calculus has a groundsequent in place of *R*&, a possibility mentioned in Gentzen's manuscript. Further, the printed thesis has this groundsequent (pp. 193–94). Bernays also discusses the possibility of a calculus with inversions of rules, similar to Gentzen's discussion in the handwritten thesis manuscript.

Section 3 of Bernays' paper is devoted to a discussion of cut elimination. First, Bernays recalls the calculus of table 9.9, obtained through a translation from natural deduction. Then, as he writes (p. 20):

> Beside the system thus obtained, we put up a second one, all rules of which are passages from one sequent to another such that the passages are pairwise inverses to each other.

> The system of intuitionistic sequent calculus used by Gentzen himself in his "Investigations into logical inference" is different from both of the mentioned forms of sequent calculus. The setting up of this calculus is directed toward a specific goal.

The goal was, naturally, a cut-free calculus that has the subformula property. The impression possibly given by the preceding passage, that Bernays is discussing his own work, gets corrected later, on page 32, where classical calculi are treated. There, as we shall see, one obtains a particularly nice calculus with just groundsequents and weakening and cut (Gentzen's calculus *LDK*, for which see von Plato 2012). Bernays writes (p. 32) that "if we consider the dual-symmetric sequent calculus without a view towards cut elimination, we can, as already emphasized by Gentzen, simplify remarkably the rules for the connectives." It is clear from Bernays' paper that he had studied the original package for a thesis given to him by Gentzen, or that the latter had explained its contents and results carefully, except for the part on normalization. It is, moreover, clear that Bernays himself had worked quite a bit with both the variant of *LIG* and the invertible calculus.

Bernays' paper reproduces with such precision Gentzen's so far unknown calculi from the handwritten thesis manuscript that I take the further development in Bernays (1965) as part of what Gentzen himself had accomplished. First, Bernays shows by an example that *LIG* is not closed with respect to cuts. Second, Bernays notes (p. 25):

> Even if the eliminability of cuts fails for this system, there remains for *suitable* derivations a subformula property.

A brief sketch of how one arrives at this result is given. Starting from a derivation in Gentzen's cut-free calculus *LI*, instances of rules are substituted by instances of groundsequents of *LIG*:

> The substitutions come out almost all simply through cuts. Only the rules for negation ... require a careful discussion.

Note that these substitutions are precisely Gentzen's steps of translation from *LI* to *LIG*, as in table 9.6. By the completeness of cut-free *LI*, a complete class of derivations in *LIG* with the subformula property follows. This indirect argument for the subformula property in suitable *LIG*-derivations, through the calculus *LI* of the printed thesis and the translation, may well be a reverse of the historical order of things.

(C) SEQUENT CALCULUS, FINAL FORMULATION.. Gentzen's analysis of "actual proofs in mathematics" led to a system of natural deduction that is intuitionistic. As mentioned, he obtained a system of classical logic *NK* by adding a rule of double negation. Another possibility that he mentions was to let instances of the law of excluded middle appear among the topformulas of derivations. Now not all formulas in a normal derivation need be subformulas of the open assumptions or of the conclusion. To obtain a proof system for classical logic with the subformula property, Gentzen devised a calculus in which the single conclusion of each rule of natural deduction is replaced by a finite number of *possible cases*. Such cases, say Θ, under the open assumptions, say Γ, are expressed formally by a sequent written as $\Gamma \rightarrow \Theta$. In Gentzen's notation Γ and Θ are lists of formulas, so we can render sequent calculus as "a calculus with lists of assumptions and cases." The German word *Sequenz* means sequence. The use of the English "sequent" as a noun stems from Kleene's book *Introduction to Metamathematics* (p. 441), the motivation being that "sequence" had another formal use in that book. This neologism has been carried over to several other languages, with such contrived nouns as the Italian "sequente."

We now present sequent calculus as it appears in Gentzen's published papers. That is how his work was received, in the almost total absence of conference and seminar talks and other contacts after 1933. Gentzen's original explanation (1934–35, I.2.4) of the meaning of a sequent $A_1, \ldots, A_m \rightarrow B_1, \ldots, B_n$ was that it expresses the same thing as the formula $A_1 \& \ldots \& A_m \supset B_1 \vee \cdots \vee B_n$ ("bedeutet inhaltlich genau dasselbe wie"). A review of that work by Gentzen's contemporary Arnold Schmidt in Göttingen instead contains the following: A sequent $A_1, \ldots, A_m \rightarrow B_1, \ldots, B_n$ means that "B_1 *or* $\ldots B_n$ depends on the assumptions A_1 *and* $\ldots A_m$." The B_j are referred to as "claims" (*Behauptungen*, Schmidt 1935, p. 145). It is clear from circumstantial evidence that this suggestion came from Gentzen.

In reference to his first paper (1936) on the consistency of arithmetic, Gentzen writes in his second paper on the same topic (1938b, 1.2):

> In the previous work I had introduced the concept of a sequent, with just one succedent formula, in its immediate connection to the natural representation of mathematical proofs (§5). It is possible to arrive at the new, symmetric concept of a sequent also from that same point of view, namely, by striving at a particularly natural representation of the *division into cases* (see §4 of the previous work, and in particular 5.26). Namely, a $\vee$-elimination can now be represented simply as: From $\rightarrow A \vee B$ one concludes $\rightarrow A, B$, to be read as: 'Both possibilities A and B obtain'.

We shall return to his last remarks in a while.

A formal calculus of sequents requires rules for the handling of lists of formulas in the antecedent (left) and succedent (right) parts of sequents. Gentzen's *structural rules* are:

Table 9.10. The structural rules of sequent calculus LK

$$\frac{\Gamma \rightarrow \Theta}{D, \Gamma \rightarrow \Theta} \; LW \qquad \frac{\Gamma \rightarrow \Theta}{\Gamma \rightarrow \Theta, D} \; RW$$

$$\frac{D, D, \Gamma \rightarrow \Theta}{D, \Gamma \rightarrow \Theta} \; LC \qquad \frac{\Gamma \rightarrow \Theta, D, D}{\Gamma \rightarrow \Theta, D} \; RC$$

$$\frac{\Delta, D, E, \Gamma \rightarrow \Theta}{\Delta, E, D, \Gamma \rightarrow \Theta} \; LE \qquad \frac{\Gamma \rightarrow \Theta, E, D, \Lambda}{\Gamma \rightarrow \Theta, D, E, \Lambda} \; RE$$

$$\frac{\Gamma \rightarrow \Theta, D \quad D, \Delta \rightarrow \Lambda}{\Gamma, \Delta \rightarrow \Theta, \Lambda} \; Cut$$

Here L and R stand for left and right (A and S in Gentzen), W for weakening (*Verdünnung*, *Vd*), C for contraction (*Zusammenziehung*,

Zz), and *E* for exchange (*Vertauschung, Vt*). Note the mirror-image arrangement of the rules, confirmed by the deviation from alphabetical order in the premiss of rule *RE*. Rule *Cut* (*Schnitt*) is self-symmetric. It permits the *composition* of two derivations, one with the case *D*, the other with the assumption *D*.

The logical rules of Gentzen's classical sequent calculus follow rather easily if instead of the groundsequent style with half axioms and half rules, only rules are given.

Table 9.11. The logical rules of sequent calculus *LK*.

$$\frac{\Gamma \to \Theta, A \quad \Gamma \to \Theta, B}{\Gamma \to \Theta, A \& B} \, R\& \qquad \frac{A, \Gamma \to \Theta \quad B, \Gamma \to \Theta}{A \lor B, \Gamma \to \Theta} \, L\lor$$

$$\frac{A, \Gamma \to \Theta}{A \& B, \Gamma \to \Theta} \, L\&_1 \quad \frac{B, \Gamma \to \Theta}{A \& B, \Gamma \to \Theta} \, L\&_2 \quad \frac{\Gamma \to \Theta, A}{\Gamma \to \Theta, A \lor B} \, R\lor_1 \quad \frac{\Gamma \to \Theta, B}{\Gamma \to \Theta, A \lor B} \, R\lor_2$$

$$\frac{\Gamma \to \Theta, A(y/x)}{\Gamma \to \Theta, \forall x A} \, R\forall \qquad \frac{A(y/x), \Gamma \to \Theta}{\exists x A, \Gamma \to \Theta} \, L\exists$$

$$\frac{A(t/x), \Gamma \to \Theta}{\forall x A, \Gamma \to \Theta} \, L\forall \qquad \frac{\Gamma \to \Theta, A(t/x)}{\Gamma \to \Theta, \exists x A} \, R\exists$$

$$\frac{A, \Gamma \to \Theta}{\Gamma \to \Theta, \neg A} \, R\neg \qquad \frac{\Gamma \to \Theta, A}{\neg A, \Gamma \to \Theta} \, L\neg$$

$$\frac{A, \Gamma \to \Theta, B}{\Gamma \to \Theta, A \supset B} \, R\supset \qquad \frac{\Gamma \to \Theta, A \quad B, \Delta \to \Lambda}{A \supset B, \Gamma, \Delta \to \Theta, \Lambda} \, L\supset$$

Rules *R∀* and *L∃* have the variable restriction that the eigenvariable *y* must not occur free in the conclusion. The rules are displayed in a mirror-image symmetry down to the last two that have no symmetry. These are separated in the German original by a horizontal line, the ones above by a vertical line. The point is to display the duality of & and ∨ resp. ∀ and ∃, and the self-duality of ¬ (1934–35, III.2.2.4). Note that the symmetric arrangement is not reproduced in the layout of the 1969 English translation of Gentzen's papers.

In the rules, the formulas displayed in the premisses are the *active* formulas of a rule (*Nebenformeln*), and the one with the connective or quantifier is the *principal* formula (*Hauptformel*). The uppercase

Greek letters indicate what are today called the *contexts* of the rules; these can be empty. In general, the antecedent or succedent of a sequent can be empty. If both are empty, we have the *empty sequent* $\to$. The assumption of a formula A in natural deduction corresponds to the initial sequent $A \to A$ by which derivations start.

Gentzen's system of intuitionistic sequent calculus *LI* is obtained from the classical calculus *LK* as a special case by requiring that the succedent consist of at most one formula. This feature is one of Gentzen's essential design principles in his putting up the calculus *LK*, and it explains some of the main aspects of the logical rules. First of all, rule $L \supset$ has single-succedent instances only if the contexts in the two premisses are independent of each other. Note that the other two-premiss rules have instead "shared," identical contexts. The explanation of this feature is that the proof of the main result for sequent calculus, Gentzen's famous *Hauptsatz*, is simplified a bit.

To display a further special feature of the calculus *LK*, consider the steps of derivation

$$\frac{\dfrac{\dfrac{\dfrac{A, B, \Gamma \to \Theta}{A\&B, B, \Gamma \to \Theta}\ L\&_1}{B, A\&B, \Gamma \to \Theta}\ LE}{A\&B, A\&B, \Gamma \to \Theta}\ L\&_2}{A\&B, \Gamma \to \Theta}\ LC$$

The result would be obtained in one step from the topsequent if instead of Gentzen's two rules one used a single left conjunction rule that concludes $A\&B, \Gamma \to \Theta$ from $A, B, \Gamma \to \Theta$. The preceding steps of derivation show that this rule is a *derivable* rule in *LK*. In the other direction, if, say, $A, \Gamma \to \Theta$ is given, a weakening with B followed by an exchange and the modified rule of left conjunction give $A\&B, \Gamma \to \Theta$, and similarly if $B, \Gamma \to \Theta$ is given. Thus, Gentzen's rules are derivable by the modified rule.

The presence of two left conjunction rules is explained by Gentzen's wish to display the duality of & and $\vee$: A dual single right disjunction rule would conclude $\Gamma \to \Theta, A \vee B$ from $\Gamma \to \Theta, A, B$, but this rule has no single-succedent instances. The intuitionistic calculus would not come out as a special case, so Gentzen's solution was to have two left conjunction rules.

Gentzen's remark at the end of the passage from 1938 quoted earlier amounts to the *invertibility* of a single right disjunction rule that concludes $\Gamma \to \Theta, A \vee B$ from $\Gamma \to \Theta, A, B$. Invertibility of logical rules as a general phenomenon was discovered by Ketonen (1944), and we shall discuss it shortly.

It should be clear that one can arrive at such choices as Gentzen's regarding the details of a logical calculus only by trying out various formulations. What they were and how Gentzen reasoned about them is difficult to say, for whatever papers and notes Gentzen had left behind in Göttingen, have disappeared. Essential new light on the discovery of sequent calculus is now shed by the manuscript version of Gentzen's thesis.

Some variants of sequent calculi can be found in Gentzen's papers. He mentions in (1934–35, III.2.2) the possibility of letting sequents of the form $A, A \supset B \to B$ be among the topsequents of derivations, and uses this (Gentzen 1936, 1943, the former natural deduction in sequent calculus style). Given the premisses of $L\supset$, so $\Gamma \to \Theta, A$ and $B, \Delta \to \Lambda$, the conclusion is obtained as follows:

$$
\cfrac{\cfrac{\Gamma \to \Theta, A \quad A, A \supset B \to B}{\Gamma, A \supset B \to \Theta, B} \; Cut \quad B, \Delta \to \Lambda}{\cfrac{\Gamma, A \supset B, \Delta \to \Theta, \Lambda}{A \supset B, \Gamma, \Delta \to \Theta, \Lambda} \; LE^*} \; Cut
$$

The asterisk indicates as many left exchanges as are needed to bring $A \supset B$ in the head of the antecedent. The two cuts are innocuous because both cut formulas are subformulas of $A \supset B$ that appears in the conclusion.

In Gentzen (1938b), implication is considered a defined notion. This possibility is mentioned also in Gentzen (1934–35, III.2.4.1), with the remark that the definition can be given only in the classical calculus *NK*. In the sequent calculi of Gentzen (1934–35) of the doctoral thesis, all the standard connectives and quantifiers were present. A sufficient reason for this is the requirement that the intuitionistic calculus be a special case of the classical one. Possibly Gentzen also had some other ideas on the matter, because he states that "not all of the rules are mutually independent, i.e., one could substitute some by others. However, if we left some out, the 'Hauptsatz' would not hold anymore" (ibid, III.2.1). If we take just some fragment of the language of predicate logic, the rules for the connectives and quantifiers are complete for that fragment. It seems that Gentzen's thought was that if one defines logical operations by others through equivalences, as in $A \vee B \supset\subset \neg(\neg A \& \neg B)$, these equivalences can be put into use only as cut formulas.

We now come to Gentzen's *Hauptsatz*. The rule of cut can be eliminated from derivations by permuting it up until it reaches initial sequents and gets removed as superfluous. A mere look at the rest of the rules shows that derivations free of the cut rule have the subformula property.

The proof of cut elimination proceeds by considering an uppermost cut, as in a typical case of

$$\frac{\dfrac{\Gamma \to \Theta, A \quad \Gamma \to \Theta, B}{\Gamma \to \Theta, A \& B} R\& \quad \dfrac{A, \Delta \to \Lambda}{A \& B, \Delta \to \Lambda} L\&_1}{\Gamma, \Delta \to \Theta, \Lambda} Cut$$

Cut is permuted up as follows:

$$\frac{\Gamma \to \Theta, A \quad A, \Delta \to \Lambda}{\Gamma, \Delta \to \Theta, \Lambda} Cut$$

Cut on $A \& B$ is replaced by a cut, or two cuts if $R\supset$ and $L\supset$ were applied, on *strictly shorter* formulas. We note in passing that the "shared" contexts of rule $R\&$ do work here, as mentioned by Gentzen. Had they been independent, there would have had to be steps of weakening that restore the original contexts of the conclusion of cut.

Another type of cut is met when the cut formula is not principal in the left premiss, as in

$$\frac{\dfrac{C, \Gamma \to \Theta, A \& B}{C \& D, \Gamma \to \Theta, A \& B} L\&_1 \quad \dfrac{A, \Delta \to \Lambda}{A \& B, \Delta \to \Lambda} L\&_1}{C \& D, \Gamma, \Delta \to \Theta, \Lambda} Cut$$

Cut is permuted up as follows:

$$\frac{\dfrac{C, \Gamma \to \Theta, A \& B \quad \dfrac{A, \Delta \to \Lambda}{A \& B, \Delta \to \Lambda} L\&_1}{C, \Gamma, \Delta \to \Theta, \Lambda} Cut}{C \& D, \Gamma, \Delta \to \Theta, \Lambda} L\&_1$$

Here the *rank* of the cut is diminished by one. Trace up the cut formula from both premisses of cut until its uppermost appearance. Rank is defined as the sum of the number of steps of derivation from these first appearances to the cut. In Gentzen's proof of cut elimination, an uppermost cut in a derivation tree is eliminated, so the proof is an induction on the number of cuts along branches of a derivation. Secondly, one considers the length of the cut formula, and thirdly the rank of the cut.

When a cut is permuted up, it eventually reaches an initial sequent. When one premiss of a cut is an initial sequent, the conclusion is identical to the other premiss and the cut is deleted.

Gentzen met with one problematic case of cut elimination, in which the cut formula in the right premiss is principal in a contraction:

$$\cfrac{\Gamma \to \Theta, D \qquad \cfrac{D, D, \Delta \to \Lambda}{D, \Delta \to \Lambda} \; LC}{\Gamma, \Delta \to \Theta, \Lambda} \; Cut$$

Two cuts on D are needed, with the result

$$\cfrac{\Gamma \to \Theta, D \qquad \cfrac{\cfrac{\Gamma \to \Theta, D \qquad D, D, \Delta \to \Lambda}{\Gamma, D, \Delta \to \Theta, \Lambda} \; Cut}{D, \Gamma, \Delta \to \Theta, \Lambda} \; LE^*}{\cfrac{\Gamma, \Gamma, \Delta \to \Theta, \Theta, \Lambda}{\Gamma, \Delta \to \Theta, \Lambda} \; LC^*, RC^*} \; Cut$$

Here LE^* indicates as many left exchanges as are needed to bring D in front of Γ, and LC^* and RC^* indicate left and right contractions that eliminate the duplications of Γ and Θ. The cut formula remains the same, but the rank of the lower cut is not reduced, even if the left exchanges LE^* are absent. Gentzen's solution was to introduce a stronger form of cut, in which any number of cut formulas can be eliminated from the premisses of cut. He called it *Mischung*, rendered as *mix* in some older English literature, and today it is often called *multicut*.

Table 9.12. Gentzen's mix rule

$$\frac{\Gamma \to \Theta \quad \Delta \to \Lambda}{\Gamma, \Delta^* \to \Theta^*, \Lambda} \; Cut^*$$

In the rule, Θ and Δ are lists of formulas in which a "mix formula" D appears at least once in both, and Θ^* and Δ^* are lists obtained from Θ and Δ through the deletion of all occurrences of D. Gentzen (1934–35, III.3.3.1) writes that the new rule is introduced "to make the proof easier." Such a remark is a sure sign that he had also a proof with the standard cut rule. Intrigued by the remark, I set out to produce a proof of the *Hauptsatz* without multicut (von Plato 2001a). The procedure involves tracing up in the right premiss the origin of the cut formula. In the critical case, it is found that cut can be permuted up in one sweep to a place where the contraction formula is principal in a left rule, with the consequence that contraction can be applied to shorter formulas.

Cut is a special case of multicut, so the elimination theorem for multicut gives also an elimination theorem for cut.

The mix rule treats the antecedents and succedents of sequents without regard to the order of formulas. This feature can be carried through systematically for all the rules of sequent calculus by using multisets of formulas instead of lists. Then the exchange rules can be dropped out. If one treated the standard rule of cut in this multiset way, the left exchanges LE^* in the problematic case of permutation of cut given earlier would disappear, but the lower cut still would not have a reduced rank.

The first applications of the cut elimination theorem, to intuitionistic propositional logic, were listed earlier. Another application in Gentzen (1934–35) was to the consistency problem of arithmetic. If the induction scheme is left out, cut elimination goes through and the underivability of the empty sequent $\to$, expressing inconsistency, follows. This consistency result was known from the work of Johann von Neumann and Jacques Herbrand, the names that Gentzen mentions. For him, this application showed the way to a future consistency proof of the whole of Peano arithmetic within the new proof-theoretical system of sequent calculus.

9.3. LOGICAL CALCULI AND THEIR APPLICATIONS

(A) RELATIONS BETWEEN GENTZEN'S LOGICAL CALCULI. As stated, the handwritten thesis contains direct translations between natural deduction and the Hilbert-Ackermann axiomatization. The printed thesis has instead a translation from natural deduction to sequent calculus. It was of course important to show the equivalence of the two calculi (or two plus two, if one considers the intuitionistic and classical variants of both), but on the whole, the matter has been considered somewhat of an obligatory business without real intrinsic interest. The translation is instead highly interesting in that it tells us something about the relations of these two forms of logical calculi.

One aspect of the difference between natural deduction and sequent calculus is notational. The latter displays all the open assumptions on a line, to the left of the sequent arrow. It is not specific to sequent calculus, and as mentioned, Gentzen used the same notation in his first published proof of the consistency of arithmetic in 1936. Natural deduction in sequent calculus style is the first stage in the passage from derivations in natural deduction to those in sequent calculus. The second step consists in replacing of E-rules by L-rules.

In the first stage, each formula C in a natural derivation is replaced by a sequent $\Gamma \to C$ in which Γ lists the open assumptions C depends on. If there was an I-rule in the natural derivation, we would have, say,

$$
\begin{array}{cc}
\Gamma & \Delta \\
\vdots & \vdots \\
\dfrac{A \quad B}{A \& B} \, {}_{\& I}
\end{array}
\quad \rightsquigarrow \quad
\dfrac{\Gamma \to A \quad \Delta \to B}{\Gamma, \Delta \to A \& B}
$$

Gentzen's translation gives the step $R\&$, modulo a second phase of weakenings in the premisses to get a rule with the shared context Γ, Δ in the premisses. The other I-rules translate similarly, and there is a direct correspondence between introductions in natural deduction and right rules in sequent calculus.

With elimination rules, the translation is different. For example, $\supset E$ is translated in the first phase as

$$
\begin{array}{cc}
\Gamma & \Delta \\
\vdots & \vdots \\
A \supset B & A \\
\hline
B
\end{array} \; {\supset E}
\qquad \rightsquigarrow \qquad
\dfrac{\Gamma \xrightarrow{\vdots} A \supset B \quad \Delta \xrightarrow{\vdots} A}{\Gamma, \Delta \rightarrow B}
$$

Observe that, as stated, we still have a step in natural deduction, with a rule that operates on the right side of the arrow. In the next phase, an instance of a left rule of sequent calculus and a cut is inserted as follows:

$$
\dfrac{\Gamma \xrightarrow{\vdots} A \supset B \qquad \dfrac{\Delta \xrightarrow{\vdots} A \quad B \rightarrow B}{A \supset B, \Delta \rightarrow B} \; {}_{L\supset}}{\Gamma, \Delta \rightarrow B} \; {}_{Cut}
$$

The derivation in natural deduction could very well have been normal, a notion absolutely clear to Gentzen, as we have known since the rediscovery of the normalization proof, yet the result of translation is a sequent calculus derivation with a cut. Gentzen keeps absolute silence here; perhaps around the time of March to May 1933 he was in a hurry to finish his thesis. That was the case with many people, with their professors being expelled by the Nazis and the department about to go to ruin. So, I think that the correspondence between natural deduction and sequent calculus remained unclear in Gentzen's thesis mainly because of lack of time.

The treatment of disjunction and existence elimination in the logical calculus of the consistency proof in Gentzen (1936), as in the Section 10.2(D), would have easily given the right correspondence between a natural and a sequent rule. By rule $\vee E$, $\Gamma \rightarrow A \vee B$ and $A, \Delta \rightarrow C$ and $B, \Theta \rightarrow C$ give $\Gamma, \Delta, \Theta \rightarrow C$, so if the first premiss is a logical groundsequent, we get that $A \vee A \rightarrow A \vee B$ and $A, \Delta \rightarrow C$ and $B, \Theta \rightarrow C$ give $A \vee B, \Delta, \Theta \rightarrow C$. Dropping the first premiss as superfluous, we have the sequent rule.

There is another translation to consider that relates to the role of the structural rules of sequent calculus. If in rule $\supset I$ several, say two, or no copies of formula A were discharged, Gentzen's translation gives, respectively,

$$
\begin{array}{c}
\overset{1\ \ 1}{A,A,\Gamma} \\
\vdots \\
\dfrac{B}{A \supset B}\ {\supset I,1}
\end{array}
\quad\rightsquigarrow\quad
\dfrac{\dfrac{A,A,\Gamma \to B}{A,\Gamma \to B}\ LC}{\Gamma \to A \supset B}\ R\supset
\qquad
\begin{array}{c}
\Gamma \\
\vdots \\
\dfrac{B}{A \supset B}\ {\supset I}
\end{array}
\quad\rightsquigarrow\quad
\dfrac{\dfrac{\Gamma \to B}{A,\Gamma \to B}\ LW}{\Gamma \to A \supset B}\ R\supset
$$

Thus, the left contraction stems from a *multiple* discharge of an assumption in natural deduction, and the left weakening from a *vacuous* discharge. The same holds for the translation of rules $\vee E$ and $\exists E$.

The problem now is to figure out the relation of structural rules in sequent calculus (*SC*) to whatever it is that corresponds to them in natural deduction (*ND*).

Problem. *What is the correspondence between derivations in ND and SC?*

There are two main solutions.

Solution A. *Change rules $\&E, \supset E, \forall E$ of NI to have a calculus with normal derivations isomorphic to cut-free derivations in Gentzen's original LI.*

Solution B. *Restrict rules $L\&, L\supset, L\forall$ of LI suitably to have derivations isomorphic to normal derivations in Gentzen's original NI.*

Solution **A** uses what is called natural deduction with general elimination rules (in von Plato 2001b). It is obtained by writing all the E-rules in the style of $\vee E, \exists E$:

$$
\dfrac{A\&B \quad \begin{array}{c}\overset{1\ \ 1}{A,B}\\ \vdots \\ C\end{array}}{C}\ \&E,1
\qquad
\dfrac{A \supset B \quad A \quad \begin{array}{c}\overset{1}{B}\\ \vdots \\ C\end{array}}{C}\ {\supset E,1}
\qquad
\dfrac{\forall x\, A \quad \begin{array}{c}\overset{1}{A(t/x)}\\ \vdots \\ C\end{array}}{C}\ \forall E,1
$$

Next, we define a root-first translation, perfectly exemplified by

$$
\dfrac{\dfrac{\begin{array}{cc}\Gamma & \Delta \\ \vdots & \vdots \\ A & B\end{array}}{A\&B}\ \&I \quad \begin{array}{c}\overset{1\ \ 1}{A,B,\Theta}\\ \vdots \\ C\end{array}}{C}\ \&E,1
\quad\rightsquigarrow\quad
\dfrac{\dfrac{\begin{array}{cc}\Gamma & \Delta \\ \vdots & \vdots \\ A & B\end{array}}{\Gamma,\Delta \to A\&B}\ R\& \qquad \dfrac{\begin{array}{c}A,B,\Theta\\ \vdots \\ C\end{array}}{A\&B,\Theta \to C}\ L\&}{\Gamma,\Delta,\Theta \to C}\ Cut
$$

The effect of the translation is that derived major premisses of E-rules give cuts. If instead they are assumptions, we get

$$
\begin{array}{cc}
\begin{array}{c}
\overset{1\quad 1}{A,B,\Theta} \\
\vdots \\
\dfrac{A\&B \qquad C}{C}\ {\scriptstyle \&E,1}
\end{array}
&
\rightsquigarrow
&
\begin{array}{c}
A,B,\Theta \\
\vdots \\
\dfrac{\quad\quad\quad C \quad}{\ } \\
\dfrac{A\&B \to A\&B \quad A\&B,\Theta \to C}{A\&B,\Theta \to C}\ {\scriptstyle Cut}
\end{array}\ {\scriptstyle L\&}
\end{array}
$$

In this case, the translation gives a *Cut* in which the left premiss is an initial sequent so that the conclusion of the cut is identical to its right premiss. Such cuts can be deleted, so we can put the following:

Definition. *A derivation in natural deduction with general elimination rules is* **normal** *if all major premisses of E-rules are assumptions.*

Overall result. *The order of logical rules is the same in normal derivations in ND with general elimination rules and in cut-free derivations in SC.*

The correspondence that Gentzen's work gives between weakening and vacuous discharge and contraction and multiple discharge goes only in one direction, from natural deduction to sequent calculus. The correspondence becomes exact through the general elimination rules.

Interpretation of weakening and contraction in ND. *If a formula in an SC derivation is active in a logical rule and has been obtained through a weakening, translation into ND with general elimination rules produces a vacuous discharge of that formula in the logical rule. Similarly, if the formula has been contracted, translation to ND produces a multiple discharge.*

The cuts produced by Gentzen's translation of rules $\&E$, $\supset E$, and $\forall E$ find an explanation. They are produced by the hidden nonnormalities of his elimination rules, those cases in which a major premiss of an E-rule has been derived by Gentzen's $\&E$, $\supset E$, or $\forall E$. Furthermore, cuts in which the cut formula is principal in a logical rule in the right premiss of cut, and therefore in a left logical rule, can be interpreted as nonnormal instances of the corresponding elimination rule of natural deduction.

Next consider solution **B**, the change of some sequent rules (as in von Plato 2011). To do this in a transparent way, we look at Gentzen's *NI*-rules as special cases of the general rules: Just make the auxiliary derivations in the general rules degenerate, and the result is

$$
\dfrac{A\&B \quad \overset{1}{A}}{A} \; {}_{\&E,1} \qquad \dfrac{A\&B \quad \overset{1}{B}}{B} \; {}_{\&E,1} \qquad \dfrac{A \supset B \quad A \quad \overset{1}{B}}{B} \; {}_{\supset E,1} \qquad \dfrac{\forall x\, A \quad \overset{1}{A(t/x)}}{A(t/x)} \; {}_{\forall E,1}
$$

The application of isomorphic translation to Gentzen's rules gives:

$$
\dfrac{\begin{array}{c}\Gamma\\ \vdots\\ A\&B \quad \overset{1}{A}\end{array}}{A} \; {}_{\&E,1} \quad \rightsquigarrow \quad \dfrac{\begin{array}{cc}\vdots & \dfrac{A \rightarrow A}{A\&B \rightarrow A}\;{}_{L\&}\\ \Gamma \rightarrow A\&B & A\&B \rightarrow A\end{array}}{\Gamma \rightarrow A} \; {}_{Cut}
$$

$$
\dfrac{\begin{array}{cc}\Gamma & \Delta\\ \vdots & \vdots\\ A \supset B \quad A & \overset{1}{B}\end{array}}{B} \; {}_{\supset E,1} \quad \rightsquigarrow \quad \dfrac{\begin{array}{cc}\vdots & \dfrac{\Delta \rightarrow A \quad B \rightarrow B}{A \supset B, \Delta \rightarrow B}\;{}_{L\supset}\\ \Gamma \rightarrow A \supset B & A \supset B, \Delta \rightarrow B\end{array}}{\Gamma, \Delta \rightarrow B} \; {}_{Cut}
$$

$$
\dfrac{\begin{array}{c}\Gamma\\ \vdots\\ \forall x\, A \quad \overset{1}{A(t/x)}\end{array}}{A(t/x)} \; {}_{\forall E,1} \quad \rightsquigarrow \quad \dfrac{\begin{array}{cc}\vdots & \dfrac{A(t/x) \rightarrow A(t/x)}{\forall x\, A \rightarrow A(t/x)}\;{}_{L\forall}\\ \Gamma \rightarrow \forall x\, A & \forall x\, A \rightarrow A(t/x)\end{array}}{\Gamma \rightarrow A(t/x)} \; {}_{Cut}
$$

The results are pretty obvious.

Result 1. *Derivations in Gentzen's NI are isomorphic to derivations in sequent calculus when the following restrictions apply:*

(i) *Cut formulas are principal in right premisses of cuts.*

(ii) *Premisses of rules L& and L∀, and the second premiss of rule L⊃, are initial sequents, and the principal formulas are cut formulas in the next rule.*

Result 2. *A derivation in Gentzen's NI is normal if and only if the following holds for its translation into sequent calculus in which L&, L⊃, and L∀ are "restricted":*

(i) Cut formulas are not principal in left premisses of cuts.
(ii) Left premisses of cuts are not derived by $L\vee$ or $L\exists$.

There remains the question of the relation of normalization to cut elimination. There are cuts in *SC* that don't have any interpretation in the preceding terms (see the remarks after the interpretation of weakening and contraction). These cuts can be interpreted as "delayed compositions" (as in von Plato 2013, section 13.4.b).

(B) PROOF SEARCH. Gentzen's doctoral thesis marked the birth of structural proof theory, in contrast to the axiomatic proof theory of Hilbert and Bernays. As mentioned, the first to use Gentzen's methods in print was Ingebrigt Johansson (1936). The correspondence between Johansson and Heyting shows that the use of Gentzen's sequent calculus was suggested by Heyting.

In 1938, Katudi Ono published a paper that used sequent calculus for the study of consistency problems of axiomatic systems. Gentzen himself wrote a review of this paper, reprinted in Menzler-Trott (2001, pp. 91–92).

A remarkable step ahead in the development of systems of sequent calculus was taken by Oiva Ketonen in his doctoral thesis of 1944. Ketonen went to Göttingen in 1938 to study proof theory with Gentzen, becoming the closest thing to a student that Gentzen may have had. Ketonen recollected later to the present author that Gentzen was a person who "did not say much of anything." In a December 1938 letter to his mother, Ketonen mentions that finally Gentzen has given him a problem to work on, but unfortunately the letter does not tell what problem it was.

Ketonen's thesis, finished in 1944 and titled *Untersuchungen zum Prädikatenkalkül*, introduced what is today called the *G3*-sequent calculus for classical logic. The propositional part has the remarkable property that all of its rules are invertible, which gives a strikingly elegant proof of completeness. Another result was a sharpening of the midsequent theorem, by which the number of quantifier inferences with eigenvariables could be minimized. The existence of a weakest possible midsequent followed, in the sense that if any midsequent is derivable, the weakest one is. Turning this into a contrapositive, Ketonen had a purely syntactic method for proofs of underivability that he applied to projective and affine geometry. To my regret,

I never asked Ketonen what specific suggestions Gentzen had made to him. Little was found after his death in 2000. One letter to Ketonen in Göttingen from the spring of 1939, by a student friend of his in Helsinki, contains the enigmatic sentence "Gentzen's proof is in the typewriting office."

Ketonen's best-known discovery is a sequent calculus for classical propositional logic in which all the logical rules are invertible. The rules that are modified, in comparison to Gentzen's calculus *LK*, are as follows:

Table 9.13. Classical sequent calculus with invertible rules

$$\frac{A,B,\Gamma \to \Theta}{A\&B,\Gamma \to \Theta}\ L\& \qquad \frac{\Gamma \to \Theta,A,B}{\Gamma \to \Theta,A \vee B}\ R\vee \qquad \frac{\Gamma \to \Theta,A \quad B,\Gamma \to \Theta}{A \supset B,\Gamma \to \Theta}\ L\supset$$

Each connective has just one left and one right rule. Further, all two-premiss rules have shared contexts, and Ketonen obtains a fully invertible classical propositional sequent calculus. Invertibility is proved after a proof of cut elimination, in the manner of the following example of the invertibility of rule $R\vee$. Assume the sequent $\Gamma \to \Theta, A \vee B$ is derivable. Then also $\Gamma \to \Theta, A, B$ is derivable:

$$\frac{\Gamma \to \Theta, A \vee B \qquad \dfrac{\dfrac{A \to A}{A \to A,B}\ RW \qquad \dfrac{\dfrac{\dfrac{B \to B}{B \to B,A}\ RW}{B \to A,B}\ RE}{A \vee B \to A,B}\ L\vee}{A \vee B \to A,B}}{\Gamma \to \Theta, A, B}\ Cut$$

Derivations in the invertible calculus are found by decomposing the endsequent in any order whatsoever. Given a sequent $\Gamma \to \Theta$ to be derived, choose from Γ or Θ any formula with a connective. The corresponding rule determines the premisses uniquely. Repeating this root-first proof search, as it is often called, formulas are decomposed into parts until there is nothing to decompose. At this stage, it can be determined if the original sequent $\Gamma \to \Theta$ is derivable, by controlling if all topsequents are initial sequents. The process terminates in the case of classical propositional logic in a bounded number of steps as determined by the number of connectives in the given sequent.

Ketonen's proof of invertibility of the logical rules of his sequent calculus used the structural rule of cut. Later, Kurt Schütte (1950) and Haskell Curry (1963) gave direct proofs of invertibility, the latter with the explicit result that the inversions are *height preserving*: If a given sequent is derivable in at most n steps, the premises in a rule that can conclude that sequent also have a derivation in at most n steps.

The second part of Ketonen's thesis contained an improvement on Gentzen's *midsequent* theorem. This result, from Gentzen (1934–35, IV.2), states that if a sequent $\Gamma \rightarrow \Theta$ is derivable in *LK* and all quantified formulas are in prenex normal form, there is a "midsequent" $\Gamma' \rightarrow \Theta'$ in the derivation such that all logical steps above the midsequent are propositional and those below it are quantificational. Ketonen shows through suitable permutabilities of rules that the number of quantificational inferences with an eigenvariable, i.e., of instances of $R\forall$ and $L\exists$, can be minimized, with the result that there is a *weakest possible* midsequent, one with the property that if any midsequent is derivable, a weakest one is. He was further able to define a recursively enumerable sequence of all possible midsequents $M_1, M_2, \ldots$ for a sequent $\Gamma \rightarrow \Theta$, by which the undecidability of predicate logic could be seen in a proper light: The derivability of any given midsequent M_i of the sequence $M_1, M_2, \ldots$ is decidable, but at no finite stage can it be known if a midsequent $\Gamma' \rightarrow \Theta'$ for the sequent $\Gamma \rightarrow \Theta$ to be derived will show up.

Gentzen (1934–35, IV.2.2) notes that Herbrand's theorem is (for formulas in prenex form) a special case of his midsequent theorem. Gentzen used the midsequent theorem for a proof of the consistency of arithmetic without the induction principle (IV.3).

Ketonen's main aim in defining his weakest possible midsequents was to use them in proofs of underivability. He writes in the introduction to his third and final chapter that "in proof theory, the most common task—the exclusive one in the applications we discuss—is to show the underivability of a given sequent. Our method is especially directed towards this task." By the defining property of a weakest possible midsequent, if it is not derivable, no midsequent is derivable, and neither is the given sequent $\Gamma \rightarrow \Theta$. The application that Ketonen made of his method of proofs of underivability was to plane projective and affine geometry. The former had been studied in Skolem (1920), a paper universally known for the Löwenheim-Skolem theorem, contained in its first part. The second part, mostly forgotten until the 1990s, contains a polynomial-time algorithm for deciding derivability of an atomic formula from a finite number of atomic formulas by the axioms of lattice theory. The third part contains a similar result for plane projective geometry, with the derivability relation between atomic formulas generalized into a finite number of possible atomic cases, a generalization analogous to that in Gentzen's sequent calculus. Skolem needed it because one of his axioms of projective geometry contains a disjunction in a positive part. To be able to analyze derivations of atomic cases from atomic assumptions purely combinatorially, without the interference of logical operations, he needed the generalization. Skolem's method was to convert the axioms into rules for the construction of formal derivations. The latter do not appear anywhere in print in Skolem's paper, but his verbal explanations amount to as much (1920, p. 116): "The axioms are in fact principles of demonstration."

In the third part of his thesis, Ketonen took up Skolem's geometrical work. He used his classical sequent calculus, with the geometrical axioms appearing through their instances in the antecedent parts of sequents. The cut elimination theorem applies in this situation, and first Ketonen repeated Skolem's syntactic proof of the independence of what is known as Desargues' theorem in projective geometry. He then went on to study affine geometry, and gave a similar proof of independence of Euclid's fifth postulate. Skolem's and Ketonen's axiomatizations corresponded to a purely universal axiom system, which leads to one essential miss in geometry: the axiom of noncollinearity that states the existence of at least three noncollinear points.

How much of Ketonen's work stems from suggestions on the part of Gentzen remains unknown because no correspondence has been found. Ketonen writes in the preface of his thesis that "Dr. G. Gentzen of Göttingen directed me towards the problem area of this dissertation during my period of study there in 1938–1939." The thesis was Ketonen's only original work in logic, saved from oblivion by a long review that Bernays wrote of it for *The Journal of Symbolic Logic* in 1945. Two somewhat earlier papers in Finnish mention results contained in the thesis, one of them about the completeness of the predicate calculus, the other about proof search with the invertible propositional rules (see Negri and von Plato 2001, pp. 60 and 86 for these).

In this list of early uses of Gentzen's sequent calculus, we note Karl Popper's paper (1947). Its 1947 review by Evert Beth in *Mathematical Reviews* attributes aspects of Popper's calculus to Gentzen, Jaśkowski, and Ketonen, even if Popper does not refer to them explicitly. The review shows that Beth knew the calculi of Gentzen as well as the modifications made by Ketonen. Later, when Beth (1955) presented the tableau calculus, he referred not to Ketonen but to Kleene (1952c). The early work of Hertz, Gentzen, and Popper is discussed at length in Bernays (1965). Invertibility of logical rules is ascribed to Popper, instead of Ketonen, even if Bernays had made the issue completely clear twenty years earlier, in his review (Bernays 1945) of Ketonen's thesis.

Ketonen's invertible rules were taken into use in Kleene (1952a), in which it is stated that the author knows Ketonen's work only through the Bernays review. Kleene further used them in his influential 1952 book that also treated intuitionistic sequent calculus in which not all rules can have invertibility. We may consider that with Kleene's book Gentzen's sequent calculi became generally known and accessible.

Kleene's work of the early 1950s also pioneered a remarkable development in sequent calculus, the "contraction-free" classical and intuitionistic calculi today denoted by *G3c* and *G3i*. These calculi have the property that no structural rules are needed. Weakening is made an admissible rule by letting initial sequents have the form $A, \Gamma \rightarrow \Delta, A$, instead of Gentzen's $A \rightarrow A$. Contraction is likewise made admissible by a suitable formulation of the rules. If we look at the invertible rules of classical propositional logic, such as in table 9.13, we see how the conclusion uniquely determines the premises, once the rule applied is specified. Thus, because of the invertibility of the logical rules, "the making of derivations consists of purely mechanical decomposition." These are the words of Ketonen from a paper (Ketonen 1943) in Finnish in which he first presented the calculus. A calculus for intuitionistic propositional logic with the same kind of termination of root-first proof search was found as late as 1992 by Roy Dyckhoff and at the same time by Jörg Hudelmaier.

As mentioned, the classical calculus has the property of height-preserving invertibility of its logical rules. Albert Dragalin (1988, Russian original 1979) refined the calculus into one in which the structural rules are moreover "height-preserving admissible," meaning that whenever the premiss of such a rule is derivable, the conclusion is derivable with at most the same size of derivation. This property has profound effects on cut elimination. In permuting cut upward, Gentzen had to restore the original contexts through weakenings and contractions. With the height-preserving admissibility of these rules, the size of a derivation does not increase when the rules are applied. Dragalin (1988) gave also an intuitionistic multisuccedent calculus with the same type of admissibility of the structural rules. Finally, Troelsta (Troelsta and Schwichtenberg 2000, first ed. 1996) gave a single-succedent intuitionistic calculus with height-preserving admissibility of weakening and contraction. The contraction-free sequent calculi are powerful tools for the analysis of formal derivations. Many difficult research results in logic, such as Harrop's theorem, become exercises through the control over the structure of proofs that the *G3*-calculi makes possible.

(C) APPLICATIONS OF SEQUENT CALCULUS. The earliest application of sequent calculus in mathematics was in the proof theory of arithmetic, in Gentzen's thesis and in a decisive way in his 1938 proof of the consistency of arithmetic, the latter to be discussed in the next chapter. Troelstra mentions Ketonen's work as "an early analysis of cutfree proofs in Gentzen calculi with axioms; but he considers the form of cutfree derivations in the pure calculus where axioms are present in the antecedent of the sequents derived" (Troelstra and Schwichtenberg 2000, p. 142). Another way to apply sequent calculus is to let topsequents in derivations have the form $\rightarrow A$, in which A is an axiom, or an instance of a universal axiom. Now, by Gentzen's "extended Hauptsatz," cuts in derivations can be permuted up until one of their premises is an axiom, but these cuts on axioms remain. Thus, it was believed that "cut elimination fails in the presence of axioms," as Girard (1987, p. 125) put it. Nevertheless, a method was found by which axioms are converted into rules of sequent calculus of a suitable form. These extend the logical rules of sequent calculus while maintaining the eliminability of cuts. The method was discovered first in the study of some intuitionistic axiom systems for apartness and order in Negri (1999). It was then shown that the method applies to any universal axiom systems (Negri and von Plato 1998). Next, Negri (2003) showed that it extends to axiom systems in which the axioms are "geometric implications," a terminology drawn from category theory. This vast class of axioms comprises the usual forms of existential axioms in mathematics. Finally, Dyckhoff and Negri (2015) were able to cover any axioms expressed in first-order logic by using a method of "geometrization."

Extension of sequent calculus by geometrization maintains cut elimination, but atomic formulas typically get removed in the application of the rules so that the subformula property of cut-free derivations is lost. Instead, an inessential modification of the subformula property often applies, in which the number of atomic formulas that can get removed is bounded. If we use a classical sequent calculus, such as *G3c*, logical rules can be permuted below the mathematical ones, so that the latter can be considered in isolation. All active formulas are atomic in the rules, and the derivability problem of a mathematical theory concerns in the first place sequents $\Gamma \to \Delta$ in which all formulas are atomic.

Consider a free-variable axiom A, given in conjunctive normal form. Each conjunct can be given as an implication

$$P_1 \& \ldots \& P_m \supset Q_1 \vee \cdots \vee Q_n$$

with $Q_1 \vee \cdots \vee Q_n$ and $\neg(P_1 \& \ldots \& P_m)$ as limiting cases when $m = 0$ and $n = 0$, respectively. These are turned into sequent calculus rules, with two possibilities:

Table 9.14. Axioms converted into sequent calculus rules

$$\frac{\Gamma \to \Delta, P_1 \quad \ldots \quad \Gamma \to \Delta, P_m}{\Gamma \to \Delta, Q_1, \ldots, Q_n} \; R\text{-rule} \qquad \frac{Q_1, \Gamma \to \Delta \quad \ldots \quad Q_n, \Gamma \to \Delta}{P_1, \ldots, P_m, \Gamma \to \Delta} \; L\text{-rule}$$

The first type of rule says that if each of the P_i follows as a case from some assumptions Γ, then the Q_j follow as cases, and the second rule is a dual to this. The latter types of rules, those that act on the left part of the sequent, are best seen in a root-first order. Assume as given a system of such left rules and that $\Gamma \rightarrow \Delta$ contains only atomic formulas. Now try matching $\Gamma \rightarrow \Delta$ as a conclusion to the rules. Whenever there is a match, we have premisses each of which get one new atomic formula in the antecedent. Thus, in the end, we obtain the deductive closure of Γ relative to the rules, with branchings into several possible closures each time there is more than one premiss. The succedent Δ remains untouched by the rules. The importance of height-preserving admissibility of contraction is that each rule has to give a new atomic formula in a premiss. In the contrary case, the conclusion would follow by the structural rules from a premiss that has a duplicated formula.

Many theories allow proof search for a sequent $\Gamma \rightarrow \Delta$ to be restricted just to atomic formulas so that all terms in the derivation are terms known from Γ, Δ. Derivations with this property are said to have the *subterm* property, in analogy to Gentzen's subformula property of cut-free derivations. There are a bounded number of distinct atomic formulas with known terms, and instances of rules that produce duplications can be excluded by the height-preserving admissibility of contraction. Therefore, the search space for a derivation is bounded, and the derivability problem is decidable. Theories with the subterm property include lattice theory, both in a relational formulation as in Skolem's early work (Skolem 1920) and in a formulation with the lattice operations (Negri and von Plato 2011). In a relational formulation, there are axioms that postulate the existence of the lattice meet (resp. join) of any two elements a, b of a lattice. For derivations with the mathematical rules, the subterm property corresponds to the conservativity of these axioms over the other lattice axioms. This conservativity was Skolem's main lemma in his early proof of the decidability of derivability with the axioms of lattice theory. In a formulation with operations, we have for any two elements a, b a meet $a \wedge b$ and a join $a \vee b$. The subterm property means that no new objects have to be constructed, by which decidability of derivability with the lattice rules follows. We note that first-order lattice theory is not decidable.

Skolem (1920) established a result for plane projective geometry analogous to that for lattices. This result was redone within sequent calculus by Ketonen (1944), who also extended it to affine geometry. As mentioned, neither was able to include the axiom of noncollinearity in their systems. The subterm property for plane projective and affine geometry with noncollinearity included, treated by extending sequent calculus with mathematical rules, was established in von Plato (2010). The subterm property for these geometries has the effect that the existence of at least three noncollinear points is conservative over the rest of the geometrical rules: A proper use of an existential axiom requires an existential theorem.

* * *

Most of Gentzen's papers from the time of his thesis work have been lost, some possibly in 1964 together with the destruction of the papers of Arnold Schmidt, or possibly later when in the 1970s papers from the attic of the Göttingen mathematics building were, it seems, hauled to the garbage dump. What is left are the two slim folders of stenographic notes and the thesis manuscript in Zurich, the latter contained in the collection *Saved from the Cellar* (Gentzen 2017). Its contents together with the published papers are the main sources for this chapter, seconded by a lot of formal work with proof systems.

As to the timing of the discovery of sequent calculus, the crucial turning point in the development of logic, I believe the earliest notes, as in the handwritten thesis manuscript, to be from late 1932. A letter to Heyting from the end of February 1933 would have mentioned the decidability of intuitionistic propositional logic, one of the corollaries to cut elimination in *LI*, had Gentzen progressed that far by then. So sequent calculus proper seems to have been developed from nothing to perfection in a miraculously short period between March and May 1933.

The idea of changing some of Gentzen's elimination rules perhaps occurred first in a 1984 work by Schroeder-Heister, who changed the conjunction elimination rule. The full system of general E-rules is found in the rare publication Dyckhoff (1988), then in works by Tennant (1992) and Lopez-Escobar (1999). Solution **A** of section 9.3(A), the isomorphic translation between cut-free derivations in sequent calculus and normal derivations in natural deduction with general elimination rules, was defined in von Plato (2001b). Solution **B**, the dual idea of restricting Gentzen's sequent calculus so as to obtain isomorphism with his original system of natural deduction, is carried through in von Plato (2011).

10

THE PROBLEM OF CONSISTENCY

The translations from classical to intuitionistic arithmetic that Gödel and Gentzen found in 1932–33 showed that the consistency of classical arithmetic reduces to that of intuitionistic arithmetic. The main aim of Hilbert's program in the 1920s had been to show that the infinitistic component of arithmetic, namely the use of quantificational inferences in indirect proofs, does not lead to contradiction. It followed from Gödel's and Gentzen's result that these classical steps are harmless. Therefore, as explained in the previous chapter, the consistency problem of arithmetic has an intuitionistic sense and, consequently, possibly also an intuitionistic solution. Bernays was quick to point out that intuitionism transcends the boundaries of strictly finitary reasoning.

Gödel seems not to have pursued the idea of an intuitionistic solution to the consistency problem, even if he reflected on his incompleteness theorems in a 1933 talk given in Boston. It is titled "The present situation in the foundations of mathematics" (Gödel 1933b) and was published from a handwritten English manuscript in the third volume of Gödel's *Collected Works* (1995). He notes (pp. 50–51) that consistency is a purely syntactic notion, so "the whole matter becomes merely a combinatorial question about the handling of symbols according to given rules." Further, "the chief point in the desired proof of freedom from contradiction is that it must be conducted by perfectly unobjectionable methods." These methods are codified in what Gödel calls "system A," of which he lists some principles. He then notes that such finitistic methods cannot lead to a proof, so the hope for a consistency proof by "Hilbert and his disciples ... has vanished entirely in view of some recently discovered facts" (p. 52). Next, Gödel notes that intuitionism clearly goes beyond what is finitistic. In particular, he refers to the interpretation of classical arithmetic in intuitionistic arithmetic as one that gives an intuitionistic proof of consistency, but later he adds that this foundation "is of doubtful value" (p. 53). Gödel ends his talk with the following remark:

> There remains the hope that in future one may find other and more satisfactory methods of construction beyond the limits of system A, which may enable us to found classical arithmetic and analysis upon them. This question promises to be a fruitful field for further investigations.

It seems that only the appearance of Gentzen's proof in 1935 made him take this possibility seriously.

10.1. WHAT DOES A CONSISTENCY PROOF PROVE?

With his thesis finished in May 1933, Gentzen had other things to worry about than the consistency of arithmetic and analysis. The mathematics department of Göttingen was in ruins after the Nazi takeover, and his professor Bernays had been fired as a "non-Aryan." Gentzen took up his research in 1934, helped by a little scholarship. One thing he tried was to use type theory as the language of mathematics. A result from the spring of 1934 is a consistency proof of Hermann Weyl's system of predicative analysis. Very little is known about the proof. One letter from Bernays to Weyl tells that Gentzen was not able to reproduce it without his notes in 1937, when he met Bernays in Paris (Menzler-Trott 2007, p. 82). The result seems to have been a by-product of the attempts at producing a proof of the consistency of arithmetic and thus not a strong result. Jean Cavaillès mentions in his book *Méthode axiomatique et formalisme* that the method of the consistency proof for arithmetic "extends without modification to mathematical theories in which the predicates and functions are decidable or calculable in finitary terms: so for the constructive part of analysis" (1938, p. 162). On 11 December 1935, Gentzen wrote to Bernays about the discussions he had had with Cavaillès, who was visiting Göttingen at the time (see Menzler-Trott 2007, p. 64).

By the end of 1934, Gentzen had found a proof of consistency of arithmetic. A letter to Bernays dated 12 May 1938 tells about a much later proof, the one that became standard through Gentzen (1938b):

"How I have obtained the consistency proof from the methods of proof in my dissertation is, I believe, now somewhat easy to see in the new version" (Menzler-Trott 2007, p. 95). As we shall see at the end of Section 10.2, the very first proof used a sequent calculus instead of the natural calculus of the 1935 proof submitted for publication in August of that year. There Gentzen (1936, p. 512) notes that the proof would be simpler, though "less natural," if a sequent calculus were used. The analogy to cut elimination that he mentions is the *Hilfssatz*, to be treated in detail in Section 10.2.

There is even a letter dated 11 April 1934 to Bernays by which a consistency proof by transfinite induction existed already at that time (Menzler-Trott, p. 54). First Gentzen writes that "the consistency of mathematics is equivalent to the carrying over of the *Hauptsatz* of my dissertation from predicate logic to type theory" (*Stufenlogik*, second-order logic with an axiom of infinity). He hopes to achieve such a consistency proof soon "by force," after which he adds: "It remains to modify the proof so that only allowed forms of inference are used. I hope to achieve this, in analogy to arithmetic only, through transfinite numbers."

It is known also through discussions that Kreisel had with Bernays that the use of transfinite induction in the published 1936 proof was, in contrast to the proof submitted for publication in 1935, a return to "an earlier idea" (as in Kreisel 1987, p. 174), discarded in favor of the 1935 proof for reasons that are at least to some extent explained in **INH**.

Bernays gave in Geneva in June 1934 a lecture in French on "some essential points of metamathematics," (see also p. 212). He notes that the statement of consistency of arithmetic is expressible in a formalism that extends the standard formalism $\mathfrak{N}$ by "certain non-elementary recursive definitions" in which existential quantification in the defining predicate can appear, as in (p. 90):

$$\Psi(k, 0) \rightleftarrows \mathfrak{B}(k)$$
$$\Psi(k, n + 1) \rightleftarrows (Ex)(\Psi(x, n) \,\&\, \mathfrak{B}(k, x, n))$$

He then notes that a similar recursive definition appears also in "the formal deduction of a principle of transfinite induction applied to an order of a limit ordinal," this ordinal being ε_0 (p. 91):

The principle mentioned is expressed, for this order, by the formula

$$(x)\{(y)(y \prec x \longrightarrow A(y)) \longrightarrow A(x)\} \longrightarrow (x)A(x)$$

$$\vdots$$

To deduce this formula, it is sufficient to use beyond the rules of the formalism $\mathfrak{N}$ a recursive definition of the sort indicated. It seems that such a definition cannot be avoided here, unless one extends the symbolism of $\mathfrak{N}$, for example, through the introduction of bound propositional variables, i.e., by ascending to the logical formalism of second order.

On the other hand, though, one can prove the principle represented by this formula in intuitionistic mathematics.

Therefore, it is apparent that the special case of the principle of transfinite induction considered is already an example of a theorem that is provable by intuitionistic mathematics, but not deducible in $\mathfrak{N}$.

Consequently one would propose, in concordance with the theorem of Gödel, that one finds an intuitionistic proof of the consistency of the formalism $\mathfrak{N}$ in which the only part non-formalizable in $\mathfrak{N}$ is the application of the said principle of transfinite induction.

Bernays adds that "at the moment, this is just one possibility" and then explains in detail Gentzen's double-negation translation from classical to intuitionistic arithmetic.

There was in 1932 no easy way to a consistency proof of arithmetic by transfinite induction. Within a week of the surfacing of the "new idea" of using such induction, Gentzen in his characteristic manner had already set out to determine what he was actually trying to do by asking in **INH** (date 16.X.32) what meaning a consistency proof can have:

Why is a consistency proof through a coarse contentful explanation,

$\mathfrak{A} \& \mathfrak{B}$ correct when $\mathfrak{A}$ correct and $\mathfrak{B}$ correct, $\mathfrak{A} \to \mathfrak{B}$ correct when from the correctness of $\mathfrak{A}$ the one of $\mathfrak{B}$ follows, $x \, \mathcal{A}x$ when $\mathcal{A}\nu$ correct for all numbers, $\neg \mathfrak{A}$ correct when $\mathfrak{A}$ not correct,

after Gödel not formal? Does it contain a circularity? One infers: The logical axioms are correct, the mathematical axioms are correct, inference scheme and substitution give correct from correct, therefore all things provable are correct.

He asks at one place: *"Where is the Gödel-point hiding?"* It took him just a few days more to come to the conclusion that the notion of correctness in arithmetic transcends what can be expressed and proved in arithmetic (**INH**, date 21.X.):

> I believe I can now see clearly why a consistency proof cannot be formalized through the giving of a coarse contentful meaning. To wit, because the meaning is not formalizable, and this naturally always: in the usual formalism, e.g., of Gödel.

Within two days, the proof strategy was clear (**INH**, date 23.X.):

> One shows now through ordinary inferences, i.e., without *CI*: There is to each proof with a numerical result a proof with a lower value and the same result. Namely, one shows existence of a peak, this peak can be reduced. The assignment of values follows according to 88.3 bottom ff.[1] So, the value of a proof is a system of transfinite numbers of the form: a polynomial in ω with natural coefficients. (To be replaced by $\omega^{a_1} + [..] + \omega^{a_v}$.)
>
> The main inference can be seen as a transfinite induction over a decidable proposition, namely the proposition: The numerical result is correct.
>
> $\vdots$
>
> There must obtain, in my opinion, some kind of a connection between the informal element in the non-formalizable definition of "correctness" and the non-formalizable (?) transfinite induction. For both of them seem to make possible a non-formalizable proof of consistency.

The attempts do not lead to any definitive result, and by early November, they peter out.

The manuscript **INH** continues with remarks that stem from February, April, and June 1933. In April, there is a clear division of proofs of consistency into three types:

1. A "purely-formal" proof
2. A "semi-contentful" proof
3. A proof through reducibility

---

[1] The numbers refer to the stenographic series **D** in which each sheet, such as 88, contains four pages, from 88.1 to 88.4. By sheet 92, this series had been renamed **INH**.

The ordinal that is needed in a purely formal proof is estimated to be $\omega^{\omega^{\omega}}$. The published proof of 1936 contains remarks about such a proof (§10.7). The third type of proof should proceed through the "peak theorem," i.e., through normalization.

There follow what Gentzen in a later addition indicated as *General thoughts about the proof of consistency*:

> The idea as a whole: Each proof has a (transfinite) *value*. Consistency of a system of proofs can be shown only through a proof that has a higher value than all of these. Therefore the theorem of Gödel.

The idea became the central one in *ordinal proof theory* that arose as a generalization of the proof theory of arithmetic. After the quoted passage, there are the added words "taken over to **WTZ**." That signum stands for something like *Widerspruchsfreiheit transfinite Zahlen* (consistency transfinite numbers) and fits well with the published 1936 consistency proof, but no pages of such a series of notes are left.

By June 1933, the consistency problem is formulated in terms of sequent calculus:

> (VI 33) *The possibility of a transfinite assignment of values* seems almost sure, more or less on the basis of the reducibility theorem. Let us take the following consideration: Proofs that become continuously smaller with reduction are assigned values according to the number of their sequents. ... One should just be able to classify each proof directly in a correct way. The best should be to begin with simple calculi.

Now there is a leap to October 1934, when the consistency proof seems to have already been finished. We read (date X.34):

> One must distinguish between the semi-contentful proof that associates to each formula resp. sequent a semi-contentful concept of correctness, and the proof by the concept of reducibility that works with reductions of a derivation. This one leads over to the purely-formal proof that considers only the reductions of a derivation of a contradiction.

It is the semi-contentful proof, or, in Kreisel's terms, the proof that is partly in terms of meaning, that would give a true insight into the significance of consistency, and that Gentzen sets out to write down towards the end of 1934.

The passages of **INH** from October 1934 already contain references to a series with the signum **WAV** that stands for *Widerspruchsfreiheit Arithmetik Veröffentlichung* (consistency arithmetic publication) and of which some pages have been preserved. They deal mainly with the production of sequents with formulas in prenex normal form and with a variant of Gentzen's reduction procedure for the classical sequent calculus *LK* of the doctoral thesis, to be discussed later. The writing proceeded chapter by chapter in the spring of 1935, each chapter being sent to Bernays as soon as it was ready. Bernays made comments concerning which only Gentzen's replies have been preserved. These comments provoked some changes, after which Gentzen submitted his long manuscript, some 100 typewritten pages, to the *Mathematische Annalen*, where it was received on 11 August 1935. A copy was sent to Weyl.

The quotation from X.34 refers to "the concept of reducibility that works with reductions of a derivation." There are two distinct notions that are called reducibility. One is the *syntactic* notion of conversion of nonnormalities in derivations, and the analogous situation with the induction rule, where the rule has as a conclusion a numerical instance, and the step is resolved into a number of instances of logical rules. This notion can be applied also to derivations in sequent calculus, because of the correspondence between natural and sequent derivations. On the other hand, Gentzen's search for a meaning to a consistency proof had led him to a general *semantic* notion of reducibility that applies to formulas and sequents. In the list of three suggested consistency proofs from April 1933, the second, "semi-contentful," type uses the semantic notion of *reducibility of sequents*, and the third uses instead the syntactic notion of *reducibility of derivations*. Confusion can be created when the reducibility of sequents in the semantic sense is applied to the sequents of a derivation. The aim with the notion of reducibility of sequents was to give a finitary interpretation to arithmetic. "Finitary" here has to be taken in broad terms, not in the way of the strict finitism of Hilbert. It turns out that by the end of 1935, Gentzen's variant of finitism encompassed the whole of the second number class, the constructive transfinite ordinals.

Gentzen's *reduction procedure* for sequents is intended as a semantic explanation of arithmetic. The reduction rules are modeled on "the mathematics of finite domains," the title of Gentzen's §7, in which the quantifiers can be replaced by conjunctions and disjunctions, and classical propositional logic dictates what the conditions of correctness for the formulas are: $A \& B$ is correct when both A and B are correct, $\neg A$ is correct when A is false, etc. The correctness of the rules of inference of propositional logic is almost immediate. Let us note that Gentzen's view of classical logic here is the same as Brouwer's: It is the logic of finite domains. This is the second of the "four insights" in Brouwer's 1928 paper *Intuitionistische Betrachtungen über den Formalismus* (Intuitionistic considerations on formalism). Brouwer makes the same point much later, in his 1955 article on Boole's logic.

Brouwer's first insight was that "the formalists" have to differentiate between the generation of theorems in formal systems and the contentful theory of these systems, and that the latter is based on "the intuitionistic theory of the set of natural numbers." The second insight was cited previously. The third insight was that excluded middle equals the assumption of the solvability of every mathematical problem. The fourth insight is most relevant for Gentzen: *"The recognition that a contentful justification of formalistic mathematics by a proof of its consistency contains a vicious circle."* This is directly the terminology of Gentzen's initial ponderings in **INH**. Brouwer's insights are also seen in action in among others, Gentzen (1936a): "I believe that, for example, in the general theory of sets a careful proof theoretic investigation will finally confirm the opinion that all powers exceeding the countable are, in a quite definite sense, only empty appearances and one will have to have the good sense to do without these concepts." All in all, a trusted disciple, from among "the formalists" to boot, had emerged as if by itself, to whom the typically Brouwerian exclamation in the beginning of the *Betrachtungen* applies:

> The acceptance of these insights is only a question of time, because they are the results of pure reflection and hence contain no disputable element, so that anyone who has once understood them must accept them.

Gentzen refers to Brouwer's paper at the very end of his long article. It is, in addition to the reference to Brouwer's 1924 paper on the continuity of real functions in Gentzen (1938a), his only reference to a work of Brouwer's. He would instead refer freely to Heyting's formalization of intuitionistic logic. I think these facts just tell us what Gentzen thought proper to refer to as a Göttingen logician whose future depends on the opinion of Hilbert, rather than what he was indebted to in his work. How could he, who in the spring of 1935 had applied for an assistantship with Hilbert, have stated the simple truth? The consistency proof of 1935 resolved, in the words of Bernays, the "temporary fiasco" of Hilbert's *Beweistheorie* by using the methods of Brouwer's intuitionistic mathematics.

10.2. GENTZEN'S ORIGINAL PROOF OF CONSISTENCY

(A) THE PLAN AND CIRCUMSTANCES OF THE ORIGINAL PROOF. The main idea of Gentzen seems to have been that the consistency of arithmetic is proved by giving a special semantic explanation of correctness in arithmetic, either of formulas or of sequents. Next, this notion is applied to formal derivations. Finally, it is shown that there is no derivation of a contradiction that would be correct in the semantic sense.

By the preceding, consistency was a by-product of the more ambitious idea of giving a constructive semantics to intuitionistic arithmetic. Syntax and semantics have to match each other, and it has to be laid down what is achieved by a consistency proof; in particular, that it does not somehow assume what it sets out to prove. In his discussion of these topics in **INH**, Gentzen carefully avoids talking about the traditionally central notion of semantics, namely truth. Like Brouwer, he talks about *correctness* (Richtigkeit) and says that a statement *holds* (gilt).

The first mention in print of Gentzen's original consistency proof seems to be a remark in Bernays' 1935 account of Hilbert's research on the foundations of arithmetic (Bernays 1935b, p. 216):

> During the printing of this account, G. Gentzen has brought forth a demonstration of consistency of the full number-theoretic formalism (to be published soon in the Math. Ann.), by a method that corresponds throughout to the basic requirements of the finitary standpoint. Thereby the assumption mentioned above about the extent of finitary methods (p. 212) finds its refutation.

The passage is somewhat ambiguous. Bernays writes earlier that the "finitary standpoint at the basis of Hilbert's proof theory could not lead to a proof of the consistency of the usual number-theoretic formalism" (p. 213). This view clearly stems from Gödel's theorem, by which, writes Bernays, any suggested proof of consistency must contain an explanation of why it is not representable in the arithmetic formalism itself. Now, on the other hand, he seems to suggest that Gentzen's proof has refuted the said limitation. The matter provoked a long exchange, as we will soon see.

When he sailed to New York in September 1935, Bernays had with him the submitted proof, which apparently had been accepted for publication before he traveled. On board was the first class passenger Gödel. I have seen a postcard in the Bernays collection of the ETH-Zurich in which Gödel requests a meeting with Bernays, for the fired professor had to travel in a tourist class and could not go and meet Gödel just like that. During the fall term, the two commented on Gentzen's proof, but only the answers of the latter have been preserved. They contain some information, though in a form that is often bound to frustrate the reader, such as the following passage from a letter of 11 December 1935 (Menzler-Trott 2007, p. 64)

> The possible changes indicated by Gödel were known to me, but are in fact inapplicable from the finite standpoint because of their impredicative character.

Gentzen answered the criticisms by changing the semantically based consistency proof into one that uses the now generally known transfinite induction principle, with essential changes of large parts of the manuscript being sent to the journal in February 1936. As mentioned, they contained a turn into an older idea, and various passages from **INH** make evident this remark of Bernays, transmitted through Kreisel's recollections (Kreisel 1987, pp. 173–175). By good luck, the proof originally submitted for publication was preserved by Bernays in the form of galleys. They were published in English translation in 1969 in the edition of Gentzen's papers by M. Szabo and in 1974 in the original German.

The net effect of the criticisms was a proof that mixed elements from the purely formal and semi-contentful approaches instead of arriving at the former through the third proof idea, that of a proof through the syntactic notion of reducibility. The presentation suffered from these changes, but Gentzen was happy with the overall result he had found during the fall of 1935, that a clear-cut transfinite induction can replace his original proof with a precise "Gödel-point," the transfinite ordinal ε_0 that characterized Peano arithmetic.

The version submitted in August 1935, referred to here as Gentzen (1935), was mutilated by the changes Gentzen made. Gentzen (1935) gives a semantics for the derivability relation in arithmetic, expressed as a sequent $A_1, \ldots, A_n \to C$. There is just a single conclusion C from the assumptions in the list $A_1, \ldots, A_n$, instead of a finite number of possible cases as in the classical sequent calculus LK of his doctoral thesis (Gentzen 1934–35).

When the sequent notation is used, there is a double sense to derivability. The arrow is like the "vertical dots" in the inference schemes of natural deduction. On the other hand, there is the notion of derivability of a sequent by the rules of sequent calculus. Thus, these rules relate derivabilities in the first sense to each other, such as in the way exemplified by the left sequent calculus rule for disjunction. Disjunction elimination becomes the sequent rule: If C is derivable from A and from B, it is derivable from $A \vee B$. With assumptions added, we have the correspondence

$$
\cfrac{A \vee B \quad \cfrac{[A], \Gamma \quad [B], \Delta}{\vdots \qquad \vdots} \\ C \qquad C}{C} \vee E
\qquad \rightsquigarrow \qquad
\cfrac{A, \Gamma \to C \quad B, \Delta \to C}{A \vee B, \Gamma, \Delta \to C} \, L\vee
$$

Above the inference line of rule $\vee E$, there are two schematic derivations that are given as two corresponding sequents above the inference line of rule $L\vee$. Its conclusion gives the final situation of derivability of rule $\vee E$.

The correspondence goes the same way for the other rules. For simplicity, I have taken the situation in which the major premiss of rule $\vee E$ in natural deduction is an assumption. The correspondence between natural deduction and sequent calculus was understood rather well by Gentzen, though not in full (see Section 9.3 (A)).

In Gentzen (1935), a semantics of derivability in the first sense, as represented by the dots or arrows, is given. Then it is applied to derivability in the second sense. Next, we look at the reduction of sequents.

(C) THE REDUCTION OF SEQUENTS. The atomic formulas of arithmetic are decidable equalities between numerical terms. It follows that the whole propositional part of arithmetic is decidable. Gentzen's *reduction procedure* is carried over from the classical propositional logic of formulas to sequents, as exemplified by the following: If $A \& B$ in the antecedent of a sequent $A \& B, \Gamma \rightarrow C$ is false, one of A and B is false, and each can be tried in turn in the place of $A \& B$. If $\neg A$ in $\neg A, \Gamma \rightarrow C$ is false, it is deleted and the sequent is changed into $\Gamma \rightarrow A$.

Gentzen's essential idea is to extend the procedure from the finitary domain to quantified formulas by applying the "transfinite sense" of $\forall x A(x)$ in a certain way. Gentzen calls it "the in-itself sense" (*der an-sich Sinn*).

A way to think of the reduction procedure is that the correctness of a sequent $\Gamma \rightarrow C$ is guaranteed if, in whatever way C may have as a consequence a false claim, it can be shown that some assumption in Γ likewise presupposes a falsity. Then, whenever the assumptions Γ hold, also C holds. To put it in figurative terms, say we have a sequent of the form $\Gamma \rightarrow \forall x A(x) \& \forall x B(x)$ and an omniscient opponent who can *reason classically* by the in-itself sense of things and to whom the infinity of the natural numbers is not an obstacle. Such a creature can decide when $\forall x A(x) \& \forall x B(x)$ is false in its eyes, with, say, $\forall x A(x)$ a false conjunct, and next take a falsifying instance $A(n)$ out of the infinitely many possibilities. Our task is to show that, even if we don't have the opponent's classical and transfinite capacities, we can make finitarily choices *after* the opponent's choices so that some assumption in Γ turns out false. It is this "finitary sense" that Gentzen is after in his semantical explanations.

The reduction of sequents is effected by suitable moves in what I, continuing to speak of Gentzen's procedure in suggestive terms, call a "falsification game" in which first certain "S-moves" are taken in the succedents of sequents, followed by "A-moves" in the antecedent.

S-move

SVar. *The sequent $\Gamma \rightarrow C$ has free variables. Numbers are chosen at will to instantiate these until there are no free variables left.*

S&. *The sequent is* $\Gamma \to A \& B$ *and either* $\Gamma \to A$ *or* $\Gamma \to B$ *is chosen at will.*

S¬. *The sequent is* $\Gamma \to \neg A$ *and the reduced sequent is* $A, \Gamma \to 0{=}1$.

S∀. *The sequent is* $\Gamma \to \forall x\, A(x)$ *and some instance* $\Gamma \to A(n)$ *is chosen at will.*

Order of precedence: *Move* **SVar** *comes before the other* **S**-*moves.*

The **S**-moves are classical, for the falsifier knows how to end up with the worst possible case; here, a false equation as a conclusion. Each **S**-step simplifies the succedent of the sequent to be reduced until an equation $m = n$ remains. If the equation is true, the attempt at falsifying the sequent failed. Otherwise, when no **S**-move is applicable and $m = n$ is false, the task is to show that some of the assumptions must contain a falsity, too. To do this, the following steps can be taken in the antecedent.

A-moves

A&. *The sequent is* $A \& B, \Gamma \to m = n$ *with* $m = n$ *false. The reduced sequent is* $A, A \& B, \Gamma \to m = n$ *or* $B, A \& B, \Gamma \to m{=}n$.

A¬. *The sequent is* $\neg A, \Gamma \to m = n$ *with* $m = n$ *false. The reduced sequent is* $\neg A, \Gamma \to A$.

A∀. *The sequent is* $\forall x\, A(x), \Gamma \to m = n$ *with* $m = n$ *false. The reduced sequent is* $A(k), \forall x\, A(x), \Gamma \to m = n$ *for some k.*

Order of precedence: **S**-*moves always come before* **A**-*moves.*

In the first of the **A**-steps, the conjunction is repeated, for it can happen that at some later stage one also needs the other conjunct. It would be possible to have a single move with A, B that replaces $A \& B$ with no repetition. The negation step seems to be classical in that $\neg A$ in the antecedent and a falsity in the succedent does not lead to the intuitionistically derivable $\neg\neg A$ in the succedent but rather to A; however, as said, the reasoning in the succedent part is classical.

The aim of the reduction procedure is to ensure that a false formula in the antecedent part of a sequent can be produced whenever a false numerical equation appears in the succedent. Note that if a negation at the left is reduced, there will be an **S**-step, unless it was a negation of an equality. Given a sequent $\Gamma \to C$, the result of reduction, provided the process terminates, is a sequent to which no reduction step applies.

Let us now check that indeed the reasoning in the succedent side is classical even if the domain is infinite.

Let the sequent be $\Gamma \to A \& B$. The **S**-steps should produce something false out of the succedent, by which the succedent itself is also false. If that is so, then $\neg(A \& B)$ is true; in other words classically $\neg A \vee \neg B$ is true. The worst case is produced by a choice, say $\neg A$, that gives a sequent $\Gamma \to A$ with a false succedent. Note that the conjuncts in $A \& B$ may very well be "transfinite," universally quantified formulas, and that it need not be decidable which of them is false. This does not matter, because **A**-steps have to be such that they apply to any choice of succedent that may have been made.

Let the sequent be $\Gamma \to \forall x A(x)$. As previously, if the succedent is false, then $\neg\forall x A(x)$ is true, so classically $\exists x \neg A(x)$ is true. There is an instance $\neg A(k)$ true "in itself," and the sequent $\Gamma \to A(k)$ with a false succedent has to be dealt with.

Finally, if the sequent is $\Gamma \to \neg A$, there are no choices, and the reduction goes on with $A, \Gamma \to 0 = 1$.

Definition 1. Irreducibility, endform, correctness.

(i) *A sequent is* **irreducible** *if no reduction move applies to it.*

(ii) *A sequent $\Gamma \to m = n$ is in* **endform** *if either $m = n$ is true or there is some false equality in Γ.*

(iii) *A sequent $\Gamma \to C$ is* **correct** *if for each choice of* **S**-*moves there are* **A**-*moves such that $\Gamma \to C$ reduces to endform.*

We often say simply that a sequent is reducible if it is reducible to endform. The aim of Gentzen's consistency proof is to show that all derivable sequents are reducible. It follows that the sequent $\to 0 = 1$ is not derivable, because it is irreducible but not in endform: No atom in the antecedent is false, because there are none.

(D) THE CALCULUS NLK. As can be seen, the reduction of sequents is an idea independent of a particular logical calculus. To emphasize this important aspect, I reversed the order of presentation of the calculus and the reduction procedure from that in Gentzen (1935). In fact, it is this aspect that made it possible for Gentzen to change the calculus into another one in the published proof instead of rewriting the whole paper (as he perhaps should have done).

The calculus in Gentzen (1935) is what is today called "natural deduction in sequent calculus style." It can be found already in Gentzen's handwritten thesis manuscript with the nomenclature *NLK*, where the letters stand for "natürlich-logistisch klassisch." *NLK* is an obvious intermediate stage in the translation from natural deduction proper into sequent calculus. The idea is simply to display for each formula occurrence in a natural derivation all the open assumptions the formula depends on. There is a fundamental difference to sequent calculus proper, because there are no left rules for conjunction, implication, and universal quantification. To finish the translation to sequent calculus, Gentzen inserts cuts (as in Section 9.3 (A)).

A further aspect of *NLK* is its classical character. Gentzen knew that classical logic would not be necessary but used it anyway. My guess is that he did it mainly for expository purposes, so that his intended general reader of the *Mathematische Annalen* would not be put off by a reliance on such esoteric things as intuitionistic logic. Table 10.1 presents the rules of *NLK*, as they are given in Gentzen's paper, except for the fraktur typeface.

Table 10.1. The rules of Gentzen's calculus NLK

&-*introduction*: The sequents $\Gamma \to A$ and $\Delta \to B$ give the sequent
$\Gamma, \Delta \to A \& B$

&-*elimination*: $\Gamma \to A \& B$ gives $\Gamma \to A$ resp. $\Gamma \to B$

$\vee$-*introduction*: $\Gamma \to A$ gives $\Gamma \to A \vee B$ resp. $\Gamma \to B \vee A$

$\vee$-*elimination*: $\Gamma \to A \vee B$ and $A, \Delta \to C$ and $B, \Theta \to C$ give
$\Gamma, \Delta, \Theta \to C$

$\forall$-*introduction*: $\Gamma \to A(a)$ gives $\Gamma \to \forall x\, A(x)$ on the condition that the free variable *a does not occur* in Γ nor in $\forall x\, A(x)$

$\forall$-*elimination*: $\Gamma \to \forall x\, A(x)$ gives $\Gamma \to A(t)$

$\exists$-*introduction*: $\Gamma \to A(t)$ gives $\Gamma \to \exists x\, A(x)$

$\exists$-*elimination*: $\Gamma \to \exists x\, A(x)$ and $A(a), \Delta \to C$ give $\Gamma, \Delta \to C$ on the condition that the free variable *a does not occur* in Γ, Δ, C nor in $\exists x\, A(x)$

$\supset$-*introduction*: $A, \Gamma \to B$ gives $\Gamma \to A \supset B$

$\supset$-*elimination*: $\Gamma \to A$ and $\Delta \to A \supset B$ give $\Gamma, \Delta \to B$

Rule of "*refutation*": $A, \Gamma \to B$ and $A, \Delta \to \neg B$ give $\Gamma, \Delta \to \neg A$

"*Elimination of double negation*": $\Gamma \to \neg\neg A$ gives $\Gamma \to A$

Rule of "*complete induction*": $\Gamma \to A(0)$ and $A(x), \Delta \to A(x+1)$ give
$\Gamma, \Delta \to A(t)$

These rules are direct translations of the rules of classical natural deduction and the induction rule into the notation of sequent calculus. In rules that have more than one premiss, the contexts $\Gamma, \Delta, \ldots$ are accordingly added up in the antecedent of the conclusion. To the assumptions of natural deduction correspond "logical groundsequents" of the form $A \rightarrow A$. *Refutation* is from the German *Widerlegung*, abbreviated *Wid* in what follows.

The rules of inference are given in a linear form of sentence. There is, in fact, not a single inference line printed in the whole work. Here again, Gentzen perhaps wanted to appeal to a general readership, to whom the notation of two-dimensional proof trees with their inference lines was completely unknown at the time. Those few specialists who had read his doctoral thesis were an exception.

Formal derivations within Gentzen's calculus consist of series of sequents, with the following definition (Gentzen 1936, p. 513):

> A derivation consists of a number of sequents in succession, such that each of these is either a "groundsequent" or results from some previous sequents through a "structural modification" or a "rule of inference."

To deal with the explicit listing of the assumptions in the antecedent parts of sequents, Gentzen adds the following "structural modifications":

1. *Exchange* of the order of assumptions in the list.
2. *Contraction* of two occurrences of an assumption into one.
3. *Weakening* of an antecedent by the addition of an assumption.
4. *Change* of a bound variable by a *fresh* one.

Gentzen writes (pp. 513–514) that these rules are "purely formal in nature and inconsequential in their content; they have to be mentioned explicitly because of the peculiarities of the formalism."

The calculus is completed by adding what Gentzen calls "mathematical groundsequents." They have the form $\rightarrow A$, with A a mathematical axiom. The right axioms are not listed; instead, Gentzen writes that for the consistency proof, it is not so essential what the mathematical axioms are. He gives as examples the following:

$$\forall x\, x = x, \quad \forall x \forall y (x = y \supset y = x),$$
$$\forall x \forall y \forall z (x = y \,\&\, y = z \supset x = z),$$
$$\forall x \, \neg x + 1 = x, \quad \forall x \forall y\, x + y = y + x,$$
$$\forall x \forall y \forall z (x + y) + z = x + (y + z).$$

Gentzen was convinced that the rule of induction was the only one that created real problems for the consistency proof. The rest of the arithmetic principles could be dealt with in whatever way was easiest. One such method was given in the doctoral thesis. It contains "as an application of the sharpened *Hauptsatz*" a consistency proof for induction-free arithmetic (IV §3). Axioms are allowed to appear in the antecedent parts of sequents in a classical calculus, and consistency is proved by the midsequent theorem. An alternative method was to formulate the axioms as groundsequents with free parameters, in the form

$$\rightarrow a = a, \quad a = b \rightarrow b = a, \quad a = b,\, b = c \rightarrow a = c,$$
$$a + 1 = a \rightarrow \quad , \quad \rightarrow a + b = b + a, \quad \rightarrow (a + b) + c = a + (b + c).$$

In Gentzen's consistency proof of 1938, after some transformations, such groundsequents contain only numerical terms, and it can be decided whether they are correct, i.e., whether an equation in the succedent is a true numerical equation or whether an equation in the antecedent is a false one.[2]

For this presentation, we grant to Gentzen what he presumes, that the arithmetical principles, except that of complete induction, will not cause problems. It will be sufficient to prove the consistency of the system of classical natural deduction augmented by the rule of complete induction.

The rules of *NLK* exhibit some strange features: Why does the classical calculus *NLK* contain a full set of connectives and quantifiers? Further, there was no normalization theorem for the classical calculus. How could the ideas about a meaning explanation through normalization be carried over to a consistency proof in terms of *NLK*?

---

[2] For a proper proof-theoretical treatment of the arithmetical axioms, see Siders (2015).

The essential difference of *NLK* with respect to a proper sequent calculus is that the elimination rules for conjunction, implication, and universal quantification operate on the right part of the sequent. The corresponding sequent calculus rules operate on the left, antecedent part of the sequent. Looking at rules $\vee E$ and $\exists E$, we notice that if the first premiss is a logical groundsequent, $A \vee B \rightarrow A \vee B$ resp. $\exists x\, A(x) \rightarrow \exists x\, A(x)$, and if it is left unwritten, the rules turn out identical to the left rules of sequent calculus. I followed this method in my (2009) article, with an intuitionistic sequent calculus for Heyting arithmetic, and gave a proof of its consistency directly along the lines of Gentzen's proof.

When Gentzen comes to the proof of consistency in his paragraph 14, he has already removed the connectives $\vee, \supset$, and $\exists$ by making the obvious translations into the fragment with just $\&, \neg$, and $\forall$. Even the inferences by the rules for the former group are transformed in the obvious way. Gentzen notes, at the end of paragraph 12, that a transformed derivation "is an essentially intuitionistically acceptable number-theoretic derivation: namely, the 'elimination of a double negation' could, where it is used, be replaced by other rules of inference." We saw earlier, in the note of January 1933, how this goes through, and there is even more reason to ask why the classical rule is kept.

(E) THE REDUCTION OF DERIVATIONS. The main part of Gentzen's original consistency proof consists of a few lemmas, that I state as follows, with some typical cases of the proofs covered.

Lemma 2. *Initial sequents $A \rightarrow A$ are correct.*

The proof is by induction on the length of A. Assume **SVar**-moves have been made, so that there are no free variables. There are four cases of which we show two:

1. A is an equality $m = n$, and we have $m = n \rightarrow m = n$. By the decidability of numerical equality, if $m = n$ is true, $A \rightarrow A$ is in endform, and likewise if $m = n$ is false.

2. A is $B \& C$. Then $B \& C \rightarrow B \& C$ reduces by **S&** to $B \& C \rightarrow B$ or to $B \& C \rightarrow C$.

Case 2.1. Consider the first time when the reduction of $B \& C \rightarrow B$ by arbitrary **S**-moves gives a sequent of the form $B \& C, \Gamma \rightarrow m = n$, namely the first time for an **A**-move. The sequent $B \to B$ is reducible by the inductive hypothesis, so the same sequence of **S**-moves as for $B \& C \to B$ gives the reducible sequent $B, \Gamma \to m = n$. Application of **A**& to $B \& C, \Gamma \to m = n$ gives $B, B \& C, \Gamma \to m = n$. When formula $B \& C$ in the antecedent is left intact, the sequent reduces exactly as $B, \Gamma \to m = n$.

Case 2.2. If $B \& C \to B \& C$ is reduced by **S**& to $B \& C \to C$, the proof is as earlier, with C in place of B.

We see here in action the method of simulating in **A**-moves the choices made in the preceding **S**-moves. The remaining two cases of $A \equiv \forall x \, B(x)$ and $A \equiv \neg B$ are treated similarly. QED.

The *rule of composition* can be given as the inference scheme:

Table 10.2. Composition of two sequents.

$$\frac{\Gamma \to D \quad D, \Delta \to C}{\Gamma, \Delta \to C} \; Comp$$

Gentzen takes it for granted that derivations can be composed in his calculus *NLK*.

Lemma 3. Closure of derivability under composition. *If the sequents* $\Gamma \to D$ *and* $D, \Delta \to C$ *are derivable in NLK and possible eigenvariables distinct, also the sequent* $\Gamma, \Delta \to C$ *obtained by composition is derivable in NLK.*

The proof would be straightforward were the calculus intuitionistic, as in von Plato (2009b, lemma 5.2). I have not tried to determine how a proof with Gentzen's rules would go through, but let's assume it does.

Next in Gentzen's article comes the crucial property of the whole proof of consistency, which he named the *Hilfssatz* in obvious analogy to his famous *Hauptsatz*, or cut elimination theorem for predicate logic. It states that composition preserves the correctness of sequents in the sense of definition 1.

Hilfssatz 4. Closure of reducibility under composition. *If the sequents* $\Gamma \to D$ *and* $D, \Delta \to C$ *are reducible to endform and possible eigenvariables distinct, their composition into* $\Gamma, \Delta \to C$ *is reducible to endform.*

The proof is by induction on the length of the composition formula D. We can assume that possible free variables have been removed by **SVar**.

1. $D \equiv m = n$. Then the first premiss of *Comp* reduces to $\Gamma^* \to 0 = 1$, or $\Gamma^* \to m = n$ if move **S¬** was never applied. Assume **S**-moves have been applied to the conclusion $\Gamma, \Delta \to C$ until $\Gamma, \Delta, \Delta^* \to k = l$ is produced, in which $k = l$ can be assumed false and Δ^* consists of those formulas, possibly none, that applications of **S¬** have brought to the antecedent. Leaving Δ, Δ^* intact, the sequence of **A**-moves that reduces $\Gamma \to m = n$ to the endform $\Gamma^* \to 0 = 1$ (or $\Gamma \to m = n$) reduces $\Gamma, \Delta, \Delta^* \to k = l$ to an endform.

We note that if $\Gamma \to m = n$ is reducible and $m = n$ is false, the equation $0 = 1$ can replace $m = n$: Compose $\Gamma \to m = n$ with the sequent in endform $m = n \to 0 = 1$ to get $\Gamma \to 0 = 1$.

2. $D \equiv A \& B$. The composition is

$$\frac{\Gamma \to A \& B \quad A \& B, \Delta \to C}{\Gamma, \Delta \to C} \ \textit{Comp}$$

By assumption, $\Gamma \to A \& B$ is reducible, so both $\Gamma \to A$ and $\Gamma \to B$ are. Consider the second premiss $A \& B, \Delta \to C$. Either there is no application of **A&** to $A \& B$ in its reduction and $A \& B$ can be removed. Then $\Delta \to C$ is reducible, and therefore also $\Gamma, \Delta \to C$. Else **A&** is applied at some stage to a reducible sequent $A \& B, \Delta^* \to 0 = 1$ with, say, the reducible sequent $A, A \& B, \Delta^* \to 0 = 1$ as the result. We now apply *Comp*:

$$\frac{\Gamma \to A \quad A, A \& B, \Delta^* \to 0 = 1}{A \& B, \Gamma, \Delta^* \to 0 = 1} \ \textit{Comp}$$

By the inductive hypothesis, *Comp* applied to shorter formulas maintains reducibility, so $A \& B, \Gamma, \Delta^* \to 0 = 1$ is reducible. The reduction of $\Gamma, \Delta \to C$ by the arbitrarily chosen **S**-moves that reduce the premiss $A \& B, \Delta \to C$ to $A \& B, \Delta^* \to 0 = 1$, gives the sequent $\Gamma, \Delta^* \to 0 = 1$ that is reducible to endform by the same **A**-moves as $A \& B, \Gamma, \Delta^* \to 0 = 1$.

3. $D \equiv \forall x \, A(x)$. The composition is

$$\frac{\Gamma \to \forall x \, A(x) \quad \forall x \, A(x), \Delta \to C}{\Gamma, \Delta \to C} \ \textit{Comp}$$

By assumption, $\Gamma \to \forall x\, A(x)$ is reducible, so $\Gamma \to A(n)$ is reducible for any choice of n. As in **2**, either there is no application of $\mathbf{A\forall}$ to $\forall x\, A(x)$ in the reduction of the second premiss and $\forall x\, A$ can be removed. Then $\Delta \to C$ is reducible, and therefore also $\Gamma, \Delta \to C$. Else $\mathbf{A\forall}$ is applied at some stage to a reducible sequent $\forall x\, A(x), \Delta^* \to 0 = 1$, with the reducible sequent $A(k), \forall x\, A(x), \Delta^* \to 0 = 1$ as the result. With the instance k also in the first premiss, application of *Comp* to the shorter formula $A(k)$ gives

$$\frac{\Gamma \to A(k) \quad A(k), \forall x\, A(x), \Delta^* \to 0 = 1}{\forall x\, A(x), \Gamma, \Delta^* \to 0 = 1}\ Comp$$

The conclusion is reducible by the inductive hypothesis. The reduction of $\Gamma, \Delta \to C$ by the arbitrarily chosen **S**-moves that reduce the premiss $\forall x\, A(x), \Delta \to C$ to $\forall x\, A(x), \Delta^* \to 0 = 1$, gives the sequent $\Gamma, \Delta^* \to 0 = 1$, which is reducible to endform by the same **A**-moves as $\forall x\, A(x), \Gamma, \Delta^* \to 0 = 1$.

4. $D \equiv \neg A$. The composition is

$$\frac{\Gamma \to \neg A \quad \neg A, \Delta \to C}{\Gamma, \Delta \to C}\ Comp$$

In the reduction of the second premiss of *Comp*, if $\mathbf{A}\neg$ is never applied to $\neg A$, it can be deleted, and what remains, the sequent $\Delta \to C$, is reducible. Then also $\Gamma, \Delta \to C$ is reducible. Otherwise there is a reducible sequent $\neg A, \Delta^* \to 0 = 1$, to which in turn $\mathbf{A}\neg$ is applied to give the reducible sequent $\neg A, \Delta^* \to A$.

The first premiss of *Comp* reduces by $\mathbf{S}\neg$ to $A, \Gamma \to 0 = 1$. Application of *Comp* to the shorter formula A gives

$$\frac{\neg A, \Delta^* \to A \quad A, \Gamma \to 0 = 1}{\neg A, \Gamma, \Delta^* \to 0 = 1}\ Comp$$

The conclusion is reducible by the inductive hypothesis.

As previously, if in the reduction of $\neg A, \Gamma, \Delta^* \to 0 = 1$ move $\mathbf{A}\neg$ is never applied to $\neg A$, it can be deleted, and the remaining sequent $\Gamma, \Delta^* \to 0 = 1$ is reducible. This is the sequent produced from $\Gamma, \Delta \to C$ by the arbitrary initial **S**-moves that gave $\neg A, \Delta^* \to 0 = 1$, so $\Gamma, \Delta \to C$ is reducible.

If instead in the reduction of $\neg A, \Gamma, \Delta^* \rightarrow 0 = 1$ move $\mathbf{A}\neg$ is applied at some stage to $\neg A$ in a reducible sequent $\neg A, \Gamma^*, \Delta^{**} \rightarrow 0 = 1$, the reducible sequent $\neg A, \Gamma^*, \Delta^{**} \rightarrow A$ is obtained. Composition with $A, \Gamma \rightarrow 0 = 1$ gives $\neg A, \Gamma, \Gamma^*, \Delta^{**} \rightarrow 0 = 1$, which is reducible. Therefore, continuing this analysis, at some stage the formula $\neg A$ in the antecedent of the result of composition must remain unreduced and can be deleted. The resulting sequent is then reducible.   QED.

The proof seems innocent enough, even if the very last steps are a bit tedious. They perhaps bring to mind methods in proofs of underivability through failed proof search.

Gentzen tries to persuade the reader of the constructive character of the reduction procedure by reformulating the *Hilfssatz* in the following terms (Gentzen 1935, §14.4 4): "If reduction procedures for $\Gamma \rightarrow \mathfrak{D}$ and $\mathfrak{D}, \Delta \rightarrow \mathfrak{C}$ are known, a reduction procedure for $\Gamma, \Delta \rightarrow \mathfrak{C}$ can also be given." These, however, are just words; there is no substantive difference from the formulation given earlier.

(F) THE CONSISTENCY THEOREM. The final component in Gentzen's consistency proof is to show that the rules of inference preserve the correctness of sequents.

Theorem 5. *If the sequent $\Gamma \rightarrow C$ is derivable, it reduces to endform.*

The proof is by induction on the last step of a derivation. If $\Gamma \rightarrow C$ is a logical groundsequent, it is correct as shown above by lemma 2. Otherwise, consider the last rule of the derivation and show that if the premisses reduce to endform, the conclusion also reduces. The cases are the structural modifications, seven logical rules, and *CI*.

Gentzen goes through the two cases for $\forall$. Then he notes that the three conjunction rules go through similarly. The cases for $\forall$ are as follows.

1. The last rule is $\forall I$. The conclusion is $\Gamma \rightarrow \forall x\, A(x)$, and it reduces by $\mathbf{S}\forall$ to $\Gamma \rightarrow A(m)$. The premiss $\Gamma \rightarrow A(y)$ is by assumption reducible, with y the eigenvariable. Rule $\mathbf{SVar}$ produces a sequent $\Gamma \rightarrow A(n)$ that is reducible for any choice of n; in particular, the choice m. Therefore, the conclusion of $\forall I$ is reducible.

2. The last rule is $\forall E$. The premiss is $\Gamma \rightarrow \forall x\, A(x)$. **S**-moves applied to the conclusion $\Gamma \rightarrow A(t)$ produce the sequent $\Gamma \rightarrow A(m)$.

The premiss is reducible for any choice of value for x, and therefore $\Gamma \to A(t)$ is reducible by the same **A**-moves as for $\Gamma \to \forall x\, A(x)$.

Next comes a peculiar turn, when Gentzen (1935, §14.4 4) writes that for the two negation rules and *CI*, the *Hilfssatz* is put into use. Namely, the question is: If we leave out rule *CI*, shouldn't we get a standard proof of the consistency of classical first-order logic as a result? Moreover, the classical rule is dispensable in *NLK*. What does a principle such as the *Hilfssatz* do in this connection? In fact, as shown in von Plato (2009b), it is not needed.

Observation. *With the intuitionistic calculus NLI, the* Hilfssatz *is needed only for showing that rule CI preserves the correctness of derivations.*

Gentzen naturally knew the preceding observation from a result of his thesis, that if induction is left out, the proof of consistency can be carried through finitistically in the tradition of Hilbert's program; some logical groundwork had simply remained undone.

Finally, we look at the crucial step of the consistency proof, the case of rule *CI*.

3. The last rule is *CI*. The premisses are the two sequents $\Gamma \to A(1)$ and $A(y), \Delta \to A(y+1)$, and the conclusion is $\Gamma, \Delta \to A(t)$. In its reduction, if t has free variables, applying **SVar** gives some numerical term n in place of t. In the second premiss, any application of rule **SVar** gives a reducible sequent, so that $A(m), \Delta \to A(m+1)$ is derivable and reducible for any m. An $n-1$-fold composition of $\Gamma \to A(1)$ with $A(1), \Delta \to A(2), \ldots, A(n-1), \Delta \to A(n)$ gives

$$\cfrac{\cfrac{\Gamma \to A(1) \quad A(1), \Delta \to A(2)}{\Gamma, \Delta \to A(2)}\; Comp \quad A(2), \Delta \to A(3)}{\cfrac{\Gamma, \Delta^2 \to A(3)}{\vdots}}\; Comp$$

$$\cfrac{\Gamma, \Delta^{n-2} \to A(n-1) \quad A(n-1), \Delta \to A(n)}{\Gamma, \Delta^{n-1} \to A(n)}\; Comp$$

Thus, the sequent $\Gamma, \Delta^{n-1} \to A(n)$ is derivable by the admissibility of composition and reducible by the *Hilfssatz*. For the conclusion $\Gamma, \Delta \to A(t)$ of *CI*, an **S**-move reduces it to $\Gamma, \Delta \to A(n)$, and $\Gamma, \Delta \to A(n)$ is reducible because $\Gamma, \Delta^{n-1} \to A(n)$ is.

By hindsight, we have one more aspect of later calculi of proof search present in the reduction procedure. Namely, it has to be shown that the rule of contraction preserves reducibility, and this is secured because there is a possible repetition of a formula for rules that are not invertible.

With the preceding lemmas and preparations, consistency can be easily concluded. As noted, the sequent $\rightarrow 0 = 1$ is irreducible but not in endform, and therefore it is not derivable.

Corollary 6. *The system NLK+CI+arithmetic axioms is consistent.*

More is achieved than the unprovability of $0 = 1$; namely, it follows that from derivability follows correctness, or soundness in more recent logical terminology.

(G) THE EARLIEST PRESERVED CONSISTENCY PROOF. The preceding account of the consistency proof is essentially based on the preserved galley proofs of Gentzen's article in its original 1935 form. We now have a background against which it is possible to understand recently transcribed stenographic manuscripts from the fall of 1934. These are, first, the last ten pages of **INH**, written in October of that year. Second, there is the manuscript **BZ**, for *Beweistheorie der Zahlentheorie* (Proof theory of arithmetic), written between August 1934 and March 1935, with pages 1–6 and 9–12 preserved. The third one is **WAV**, already mentioned, and written around October 1934 but without dates and with the pages 55–56, 77–80, and 83–86 preserved. It consists of preliminary notes for the preparation of the final manuscript, judging from the pages that have been preserved as well as from occasional references to it in the other manuscripts. These notes have direct connections to the article that Gentzen prepared in the spring of 1935.

There are parts in **BZ** and **WAV** that treat the same topic, the preparation of sequent derivations in which all formulas are in prenex normal form. The propositional part of arithmetic is decidable, and Gentzen wanted to delimit propositional steps in derivations to a "finitary" part, above a "transfinite part" that contains steps of inference with the quantifiers, a separation that follows from the midsequent theorem for derivations in the classical sequent calculus *LK* that he used at this stage. The aim was to have a consistency proof that is "more concentrated on what is essential" (**WAV**, p. 78). One idea in **WAV** is to minimize the number of proper rules of inference, by using groundsequents, such as $A \& B \rightarrow A, A \rightarrow A \vee B$, and $\forall x A(x) \rightarrow A(t)$. The reducibility of such sequents follows easily from the reducibility of initial sequents, say when an **A**-move is met with the first one, $A \& B$ is replaced by A, and then reduction steps can be applied in the antecedent as in the reduction of $A \rightarrow A$.

WAV contains the earliest preserved proof of consistency of arithmetic, detailed in three pages and based on the reduction procedure (pp. 78–80). It is titled "the second proof of correctness (*LK* consistency proof)," and by this proof, it becomes clearer that the first proof was also based on the reduction procedure but with the intuitionistic sequent calculus *LI* augmented by the classical sequent $\neg\neg A \rightarrow A$. When the classical "symmetric calculus" is used, as Gentzen calls it, the reduction procedure has to be defined also for disjunction and existence. (He prefers to leave implication out because it breaks the symmetry of *LK*.) The details of the reduction procedure for symmetric sequents are not spelled out, but it is clear how they are to be taken: The arbitrary choices (moves by the opponent in my terminology) now extend to the antecedent part, with the aim of producing a true numerical equation at the left. Thus, the opponent is able to make the best possible choice in the case of an antecedent formula $\exists x A(x)$ for a true instance $A(t)$. Later, the respondent can reply to such a choice in the succedent by choosing the same instance $A(t)$. Analogously, the opponent chooses one of the disjuncts in $A \vee B$ in the antecedent, and the respondent does so in the succedent. Whenever the opponent has produced a false equation in the succedent or a true equation in the antecedent, the respondent does likewise, with the aim of producing a false equation in the antecedent or a true one in the succedent. Whenever this is the case, a sequent is in endform, a notion that coincides with the earlier one for single-succedent sequents.

As before, the proof of consistency proceeds by showing that derivable sequents reduce to endform. Propositional connectives are handled by logical groundsequents, as previously, and for conjunction in the succedent by $A, B \rightarrow A \& B$ and disjunction in the antecedent by the dual $A \vee B \rightarrow A, B$. The quantifier rules are straightforward. There remain *CI* and the crux of the proof, namely that the composition of sequents in the form of a *mix rule* (Mischung), or *multicut* in more recent terminology, maintains reducibility (**WAV**, p. 79):

> Let the reducibility of both upper sequents be already shown. That for the lower sequent to be shown. We do a complete induction after the grade of the mix. That is now: The number of $\forall$ and $\exists$ at the head of the mix formula $\mathfrak{M}$.

$$\frac{\Gamma \to \Delta(\mathfrak{M}) \quad \Theta(\mathfrak{M}) \to \Lambda}{\Gamma \Theta^* \to \Delta^* \Lambda}$$

The notion of grade indicates that the formulas are in prenex normal form. The proof that the grade of the mix formula can be lowered ends with these words (**WAV**, p. 80): "This somewhat peculiar inference is subjected to detailed criticism in Section IV," clearly a reference to the paper Gentzen was writing. In that paper, the proof through a reduction procedure obtained a third form, through the classical natural calculus *NLK* that uses the sequent notation. Thus, what I have called the original proof was by Gentzen's count in **WAV** actually the third one. Moreover, **INH** and **BZ** contain references to a lost series **WTZ**, clearly for "consistency transfinite numbers," but the few indications of ordinals in that attempted proof do not yet contain the Gentzen ordinal ε_0 of 1936.

Gentzen was obviously happy and content with his original proof. A lot of work had gone into it, both formal and conceptual. The detailed discussions in **INH** especially give an indication of its conceptual importance to Gentzen. Others, however, felt that something was missing. Gentzen (1935) contains general discussions about the significance of consistency proofs and even singles out the *Hilfssatz* as central, but it does not indicate clearly what the crucial points in the proof of the *Hilfssatz* are. Specifically, the termination of the reduction process is not treated in precise terms.

10.3. BAR INDUCTION: A HIDDEN ELEMENT IN THE CONSISTENCY PROOF

(A) THE PROBLEM OF TERMINATION OF THE REDUCTION PROCEDURE. There was extensive correspondence between Gentzen and Bernays about the consistency proof, as well as some letters between Gentzen and Weyl and between Gentzen and Van der Waerden. Only the letters of Gentzen to Bernays have been preserved. The first of these letters, dated 23 June 1935, was sent from Gentzen's hometown Stralsund by the Baltic Sea and included the "final part" of the consistency paper. It then goes on to discuss the suggestions made by Bernays and notes, among other things, that the existence property of arithmetic follows for formulas $\exists x\, A(x)$, "in case $A(x)$ is not transfinite." Toward the end, Gentzen writes that in the final chapter he wanted to discuss transfinite ordinal numbers and their relation to reduction procedures and construction procedures. He then continues: "In the end, these things did not seem ripe for a presentation yet but could perhaps find place in a later separate publication."

A second letter written three weeks later, on 14 July, contains the following passage:

> I have written in fact nonsense on pp. 75–76; I held my eye on an older form of the notion of reduction, in which the reduction steps are uniquely determined. The passages could be corrected more or less as follows: At 15.21, reducibility should be replaced by: 'There is a number ν so that for each series $\mathcal{R}_\nu$ of ν numbers, a series of at most ν sequents can be given such that the first one is $\mathfrak{S}$q, and each of these is formed from the preceding one through a reduction step, and the last one has endform, and further, *the possible choices are determined through the associated numbers from the series $\mathcal{R}_\nu$.*' Correspondingly under 15.23: "For each *infinite* series $\mathcal{R}$ of numbers, a finite series of sequents can be given, the first of which …" as before. – I have, however, cancelled these passages completely, because they are not fully necessary; perhaps I could give sometime later complete proofs to both theorems in a special publication.

The uniquely determined sequence of reduction steps should refer to a reduction procedure for derivations of the false formula $0 = 1$.

The preceding passage is reminiscent of Brouwer's explanation of bar induction in his (1924b), where the connection to transfinite induction is also made. It is a pity we don't have Gentzen's proof of his 15.23 preserved. He states quite clearly that the choice sequences in steps of reduction, represented as sequences of natural numbers, lead to endform in a finite number of steps. Gentzen's use of natural numbers in the description of the reduction procedure brings him very close to Brouwer, who in 1924 formulated the bar theorem as follows:

> If to each element of a set M a natural number β is associated, M is decomposed by this association into a well-ordered species S of subsets M_α, such that each of these is determined by a finite initial segment of choices. To each element of the same M_α is associated the same natural number β_α.

In Gentzen, M consists of the collection of reduction sequences of sequents and the choices are single reduction steps.

Now there is a big gap in the correspondence, until 4 November, with a four-page tightly and very orderly written letter sent to Bernays in Princeton, where also Gödel was. Bernays, possibly with the help of Gödel, had taken up the central problem of the proof, as can be gathered from Gentzen's answer:

> I have considered all these aspects already myself, including the geometrical image of branching line segments. You are quite right that the finiteness of even a single reduction path for the sequent $\Gamma, \Delta \rightarrow \mathfrak{C}$ can get grounded on the finiteness of a whole series of different reduction paths for $\mathfrak{D}, \Delta \rightarrow \mathfrak{C}$. But this does nothing for my proof idea!

He then writes that he had added an explanation to the article, one that would be in the end of section 14 (p. 112 of the German version of 1974), the proofs of which he had sent back a few days earlier:

> Let the following be remarked to avoid misunderstandings: The type of reduction of $\mathfrak{D}, \Delta \rightarrow \mathfrak{C}$ into $\mathfrak{D}, \Delta^* \rightarrow \mathfrak{C}^*$ can eventually depend on a <u>choice</u> (14.6 2 1) that takes place in the reduction of the mix-sequent $\Gamma, \Delta \rightarrow \mathfrak{C}$.[3] The same holds of each further step of reducing back, and, it can be added, the new <u>mix</u>-sequent $\Gamma, \Delta^* \rightarrow \mathfrak{C}^*$ etc. need in no way always be the reduced one of the preceding sequent (14.6 2 3). So, the number of steps of proof can be very different, according to the result of the individual choices; the only thing that is certain is that it is in every case <u>finite</u>. To prove a claim for <u>every</u> possible choice, it is sufficient to prove it for <u>one</u> specific, arbitrary choice. Therefore it is sufficient in the entire proof to keep an eye on just

---

[3] We saw this situation in the earlier consistency proof, in the case of rule CI, in which $A(t)$ in the succedent was reduced to $A(n)$.

one <u>single specific</u> sequence of reductions of the sequent $\mathfrak{D}, \Delta \to \mathfrak{C}$, and thereby on just one <u>single specific</u> finite series of steps of proof.

No second round of proofs is known that would contain this passage. The terminology of mix-sequents is that of the doctoral thesis, where cut formulas were called mix-formulas. This terminology is used also in the consistency proof of October 1934 mentioned earlier. As to why the termination is not addressed in the paper, Gentzen writes that "since you don't seem so far to have said anything concerning the recognition of the finiteness of the forms of inference, I have left them out of the consistency proof; also because there would be still one thing and another to clarify." Gentzen obviously had a great desire to publish what he had to offer so far.

(B) THE ESSENCE OF GENTZEN'S HILFSSATZ. Gentzen's letter of 4 November contains a description of what he calls "the essence of the somewhat peculiar inductive inference" in the *Hilfssatz*, namely why the reduction procedure should terminate:

> A proposition $\forall x\, F(x)$ is proved if each of the infinitely many special cases $F(v)$ is proved. Let each of these again be equivalent to a proposition $\forall x\, F_v(x)$, each special case $F_v(\mu)$ of these propositions again equivalent to a proposition $\forall x\, F_{v,\mu}(x)$, etc. Let the following be known: Each arbitrary series of specializations $\forall x\, F(x), F(v) \supset\subset \forall x\, F_v(x), F_v(\mu) \supset\subset \forall x\, F_{v,\mu}(x), \ldots$ ends after a finite number of components in a formula $F_{v\mu\ldots}(\varrho)$, the <u>correctness of which is known</u>. To be proved now: $\forall x\, F(x)$ is correct. To this end, I infer as follows: The correctness of $\forall x\, F(x)$ is secured if $F(v)$ holds for whichever arbitrarily chosen v. So let us assume that we had chosen a specific number v, and it remains just to prove $F(v)$. This is $\supset\subset \forall x\, F_v(x)$. Now I infer just as before, namely, that to show that this proposition holds, it suffices to take whichever arbitrarily determined special case, say $F_v(\mu)$, etc. This chain of inferences must end after a finite number of steps, because each arbitrary sequence $\forall x\, F(x), F(v), F_v(\mu), \ldots$ had to be finite. Thereby $\forall x\, F(x)$ is proved.

He says that this is "an analogy" that should be compared to "the image of the branching sequence of line segments." The latter can be depicted as follows, with Gentzen's example.

Table 10.3. The branching figure of the reduction procedure

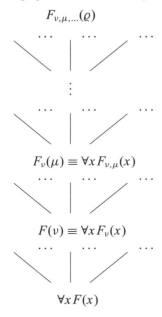

$$F_{\nu,\mu,\dots}(\varrho)$$

$$F_{\nu}(\mu) \equiv \forall x\, F_{\nu,\mu}(x)$$

$$F(\nu) \equiv \forall x\, F_{\nu}(x)$$

$$\forall x\, F(x)$$

There is no bound on how many universal quantifiers can occur in a formula, and therefore a denumerably branching tree of any finite height can occur. Moreover, the "analogy" begins with a peculiar requirement that each of the infinitely many instances of $\forall x\, F(x)$ be proved. Is the analogy an appeal to infinitary proof theory, or to the infinite capacities of the classical reasoner? "What do you think, now, about this way of inference? Shouldn't it be finite?" These are his questions to Bernays, but he adds at once the parenthetical remark:

> If one turns the proof into an indirect one, i.e., begins like this: Assume that $\forall x\, F(x)$ does not hold, then there is a counterexample ν so that $F(\nu)$ does not hold, so neither $\supset\subset \forall x\, F_{\nu}(x)$, etc, then the *tertium non datur* enters.

Now we can read the suggestion as the choice of a path in a reduction tree that has a denumerable branching at each node. If there is at least one sequence of choices such that the topformula $F_{\nu\mu\dots}(\varrho)$ gets falsified, we have established $\neg F_{\nu\mu\dots}(\varrho)$. If not, i.e., if no counterexample was found, proceeding all the way down to the root of the reduction tree, we get that the assumption that $\forall x\, F(x)$ does not hold is false, and a classical step of double-negation elimination (the *tertium non datur*) gives $\forall x\, F(x)$.

Gentzen is well aware that a new type of proof is about to surface here. One reason for expecting something new is, naturally, that the proof must go beyond those that can be justified in arithmetic. Turning now to the reduction rules, we notice that sequences of moves in the succedent and antecedent can alternate any number of times, and each block of succedent moves can produce an initial segment in the Baire space of a denumerably branching tree.

(C) A LOST CONNECTION: CONSISTENCY PROOFS AND BAR INDUCTION. Gentzen's 1935 paper was received, in a literal sense, by Bernays and Weyl. Parts of the paper were changed in February 1936, by which time the galleys of the original version had been prepared, and clearly the paper must have gone to print earlier. In fact, Menzler-Trott (2007, p. 61) reproduces a letter from Gentzen to Hellmuth Kneser, written 27 October 1935, in which it is stated that the first galley proofs have already arrived. He also wrote there that Van der Waerden, then a professor at Leipzig, had commented very positively on the proof.

As mentioned, Cavaillès was in Göttingen in the fall of 1935. His book contains a discussion of Gentzen's proof, with a description of the reduction procedure and the problem of its termination, but along the lines of the treatment by transfinite induction of the published version (Cavaillès 1938, pp. 165–170). A letter from Cavaillès to Albert Lautmann indicates that Gentzen had read the text and "repaired passages where to him I had oversimplified" (Menzler-Trott 2007, p. 82).

Weyl gave his copy to Stephen Kleene, who, by his own telling, got a job in Wisconsin and gave the copy back after only two days. That was very unfortunate for the development of proof theory and foundational study in general. It took another fifteen years before Kleene took up Gentzen's work, in an article about sequent calculus (Kleene 1952a) and in the *Introduction to Metamathematics*. In the latter, the *Hauptsatz* is presented in detail and applied to a consistency proof of arithmetic without the induction rule (p. 463). For the full consistency proof, there is just a "brief heuristic account of the method used by Gentzen" (p. 476). It is all based on the published proof. Richard Vesley worked with Kleene on the constructive theory of ordinals, and together they

studied Brouwer's work. Vesley has told me (in an e-mail on 3 March 2011) that he is sure that they never discussed the extent to which Gentzen had been influenced by Brouwer's intuitionistic theory of ordinals.

Bernays (1970) recalled that the main point of criticism was Gentzen's implicit use of the *fan theorem*, a principle of Brouwer's intuitionistic mathematics by which if all branches of a finitely branching tree are finite, the tree consists of a finite number of nodes. Bernays gives the same explaination in his prefatory words to the publication of Gentzen (1935) in 1974 (p. 97):

> A methodical objection was made against the original proof, namely that it used implicitly a principle usually described today as the "fan theorem," by which each branching figure that branches only finitely at each point and in which each thread ends after a finite number of component parts, can on the whole have only a finite extension.

The fan theorem is a special case of the *bar theorem* in which the latter branchings are denumerably infinite. These terminologies are much later than the results, but it is still a bit strange that Bernays explicitly describes the finite branching when Gentzen's proof clearly has denumerable branching. A detailed proof of Gentzen's *Hilfssatz* can indeed be given by the use of bar induction. It makes Gentzen's "peculiar inductive inference" of termination of reduction crystal clear (Siders and von Plato 2015). As we saw, Bernays writes that Gentzen's use of bar induction was "implicit," and if so, then he came to use that principle independently of Brouwer, which would be remarkable.

In Brouwer (1924b), to which Gentzen (1938a) refers, the bar theorem is called "the main theorem on well-ordered sets." The additional remarks in Brouwer (1924a) make quite explicit the associated principle of transfinite induction on well-founded trees on which the bar theorem rests (p. 645). Bernays alludes to the connection in his remark about the intuitionistic provability of transfinite induction up to the ordinal ε_0 cited in Section 10.1.

The bar theorem was known to Gödel and also to von Neumann, who also was in Princeton at that time and must have heard discussions

about Gentzen's result.[4] Kreisel had had extensive discussions with Bernays about Gentzen's original proof, and he writes (1987, p. 173) that "Gödel and von Neumann criticized the original—posthumously published—version." There is a more general principle behind the fan and the bar theorem. Kreisel (1976, p. 201) states that both Gödel and von Neumann "naturally knew the theory of choice sequences that Brouwer had developed systematically, and especially the problematic assumption (of which Brouwer was particularly proud), namely that all functions F with arbitrary choice sequences of natural numbers as arguments and natural numbers as values ... can be produced inductively. The best-known corollary is the *fan theorem*."

During and after the criticisms by Bernays and Gödel, seconded by von Neumann and possibly even Weyl (as suggested by a letter of Weyl's in Menzler-Trott 2007, p. 58), Gentzen laid the foundation of today's *ordinal proof theory*. It can be seen clearly from his letters how this topic emerged in a few months' time, with the consequence that the semantical explanation of sequents through a notion of reducibility and the consistency proof by induction on well-founded trees receded into the background. Gentzen (1938b), having closed his new proof of consistency with a presentation of transfinite induction, writes that he puts no specific weight on the notion of reducibility of derivable sequents and ends up with what seems almost a contradiction in terms: "I resorted to it at the time as *one* argument against radical intuitionism." This paper was the second part of an issue in Heinrich Scholz' publication series on logic and foundations. The first part was Gentzen's essay on *The present situation in mathematical foundational research*. There he contrasts the lessons from intuitionism, presumably those of Brouwer's four insights, against "radical intuitionism, that rejects as senseless everything in mathematics that does not correspond to the constructive point of view." Gentzen followed the path of Brouwer's choice sequences in 1932, but by 1936 it seems he had found that constructive ordinals codify intuitionistic principles in more conventional terms.

---

[4] I owe the information about von Neumann's knowledge of Brouwer's "fundamental theorem on finite sets" to Dirk van Dalen. He kindly sent me a copy of an April 1929 letter from von Neumann to Brouwer, that contains a constructive proof of the existence of a winning strategy in chess using the fan theorem. The letter is found in Van Dalen (2011).

The fate of the original proof was that it was simply put aside, just like Gentzen had put aside his detailed proof of normalization for natural deduction, the former saved only because Bernays had kept the galley proofs, the latter only because he had kept Gentzen's handwritten notes. Gentzen's use of induction on well-founded trees had been saved also in another sense, the extent of which is yet to be fully determined; namely, as shown by the titles of topics in Gödel's stenographic notes in his *Arbeitshefte*, there are at least 150 pages Gödel's of work on Gentzen's proof, with such suggestive titles as *Principal lemma of Gentzen's consistency proof with choice sequences* (*Arbeitsheft* 11, p. 28). In the earlier *Arbeitsheft* 4 (p. 39), there is the title *Gentzen with choice sequences*. A general theorem is formulated and the proof ends on p. 50 with "**Theorem**. Induction principle for functions of the finite type: $[(n)\mathfrak{A}(\Phi_n)] \supset \mathfrak{A}(\Phi)$). $\mathfrak{A}(\mathrm{const.}) \supset (\Phi)\mathfrak{A}(\Phi)$." The application to Gentzen's consistency proof is that if from the assumption that every one-step continuation Φ_n of a reduction sequence Φ has the property $\mathfrak{A}$ it follows that Φ has the property $\mathfrak{A}$, then from the base case $\mathfrak{A}(\mathrm{const.})$ it follows that all reduction sequences have the property $\mathfrak{A}$.

A picture starts emerging from a study of Gentzen's original proof, the letters he wrote to Bernays, Gödel's titles in the *Arbeitshefte*, his "Zilsel" lecture of 1938 and the Yale lecture of 1941, and Kreisel's recollections: Namely, Gödel's *no-counterexample* interpretation of the Zilsel lecture derives from Gentzen's original proof (cf. also Tait 2005). Secondly, concerning the *Dialectica-interpretation*, Kreisel (1987, p. 175) writes: "At first Gödel, like von Neumann, was ill at ease with Gentzen's use of functionals, albeit of *lowest* type. But when Gödel returned to the subject, about 5 years later, he used *all finite* types." The connections between Gentzen's proof, Gödel's functional interpretation, and bar induction and transfinite induction are suggestive enough, but the source materials are at present not sufficiently known for these matters to be discussed in any conclusive way-so here is where the story must rest for now.

REFERENCES

A great disservice is being done to scholarship by the reference system prevalent today that has running numbers, usually in square brackets, for the items in the references. The defects of this system are twofold. First, it is enormously disturbing for the reader to be constantly checking the list of references to see what article or book is being referred to. The reader's memory is burdened with information that has no meaning elsewhere. Second, the awareness of who did what and when is eroded little by little. If we read Gödel (1931) or Gentzen (1936), we know what that is, contrary to a plain [104] and [90], say, and similarly with hundreds of other works. Such couplings of names and years give us a timeline that is indispensable for an awareness of the development of logic or any other part of science. The thoughtless "bibtex" square bracket numbering system of references is destroying such awareness and should therefore be universally abandoned. It has just one, totally inessential advantage: that it saves some space. In a standard article, that may be a few lines, and in a book, a page or two.

Acerbi, F. (2007) *Euclide: Tutte le opere*. Bompiani, Milan.
———. (2010) *Il silenzio delle sirene: la matematica greca antica*. Carocci editore, Rome.
Ackermann, W. (1924) Begründung des "tertium non datur" mittels der Hilbertschen Theorie der Widerspruchsfreiheit. *Mathematische Annalen*, vol. 93, pp. 1–36.
———. (1928) Zum Hilbertschen Theorie der reellen Zahlen. *Mathematische Annalen*, vol. 99, pp. 118–133. English translation in Van Heijenoort (1967).
———. (1940) Zur Widerspruchsfreiheit der Zahlentheorie. *Mathematische Annalen*, vol. 117, pp. 162–194.

Ambrose, A. (1935) Finitism in mathematics I–II. *Mind*, vol. 44, pp. 186–203 and 317–340.

Aristotle. *Prior Analytics*.

van Atten, M. (2005) The correspondence between Oskar Becker and Arend Heyting. In V. Peckhaus, ed., *Oskar Becker und die Philosophie der Mathematik*, pp. 119–142. Fink Verlag, Munich.

Beltrami, E. (1868) *Saggio di interpretazione della geometria non-euclidea*. De Angelis, Naples.

Bernays, P. (1918) *Beiträge zur axiomatischen Behandlung des Logik-Kalküls*. Manuscript Hs. 973:193, Bernays collection, ETH-Zurich. Published in Hilbert (2013).

———. (1926) Axiomatische Untersuchung des Aussagen-Kalkuls der "Principia mathematica." *Mathematische Zeitschrift*, vol. 25, pp. 305–320.

———. (1927a) Probleme der theoretischen Logik. As reprinted in Bernays (1976), pp. 1–16.

———. (1927b) Zusatz zu Hilberts Vortrag über "Die Grundlagen der Mathematik." *Abhandlungen aus dem mathematischen Seminar der Hamburgischen Universität*, vol. 6, pp. 89–92.

———. (1932) Methoden des nachweises von Widerspruchsfreiheit und ihre Grenzen. In *Verhandlungen des Internationalen Mathematiker-Kongresses Zürich 1932*, vol. 2, pp. 342–343, Orell Füssli, Zurich.

———. (1935a) Quelques points essentiels de la métamathématique. *L'Enseignement Mathématique*, vol. 34, pp. 70–95.

———. (1935b) Hilberts Untersuchungen über die Grundlagen der Arithmetik. *Gesammelte Abhandlungen*, vol. 3, pp. 196–216.

———. (1936) *Logical Calculus*. Lectures at the Institute for Advanced Study, Princeton, 1935–36.

———. (1945) Review of Ketonen (1944). *The Journal of Symbolic Logic*, vol. 10, pp. 127–130.

———. (1951) Über das Induktionsschema in der rekursiven Zahlentheorie. In *Kontrolliertes Denken*, ed. A. Menne et al., pp. 10–17. Kommissions-Verlag Karl Alber, Freiburg Brsg.-Munich.

———. (1965) Betrachtungen zum Sequenzen-kalkul. *Contributions to Logic and Methodology in Honor of J. M. Bochenski*, pp. 1–44. North-Holland, Amsterdam.

———. (1970) On the original Gentzen consistency proof for number theory. In J. Myhill et al., eds., *Intuitionism and Proof Theory*, pp. 409–417. North-Holland, Amsterdam.

———. (1976) *Abhandlungen zur Philosophie der Mathematik*. Wissenschaftliche Buchgesellschaft, Darmstadt.

Bernays, P. and M. Schönfinkel (1928) Zum Entscheidungsproblem der mathematischen Logik. *Mathematische Annalen*, vol. 99, pp. 342–372.

Beth, E. (1947) Review of Popper (1947). *Mathematical Reviews*, MR0021924.

———. (1955) Semantic entailment and formal derivability. *Mededelingen der Koninklijke Nederlandse Akademie van Wetenschappen, Afd. Letterkunde*, vol. 18, pp. 309–342.

Boole, G. (1847) *The Mathematical Analysis of Logic*.

———. (1854) *An Investigation of the Laws of Thought*. Walton and Maberly, London.

Boolos, G. (1985) Reading the *Begriffsschrift*. *Mind*, vol. 94, pp. 331–344.

Borel, E. (1903) Contribution à l'analyse arithmétique du continu. As reprinted in Borel (1972), pp. 1439–1485.

———. (1908) Les paradoxes de la theorie des ensembles. As reprinted in Borel (1972), pp. 1271–1276.

———. (1912a) *Notice sur les travaux scientifiques*. Gauthier-Villars, Paris. Also in Borel (1972), pp. 119–190.

———. (1912b) Le calcul des intégrales définies. As reprinted in Borel (1972), pp. 827–878.

———. (1972) *Oeuvres de Emile Borel*. 4 vols. Editions du CNRS, Paris.

Brouwer, L. (1921) Besitzt jede reelle Zahl eine Dezimalbruchentwicklung? As reprinted in Brouwer (1975), pp. 236–245.

———. (1924a) Intuitionistische Zerlegung mathematischer Grundbegriffe. As reprinted in Brouwer (1975), pp. 275–280.

———. (1924b) Beweis, dass jede volle Funktion gleichmässig stetig ist. *Koninklijke Akademie van Wetenschappen te Amsterdam, Proceedings of the Section of Sciences*, vol. 27, pp. 189–193.

———. (1924c) Bemerkungen zum Beweise der gleichmässigen Stetigkeit voller Funktionen. *Koninklijke Akademie van*

Wetenschappen te Amsterdam, Proceedings of the Section of Sciences, vol. 27, pp. 644–646.

———. (1925) Zur Begründung der intuitionistischen Mathematik. I. *Mathematische Annalen*, vol. 93, pp. 244–257.

———. (1928) Intuitionistische Betrachtungen über den Formalismus. *Sitzungsberichte der Preussischen Akademie der Wissenschaften*, pp. 48–52.

———. (1955) The effect of intuitionism on the classical algebra of logic. As reprinted in Brouwer (1975), pp. 551–554.

———. (1975) *Collected Works*, vol. 1. North-Holland, Amsterdam.

Burris, S. (1995) Polynomial time uniform word problems. *Mathematical Logic Quarterly*, vol. 41, pp. 173–182.

Cantor, G. (1892) Über eine elementare Frage der Mannigfaltigkeitslehre. *Jahresbericht der Deutschen Mathematiker-Vereinigung*, vol. 1, pp. 75–78.

Carnap, R. (1929) *Abriss der Logistik*. Springer, Vienna.

———. (1931) Die logizistische Grundlegung der Mathematik. *Erkenntnis*, vol. 2, pp. 91–105.

Cavaillès, J. (1938) *Méthode axiomatique et formalisme*. Reprinted in *Oeuvres completés de Philosophie des sciences*, Hermann, Paris, 1994.

Church, A. (1932) A set of postulates for the foundation of logic. *Annals of Mathematics*, vol. 33, pp. 346–366.

———. (1936) A note on the Entscheidungsproblem. *The Journal of Symbolic Logic*, vol. 1, pp. 101–102.

Cohen, P. (1963) The independence of the continuum hypothesis. *Proceedings of the National Academy of Sciences of the United States of America*, vol. 50, pp. 1143–1148 and vol. 51, pp. 105–110.

Cooper, B. and J. van Leeuwen (2013) *Alan Turing: His Work and Impact*. Elsevier, Amsterdam.

Copeland, J., et al. (2013) *Computability: Turing, Gödel, Church, and Beyond*. MIT Press, Cambridge, Massachusetts.

Coquand, T. (2004) About Brouwer's fan theorem. *Revue internationale de philosophie*, no. 240, pp. 483–489.

Cosmadakis, S. (1988) The word and generator problem for lattices. *Information and Computation*, vol. 77, pp. 192–217.

Courant, R. (1981) Reminiscences of Hilbert's Göttingen. *The Mathematical Intelligencer*, vol. 3, pp. 154–64.

Curry, H. (1941) A formalization of recursive arithmetic. *American Journal of Mathematics*, vol. 63, pp. 263–282.

———. (1953) Review of D. Hilbert and W. Ackermann, *Grundzüge der theoretischen Logik*, third edition of 1949. *Bulletin of the American Mathematical Society*, vol. 59, pp. 263–267.

———. (1963) *Foundations of Mathematical Logic.* Page references are to the reprint: Dover, New York, 1977.

van Dalen, D. (2011) *The Selected Correspondence of L.E.J. Brouwer.* Springer, Berlin.

———. (2013) *L.E.J. Brouwer: Topologist, Intuitionist, Philosopher.* Springer, Berlin.

Davis, M. (1965) *The Undecidable. Basic Papers on Undecidable Propositions, Unsolvable Problems and Computable Functions.* Raven Press, New York. Corrected reprint Dover, New York, 2004.

———. (2013) Three proofs of the unsolvability of the Entscheidungsproblem. In Cooper and van Leeuwen (2013), pp. 49–52.

Dawson, J. (1997) *Logical Dilemmas: The Life and Work of Kurt Gödel.* A. K. Peters, Wellesley, Massachusetts.

Dawson, J. and C. Dawson (2005) Future tasks for Gödel scholars. *The Bulletin of Symbolic Logic*, vol. 11, pp. 150–171.

Dedekind, R. (1872) *Stetigkeit und Irrationale Zahlen.* Vieweg reprint, Wiesbaden, 1969.

———. (1888) *Was sind und was sollen die Zahlen?* Vieweg reprint, Wiesbaden, 1969.

Dragalin, A. (1988) *Mathematical Intuitionism: Introduction to Proof Theory.* American Mathematical Society, Providence, Rhode Island. Russian original 1979.

Dummett, M. (1959) A propositional calculus with denumerable matrix. *The Journal of Symbolic Logic*, vol. 24, pp. 97–106.

Dyckhoff, R. (1988) Implementing a simple proof assistant. In *Workshop on Programming for Logic Teaching*, Proceedings 23, pp. 49–59. University of Leeds, Centre for Theoretical Computer Science.

———. (1992) Contraction-free sequent calculi for intuitionistic logic. *The Journal of Symbolic Logic*, vol. 57, pp. 795–807.

Dyckhoff, R. and S. Negri (2015) Geometrization of first-order logic. *The Bulletin of Symbolic Logic*, vol. 21, pp. 123–163.

von Ettingshausen, A. (1826) *Die combinatorische Analysis als Vorbereitungslehre zum Studium der theoretischen höhern Mathematik*. Wallishauser, Vienna.

Frege, G. (1879) *Begriffsschrift, eine nach der arithmetischen nachgebildete Formelsprache des reinen Denkens*. Nebert, Halle. Reprinted in I. Angelelli, ed., *Begriffsschrift und andere Aufsätze*. Olms, Hildesheim, 1964. English translation in Van Heijenoort (1967).

———. (1881) Booles rechnende Logik und die Begriffsschrift. In Frege (1969), pp. 9–52.

———. (1882) Über die wissenschaftliche Berechtigung einer Begriffsschrift. *Zeitschrift für Philosophie und philosophische Kritik*, vol. 81, pp. 48–56.

———. (1884) *Grundlagen der Arithmetik*. Felix Meiner, Hamburg.

———. (1885a) *Über formale Theorien der Arithmetik*. As reprinted in Frege (1967), pp. 103–111.

———. (1885b) Review of H. Cohen's "Das Prinzip der Infinitesimal-Methode und seine Geschichte." *Zeitschrift für Philosophie und philosophische Kritik*, vol. 87, pp. 324–329. Reprinted in Frege (1967).

———. (1893) *Grundgesetze der Arithmetik. Begriffsschriftlich abgeleitet*, vol. 1. Pohle, Jena.

———. (1967) *Kleine Schriften*. Ed. I. Angelelli. Olms, Hildesheim.

———. (1969) *Nachgelassene Schriften*. Ed. H. Hermes et al. Felix Meiner, Hamburg.

———. (1976) *Wissenschaftlicher Briefwechsel*. Ed. G. Gabriel et al. Felix Meiner, Hamburg.

———. (1994) Gottlob Freges politisches Tagebuch. *Deutsche Zeitschrift für Philosophie*, vol. 42, pp. 1057–1066.

———. (2004) *Frege's Lectures on Logic: Carnap's Student Notes, 1910–1914*. Tr. and ed. E. Reck and S. Awodey. Chicago, Open Court.

Gentzen, G. (1932) Über die Existenz unabhängiger Axiomensysteme zu unendlichen Satzsysteme. *Mathematische Annalen*, vol. 107, pp. 329–350.

———. (1933) Über das Verhältnis zwischen intuitionistischer und klassischer Arithmetik. Submitted for publication March 15, 1933, but withdrawn, published in *Archiv für mathematische Logik*, vol. 16 (1974), pp. 119–132.

———. (1934–35) Untersuchungen über das logische Schliessen. *Mathematische Zeitschrift*, vol. 39, pp. 176–210 and 405–431.

———. (1935) Der erste Widerspruchsfreiheitsbeweis für die klassische Zahlentheorie. First printed in *Archiv für mathematische Logik*, vol. 16 (1974), pp. 97–118.

———. (1936) Die Widerspruchsfreiheit der reinen Zahlentheorie. *Mathematische Annalen*, vol. 112, pp. 493–565.

———. (1936a) Der Unendlichkeitsbegriff in der Mathematik. *Semester-Berichte Münster, WS 1936–37*, pp. 65–80. English translation in Menzler-Trott (2007), pp. 343–350.

———. (1938a) Die gegenwärtige Lage in der mathematischen Grundlagenforschung. *Forschungen zur Logik und zur Grundlegung der exakten Wissenschaften*, vol. 4, pp. 1–18.

———. (1938b) Neue Fassung des Widerspruchsfreiheitsbeweises für die reine Zahlentheorie. *Forschungen zur Logik und zur Grundlegung der exakten Wissenschaften*, vol. 4, pp. 19–44.

———. (1943) Beweisbarkeit und Unbeweisbarkeit von Anfangsfällen der transfiniten Induktion in der reinen Zahlentheorie. *Mathematische Annalen*, vol. 120, pp. 140–161.

———. (1964–65) Investigations into logical deduction. *American Philosophical Quarterly*, vol. 1, pp. 288–306 and vol. 2, pp. 204–218.

———. (1969) *The Collected Papers of Gerhard Gentzen*, ed. M. Szabo, North-Holland, Amsterdam.

———. (1974) Der erste Widerspruchsfreiheitsbeweis für die klassische Zahlentheorie. *Archiv für mathematische Logik*, vol. 16, pp. 97–118.

———. (2008) The normalization of derivations. *The Bulletin of Symbolic Logic*, vol. 14, pp. 245–257.

———. (2017) *Saved from the Cellar: Gerhard Gentzen's Shorthand Notes on Logic and Foundations of Mathematics*. With an introduction and translation by Jan von Plato. Studies in the History of Mathematics and Physical Sciences, Springer, Berlin.

Girard, J.-Y. (1987) *Proof Theory and Logical Complexity*, vol. 1. Bibliopolis, Naples.

Glivenko, V. (1928) Sur la logique de M. Brouwer. *Academie Royale de Belgique, Bulletin de la Classe des Sciences*, vol. 5, pp. 225–228.

Gödel, K. (1929) *Über die Vollständigkeit des Logikkalküls.* Doctoral dissertation published in Gödel (1986), pp. 60–101.

———. (1930) Die Vollständigkeit der Axiome des logischen Funktionenkalküls. *Monatshefte für Mathematik und Physik*, vol. 37, pp. 349–360. Reprinted in Gödel (1986), pp. 102–123.

———. (1931) Über formal unentscheidbare Sätze der Principia Mathematica und verwandter Systeme I. *Monatshefte für Mathematik und Physik*, vol. 38, pp. 173–198. Also in Gödel (1986)

———. (1933a) Zur intuitionistischen Arithmetik und Zahlentheorie. As reprinted in Gödel (1986), pp. 286–295.

———. (1933b) The present situation in the foundations of mathematics. In Gödel (1995), pp. 45–53.

———. (1934) On undecidable propositions of formal mathematical systems. Mimeographed lectures in Princeton, reprinted in Gödel (1986), pp. 346–372.

———. (1938a) The consistency of the axiom of choice and of the generalized continuum hypothesis. *Proceedings of the National Academy of Sciences of the United States of America*, vol. 24, pp. 556–557.

———. (1938b) Lecture at Zilsel's. In Gödel (1995), pp. 86–113.

———. (1941) In what sense is intuitionistic logic constructive? In Gödel (1995), pp. 189–200.

———. (1949) An example of a new type of cosmological solution of Einstein's field equations of gravitation. *Reviews of Modern Physics*, vol. 21, pp. 447–450.

———. (1958) Über eine noch nicht benutzte Erweiterung des finiten Standpunktes. *Dialectica*, vol. 12, pp. 280–287. Also in Gödel (1986).

———. (1986) *Collected Works*, vol. 1. Oxford University Press.

———. (1995) *Collected Works*, vol. 3. Oxford University Press.

———. (2003) *Collected Works*, vols. 4–5. Oxford University Press.

Goodstein, R. (1939) Mathematical systems. *Mind*, vol. 48, pp. 58–73.

———. (1945) Function theory in an axiom-free equation calculus. *Proceedings of the London Mathematical Society*, vol. 48, pp. 401–434.

———. (1951) *Constructive Formalism*. Leicester University Press, Leicester.

———. (1957) *Recursive Number Theory*. North-Holland, Amsterdam.

———. (1958) On the nature of mathematical systems. *Dialectica*, vol. 12, pp. 296–316.

———. (1972) Wittgenstein's philosophy of mathematics. In A. Ambrose and M. Lazerowitz, eds., *Ludwig Wittgenstein: Philosophy and Language*, pp. 271–286. Allen and Unwin, London.

Grassmann, H. (1844) *Die Wissenschaft der extensiven Grösse oder die Ausdehnungslehre*. Wigand, Leipzig.

———. (1861) *Lehrbuch der Arithmetik für höhere Lehranstalten*. Enslin, Berlin.

Hankel, H. (1867) *Vorlesungen über die complexen Zahlen und ihre Functionen I*. Leopold Voss, Leipzig.

Hashagen, U. (2008) Kein Platz für das "Genie": John von Neumann und das deutsche Wissenschaftssystem. *Acta Historica Leopoldina*, vol. 54, pp. 571–587.

Heck, R. (2011) *Frege's Theorem*. Oxford Universtiy Press.

van Heijenoort, J., ed. (1967) *From Frege to Gödel, a Source Book in Mathematical Logic, 1879–1931*. Harvard University Press, Cambridge, Massachusetts.

von Helmholtz, H. (1887) Über Zählen und Messen, erkenntnistheoretisch betrachtet. In *Philosophische Aufsätze, Eduard Zeller zu seinem fünfzigjährigen Doctorjubiläum gewidmet*, pp. 17–52. Fues' Verlag, Leipzig.

Hempel, C. (2000) An intellectual autobiography. In *Science, Explanation, and Rationality*. Ed. J. Fetzer, pp. 3–35. Oxford University Press.

Herbrand, J. (1930a) Les bases de la logique hilbertienne. *Revue de métaphysique et de morale*, vol. 37, pp. 243–255. English translation in Herbrand (1971).

———. (1930b) *Recherches sur la théorie de la démonstration*. References are to the English translation in Herbrand (1971).

———. (1931) Sur la non-contradiction de l'arithmétique. *Journal für die reine und angewandte Mathematik*, vol. 166, pp. 1–8. English translation in Van Heijenoort (1967).

———. (1971) *Logical Writings*. Ed. W. Goldfarb. Reidel, Dordrecht.

Hertz, P. (1923) Über Axiomensysteme für beliebige Satzsysteme. Teil II. *Mathematische Annalen*, vol. 89, pp. 76–102.

Heyting, A. (1925) *Intuitionistische Axiomatiek der Projektieve Meetkunde*. Noordhoff, Groningen.

———. (1927) Zur intuitionistischen Axiomatik der projektiven Geometrie. *Mathematische Annalen*, vol. 98, pp. 491–538.

———. (1930a) Die formalen Regeln der intuitionistischen Logik. *Sitzungsberichte der Preussischen Akademie von Wissenschaften, Physikalisch-mathematische Klasse*, pp. 42–56.

———. (1930b) Die formalen Regeln der intuitionistischen Mathematik. *Sitzungsberichte der Preussischen Akademie von Wissenschaften, Physikalisch-mathematische Klasse*, pp. 57–71.

———. (1931) Die intuitionistische Grundlegung der Mathematik. *Erkenntnis*, vol. 2, pp. 106–115.

———. (1934) *Mathematische Grundlagenforschung, Intuitionismus, Beweistheorie*. Springer, Berlin.

———. (1935) Intuitionistische wiskunde. *Mathematica B*, vol. 4, pp. 72–83.

———. (1956) *Intuitionism: An Introduction*. North-Holland, Amsterdam.

Hilbert, D. (1899) *Grundlagen der Geometrie*. Teubner, Leipzig. Several later editions.

———. (1900a) Mathematische Probleme. *Nachrichten der Königlichen Gesellschaft der Wissenschaften zu Göttingen, mathematisch-physikalische Klasse*, pp. 253–297.

———. (1900b) Über den Zahlbegriff. *Jahresbericht der Deutschen Mathematiker-Vereinigung*, vol. 8, pp. 180–183.

———. (1905) Über die Grundlagen der Logik und Arithmetik. *Verhandlungen des Dritten Internationalen Mathematiker-Kongresses*, pp. 174–185. English translation in Van Heijenoort (1967).

———. (1918) Axiomatisches Denken. *Mathematische Annalen*, vol. 78, pp. 405–415.

———. (1922) Neubegründung der Mathematik (Erste Mitteilung). *Abhandlungen aus dem mathematischen Seminar der Hamburgischen Universität*, vol. 1, pp. 157–177.

———. (1923) Die logischen Grundlagen der Mathematik. *Mathematische Annalen*, vol. 88, pp. 151–165.

———. (1925–26) Über das Unendliche. *Jahresbericht der Deutschen Mathematiker-Vereinigung*, vol. 36, pp. 201–215. Longer version of 1926 in *Mathematische Annalen*, vol. 95, pp. 161–190.

———. (1927) Die Grundlagen der Mathematik. *Abhandlungen aus dem mathematischen Seminar der Hamburgischen Universität*, vol. 6, pp. 65–85. Page references are to the English translation in Mancosu (1998).

———. (1928) Probleme der Grundlegung der Mathematik. As reprinted in 1929 in *Mathematische Annalen*, vol. 102, pp. 1–9.

———. (1931a) Die Grundlegung der elementaren Zahlenlehre. *Mathematische Annalen*, vol. 104, pp. 485–494.

———. (1931b) Beweis des tertium non datur. *Nachrichten von der Gesellschaft der Wissenschaften zu Göttingen, mathematisch-philosophische Klasse*, pp. 120–125.

———. (2004) *David Hilbert's Lectures on the Foundations of Geometry, 1891–1902*. Ed. M. Hallett and U. Majer. Springer, Berlin.

———. (2013) *David Hilbert's Lectures on the Foundations of Arithmetic and Logic, 1917–1933*. Ed. W. Ewald and W. Sieg. Springer, Berlin.

Hilbert, D. and W. Ackermann (1928) *Grundzüge der theoretischen Logik*. Springer, Berlin. Second edition 1938, third edition 1949, fourth edition 1959.

Hilbert, D. and P. Bernays (1934) *Grundlagen der Mathematik I*. Springer, Berlin.

———. (1939) *Grundlagen der Mathematik II*. Springer, Berlin.

Hilbert, D. and S. Cohn-Vossen (1932). *Anschauliche Geometrie*, Springer, Berlin.

Hudelmaier, J. (1992) Bounds of cut elimination in intuitionistic propositional logic. *Archive for Mathematical Logic*, vol. 31, pp. 331–354.

Jaśkowski, S. (1934) On the rules of supposition in formal logic. As reprinted in S. McCall, ed., *Polish Logic 1920–1939*, pp. 232–258. University Press, Oxford, 1967.

Johansson, I. (1936) Der Minimalkalkül, ein reduzierter intuitionistischer Formalismus. *Compositio Mathematica*, vol. 4, pp. 119–136.

Kahle, R. (2013) David Hilbert and the *Principia Mathematica*. In N. Griffin and B. Linsky, eds., *The Palgrave Centenary Companion to Principia Mathematica*, pp. 21–34. Palgrave, London.

Kahn G. (1995) Constructive geometry according to Jan von Plato. Available at the Coq proof editor library.

Kanckos, A. (2010) Consistency of Heyting arithmetic in natural deduction. *Mathematical Logic Quarterly*, vol. 56, pp. 611–624.

Ketonen, O. (1943) "Luonnollisen päättelyn" kalkyylista. (On the calculus of "natural deduction"). *Ajatus* (Yearbook of the Philosophical Society of Finland), vol. 12, pp. 128–140. (In Finnish.)

———. (1944) *Untersuchungen zum Prädikatenkalkül*, Annales Academiae Scientiarum Fennicae, Ser. A.I. 23, Helsinki.

Khintchine, A. (1926) Ideas of intuitionism and the struggle for content in contemporary mathematics. *Vestnik Kommunisticheskaya Akademiya*, vol. 16, pp. 184–192. (In Russian.)

Kleene, S. (1936) General recursive functions of natural numbers. *Mathematische Annalen*, vol. 112, pp. 727–742.

Kleene, S. (1952a) Permutability of inferences in Gentzen's calculi LK and LJ. *Memoirs of the American Mathematical Society*, vol. 10, pp. 1–26.

———. (1952b) Review of Péter (1951). *Bulletin of the American Mathematical Society*, vol. 58, pp. 270–272.

———. (1952c) *Introduction to Metamathematics*, North-Holland, Amsterdam.

Kolmogorov, A. (1925) On the principle of excluded middle. Translation of Russian original in Van Heijenoort (1967), pp. 416–437.

Kreisel, G. (1976) Wie die Beweistheorie zu ihren Ordinalzahlen kam und kommt? *Jahresbericht der Deutschen Mathematiker-Vereinung*, vol. 78, pp. 177–223.

———. (1987) Gödel's excursions into intuitionistic logic. In P. Weingartner and L. Schmetterer, eds., *Gödel Remembered*, pp. 65–186. Bibliopolis, Naples.

Kronecker, L. (1901) *Vorlesungen über Zahlentheorie*, vol. 1. Teubner, Leipzig.

Leibniz, G. (1704) *Nouveaux essais sur l'entendement humain*. Page reference to the edition by J. Erdmann, 1840. Eichler, Berlin.

Lopez-Escobar, E. (1999) Standardizing the N systems of Gentzen. In X. Caicedo and C. Montenegro, eds., *Models, Algebras, and Proofs*, pp. 411–434. Dekker, New York.

Löwenheim, L. (1915) Über Möglichkeiten im Relativkalkül. *Mathematische Annalen*, vol. 76, pp. 447–470.

Mac Lane, S. (2005) *A Mathematical Autobiography*. A. K. Peters, Wellesley, Massachusetts.

Mancosu, P. (1998) *From Brouwer to Hilbert: The Debate on the Foundations of Mathematics in the 1920s*. Oxford University Press.

———. (1999) Between Vienna and Berlin: The immediate reception of Gödel's incompleteness theorems. *History and Philosophy of Logic*, vol. 20, pp. 33–45. Reprinted in Mancosu (2010).

———. (2003) The Russellian influence on Hilbert and his school. *Synthese*, vol. 137, pp. 59–101. Reprinted in Mancosu (2010).

———. (2010) *The Adventure of Reason: Interplay between Philosophy of Mathematics and Mathematical Logic, 1900–1940*. University Press, Oxford.

Mancosu, P. and M. Marion (2003) Wittgenstein's constructivization of Euler's proof of the infinity of primes. In F. Stadler, ed., *The Vienna Circle and Logical Empiricism*, pp. 171–188, Kluwer, Dordrecht. Reprinted in Mancosu (2010).

Marion, M. and M. Okada (2012) Wittgenstein and Goodstein on the equation calculus and the uniqueness rule. Talk presented at the Goodstein centenary symposium, Leicester, 2012.

Martin, G. (1938) *Arithmetik und Kombinatorik bei Kant*. Cited from the second enlarged edition published by de Gruyter, Berlin 1972.

McGuinness, B., ed. (1967) *Ludwig Wittgenstein und der Wiener Kreis*. Blackwell, London. English translation *Ludwig Wittgenstein and the Vienna Circle*.

McGuinness, B. (1995) *Wittgenstein in Cambridge: Letters and Documents 1911–1951*. Blackwell, Oxford.

Menzler-Trott, E. (2007) *Logic's Lost Genius: The Life of Gerhard Gentzen*. American Mathematical Society, Providence, Rhode Island.

Moschovakis, Y. (1993) Sense and denotation as algorithm and value. *Lecture Notes in Logic*, vol. 2, pp. 210–249. Springer, Berlin.

Negri, S. (1999) Sequent calculus proof theory of intuitionistic apartness and order relations. *Archive for Mathematical Logic*, vol. 38, pp. 521–547.

———. (2003) Contraction-free sequent calculi for geometric theories, with an application to Barr's theorem. *Archive for Mathematical Logic*, vol. 42, pp. 389–401.

Negri, S. and J. von Plato (1998) Cut elimination in the presence of axioms. *The Bulletin of Symbolic Logic*, vol. 4, pp. 418–435.

———. (2001) *Structural Proof Theory*. Cambridge University Press, Cambridge.

———. (2011) *Proof Analysis: A Contribution to Hilbert's Last Problem*. Cambridge University Press, Cambridge.

von Neumann, J. (1927) Zur Hilbertschen Beweistheorie. *Mathematische Zeitschrift*, vol. 26, pp. 1–46.

———. (1928) Zur Theorie der Gesellschafsspiele. *Mathematische Annalen*, vol. 100, pp. 295–320.

Ohm, M. (1816) *Elementar-Zahlenlehre zum Gebrauch von Schulen und Selbstlernen auch als Leitfaden bey akademischen Vorlesungen*. Palm and Enke, Erlangen.

Ono, K. (1938) Logische Untersuchungen über die Grundlagen der Mathematik. *Journal of the Faculty of Science, Imperial University of Tokyo, Section I*, vol. 3, pp. 329–389.

Pasch, M. (1882) *Vorlesungen über neuere Geometrie*, Teubner, Leipzig.

Peano, G. (1889) *Arithmetices Principia, Nova Methodo Exposita*. Partial English translation in Van Heijenoort (1967).

———. (1901) *Formulaire de Mathématiques*. Carré and Naud, Paris.

———. (1908) *Formulario de Mathematico*. Fratres Bocca, Torino.

Péter (Politzer), R. (1934) Über den Zusammenhang der verschiedenen Begriffe der rekursiven Funktion. *Mathematische Annalen*, vol. 110, pp. 612–632.

———. (1935) Konstruktion nichtrekursiver Funktionen. *Mathematische Annalen*, vol. 111, pp. 42–60.

———. (1936) Über die mehrfache Rekursion. *Mathematische Annalen*, vol. 113, pp. 489–527.

———. (1951) *Rekursive Funktionen*. Akademiai Kiado, Budapest. Second edition 1957.

———. (1959) Rekursivität und Konstruktivität. In A. Heyting, ed., *Constructivity in Mathematics*, pp. 226–233. North-Holland, Amsterdam.

von Plato, J. (1997) Formalization of Hilbert's geometry of incidence and parallelism. *Synthese*, vol. 110, pp. 127–141.

———. (2001a) A proof of Gentzen's *Hauptsatz* without multicut. *Archive for Mathematical Logic*, vol. 40, pp. 9–18.

———. (2001b) Natural deduction with general elimination rules. *Archive for Mathematical Logic*, vol. 40, pp. 541–567.

———. (2006) Normal derivability and existence property in Heyting arithmetic. *Acta Philosophica Fennica*, vol. 78, pp. 159–163.

———. (2007) In the shadows of the Löwenheim-Skolem theorem: early combinatorial analyses of mathematical proofs. *The Bulletin of Symbolic Logic*, vol. 13, pp. 189–225.

———. (2008) Gentzen's proof of normalization for natural deduction. *The Bulletin of Symbolic Logic*, vol. 14, pp. 240–244.

———. (2009a) Gentzen's logic. *Handbook of the History of Logic*, vol. 5, pp. 667–721. North-Holland, Amsterdam.

———. (2009b) Gentzen's original proof of the consistency of arithmetic revisited. In G. Primiero and S. Rahman, eds., *Acts of Knowledge—History, Philosophy and Logic*, pp. 151–171. College Publications, London.

———. (2010) Combinatorial analysis of proofs in projective and affine geometry. *Annals of Pure and Applied Logic*, vol. 162, pp. 144–161.

———. (2011) A sequent calculus isomorphic to Gentzen's natural deduction. *The Review of Symbolic Logic*, vol. 4, pp. 43–53.

———. (2012) Gentzen's proof systems: by-products in a work of genius. *The Bulletin of Symbolic Logic*, vol. 18, pp. 313–367.

———. (2013) *Elements of Logical Reasoning*. University Press, Cambridge.

———. (2014) Generality and existence: quantificational logic in historical perspective. *The Bulletin of Symbolic Logic*, vol. 20, pp. 417–448.

———. (2016) Aristotle's deductive logic: A proof-theoretical study. In D. Probst and P. Schuster, eds., *Concepts of Proof in Mathematics, Philosophy, and Computer Science*, pp. 323–346. De Gruyter, Berlin.

Poincaré, H. (1902) Les fondements de la géométrie—Grundlagen der Geometrie par M. Hilbert. *Bulletin des sciences mathématiques*, vol. 26, pp. 249–272.

Politzer, R. (1932) Rekursive Funktionen. In *Verhandlungen des Internationalen Mathematiker-Kongresses Zürich 1932*, vol. II, pp. 336–337, Orell Füssli, Zurich.

Popper, K. (1947) Logic without assumptions. *Proceedings of the Aristotelian Society*, vol. 47, pp. 251–292.

Post, E. (1936) Finite combinatory processes. Formulation I. *The Journal of Symbolic Logic*, vol. 1, pp. 103–105.

Prawitz, D. (1965) *Natural Deduction: A Proof-Theoretical Study*. Almqvist and Wicksells. Reprint Dover, New York, 2006.

Presburger, M. (1930) Über die Vollständigkeit eines gewissen Systems der Arithmetik ganzer Zahlen, in welcher die Addition als einzige Operation hervortritt. *Comptes Rendus Premier Congrès des Mathématiciens des Pays Slaves*, pp. 92–101.

Raggio, A. (1965) Gentzen's Hauptsatz for the systems NI and NK. *Logique et analyse*, vol. 8, pp. 91–100.

Russell, B. (1903) *The Principles of Mathematics*. Cambridge University Press, Cambridge.

———. (1906) The theory of implication. *American Journal of Mathematics*, vol. 28, pp. 159–202.

———. (1908) Mathematical logic as based on the theory of types. *American Journal of Mathematics*, vol. 30, pp. 222–262. Page references are to the reprint in Van Heijenoort (1967).

———. (1919) *Introduction to Mathematical Philosophy*. Allen and Unwin, London.

Schmidt, A. (1935) Review of Gentzen (1934–35). *Zentralblatt für Mathematik und ihre Grenzgebiete*, vol. 10, pp. 145–146.

Scholz, H. (1941) Gottlob Frege. As reprinted in Scholz (1961), pp. 268–278.

———. (1942) David Hilbert, der Altmeister der mathematischen Grundlagenforschung. As reprinted in Scholz (1961), pp. 279–290.

———. (1961) *Mathesis Universalis: Abhandlungen zur Philosophie als strenge Wissenschaft*. Ed. H. Hermes et al. Benno Schwabe Verlag, Basel.

Scholz, H. and F. Bachmann (1936) Der wissenschaftliche Nachlass von Gottlob Frege. *Actes du Congrès international de philosophie scientifique Paris 1935*, pp. 24–30.

Schröder, E. (1873) *Lehrbuch der Arithmetik und Algebra für Lehrer und Studirende. Erster band. Die Sieben algebraischen Operationen.* Teubner, Leipzig.

———. (1890–1905) *Vorlesungen über die Algebra der Logik,* vols. 1–3. Teubner, Leipzig.

Schroeder-Heister, P. (1984) A natural extension of natural deduction. *The Journal of Symbolic Logic,* vol. 49, pp. 1284–1300.

———. (2002) Resolution and the origins of structural reasoning: early proof-theoretic ideas of hertz and Gentzen. *The Bulletin of Symbolic Logic,* vol. 8, pp. 246–265.

Schultz, J. (1789) *Prüfung der Kantischen Critik der reinen Vernunft,* vol. 1. Hartung, Königsberg.

———. (1790) *Anfangsgründe der reinen Mathesis.* Hartung, Königsberg.

———. (1792) *Prüfung der Kantischen Critik der reinen Vernunft,* vol. 2. Hartung, Königsberg.

———. (1797) *Kurzer Lehrbegiff der Mathematik.* Nicolovius, Königsberg.

Schütte, K. (1950) Schlussweisen-Kalküle der Prädikatenlogik. *Mathematische Annalen,* vol. 122, pp. 47–65.

Siders, A. (2015) A direct Gentzen-style consistency proof for Heyting arithmetic. In R. Kahle and M. Rathjen, eds., *Gentzen's Centenary: The Quest for Consistency,* pp. 177–211. Springer, Berlin.

Siders, A. and J. von Plato (2015) Bar induction in the proof of termination of Gentzen's reduction procedure. In R. Kahle and M. Rathjen, eds., *Gentzen's Centenary: The Quest for Consistency* pp. 127–130. Springer, Berlin.

Skolem, T. (1913) Om konstitutionen av den identiske kalkuls grupper. *Proceedings of the 3rd Scandinavian Mathematical congress, Kristiania,* pp. 149–163. English translation in Skolem (1970), pp. 53–65.

———. (1919) Untersuchungen über die Axiome des Klassenkalküls und über Produktations- und Summationsprobleme, welche gewisse Klassen von Aussagen betreffen. As reprinted in Skolem (1970), pp. 67–101.

———. (1920) Logisch-kombinatorische Untersuchungen über die Erfüllbareit oder Beweisbarkeit mathematischer Sätze, nebst einem Theoreme über dichte Mengen. As reprinted in Skolem (1970), pp. 103–136. Section 1 translated into English in Van Heijenoort (1967).

———. (1922) Einige Bemerkungen zur axiomatischen Begründung der Mengenlehre. As reprinted in Skolem (1970), pp. 137–152. English translation in Van Heijenoort (1967).

———. (1923) Begrüngung der elementaren Arithmetik durch die rekurrierende Denkweise ohne Anwendung scheinbarer Veränderliche mit unendlichem Ausdehnungsbereich. As reprinted in Skolem (1970), pp. 153–188. English translation in Van Heijenoort (1967).

———. Über die mathematische Logik. As reprinted in Skolem (1970), pp. 189–206.

———. (1930) Über einige Satzfunktionen in der Arithmetik. As reprinted in Skolem (1970), pp. 281–306.

———. (1934) Den matematiske grunnlagsforskning. *Nordisk matematisk tidskrift*, vol. 16, pp. 75–92.

———. (1937) Über die Zurückführbarkeit einiger durch Rekursionen definierter Relationen auf "arithmetische." As reprinted in Skolem (1970), pp. 425–440.

———. (1947) The development of recursive arithmetic. As reprinted in Skolem (1970), pp. 499–514.

———. (1955a) The logical background of arithmetic. As reprinted in Skolem (1970), pp. 541–552.

———. (1955b) A critical remark on foundational research. As reprinted in Skolem (1970), pp. 581–586.

———. (1970) *Selected Works in Logic*, ed. J. E. Fenstad. Universitetsforlaget, Oslo.

Spehr, F. (1840) *Vollständiger Lehrbegriff der reinen Combinationslehre mit Anwendungen derselben auf Analysis und Wahrscheinlichkeitsrechnung*, second edition. Leibrock, Braunschweig.

Sudan, G. (1927) Sur le nombre transfini ω^ω. *Bulletin mathématique de la Société roumaine des sciences*, vol. 30, pp. 11–30.

Tait, B. (2005) Gödel's reformulation of Gentzen's first consistency proof of arithmetic for arithmetic: The no-counterexample interpretation. *The Bulletin of Symbolic Logic*, vol. 11, pp. 225–238.

Tennant, N. (1992) *Autologic*. Edinburgh University Press.

Toledo, S. (2011) Sue Toledo's notes of her conversations with Gödel in 1972–5. In J. Kennedy and R. Kossak, eds., *Set Theory, Arithmetic and Foundations of Mathematics*, pp. 200–207. ASL, Cambridge University Press.

Troelstra, A. (1990) On the early history of intuitionistic logic. In P. Petkov, ed., *Mathematical Logic*, pp. 3–17. Plenum Press, New York.

Troelstra, A. and Schwichtenberg (2000) *Basic Proof Theory*, second edition, Cambridge University Press.

Turing, A. (1936) On Computable Numbers, with an Application to the Entscheidungsproblem. *Proceedings of the London Mathematical Society*, vol. 42, pp. 230–265.

———. (1944) The reform of mathematical notation and phraseology. Manuscript in the Turing Archives. As printed in Cooper and van Leeuwen (2013), pp. 245–249.

———. (1948) Practical forms of type theory. *The Journal of Symbolic Logic*, vol. 13, pp. 80–94. Also in Cooper and van Leeuwen (2013), pp. 213–225.

Waismann, F. (1936) *Einführung in das mathematische Denken*. Cited from the English translation *Introduction to Mathematical Thinking*. Harper, New York, several editions.

Weyl, H. (1918) *Das Kontinuum. Kritische Untersuchungen über die Grundlagen der Analysis*. Veit, Leipzig.

———. (1921) Über die neue Grundlagenkrise der Mathematik. *Mathematische Zeitschrift*, vol. 10, pp. 39–79.

———. (1944) David Hilbert and his mathematical Work. *Bulletin of the American Mathematical Society*, vol. 50, pp. 612–654.

Whitehead, A. and B. Russell (1910–13) *Principia Mathematica*, vols. I–III. Cambridge University Press, Cambridge. Second edition 1927.

Wittgenstein, L. (1922) *Tractatus Logico-Philosophicus*. Routledge, London.

———. (1964) *Philosophische Bemerkungen*. Blackwell, London.

———. (1969) *Philosophische Grammatik*. Blackwell, London.

———. (1975) *Wittgenstein's Lectures on the Foundations of Mathematics Cambridge, 1939*, ed. C. Diamond. Cornell University Press, Ithaca, New York.

———. (1979) *Wittgenstein's Lectures, Cambridge 1932–1935*, ed. A. Ambrose. Blackwell, Oxford.

Zach, R. (1999) Completeness before Post: Bernays, Hilbert, and the development of propositional logic. *The Bulletin of Symbolic Logic*, vol. 3, pp. 331–366.

———. (2003) The practice of finitism. *Synthese*, vol. 137, pp. 211–259.

INDEX

Acerbi, F., 16, 136
Ackermann, W., 2, 121, 126, 133, 138, 174,
    177, 188, 193–199, 201–205, 207–209,
    212, 213, 223–225, 229, 234–236, 248,
    249, 253, 261, 263, 282, 283, 285, 303
algebraic logic, 82
algorithm, 25, 115
Ambrose, A., 153–158
analysis, 69, 73, 79, 155, 183, 185, 212,
    233, 234
  predicative 319
Anzahl, 71, 121, 189, 199
Aristotle, 6, 7–16, 28, 81, 83, 111, 119
arithmetic, 29, 31, 36, 44, 47
  first-order, 225
  Heyting, 27, 143, 233, 318, 335
  Peano, 50, 138, 143
  Presburger, 225
  primitive recursive, 2, 138, 143, 153, 158,
    160, 208, 218, 224
Artin, E., 250, 254
van Atten, M., 158
Awodey, S., 127
axiom
  choice, 76, 233
  existential, 58, 171
  geometry, 5, 57, 168
  Peano, 52, 208, 220, 328

Bachmann, F., 125
bar, 177, 182, 343, 344, 348–351
Becker, O., 158
Beltrami, E., 29
Bernays, P., 2–4, 25, 36, 37, 111, 112, 126,
    130, 152, 154, 155, 159, 160, 164, 174,
    175, 187, 188, 191–193, 195–196, 198,
    199, 203–229, 234, 239, 252, 253, 255,
    256, 259, 260, 262, 263, 272, 273, 280,
    286, 293, 294, 308, 311, 312, 318–321,
    324, 326, 327, 341, 344–346, 348–351
Beth, E., 312
BHK, 262
Bohr, N., 3
Boole, G., 51, 81–87, 110, 127, 141, 325
Boolos, G., 127
Borel, E., 26, 77–80
Born, M., 2, 126
Brouwer, L., 3, 26–28, 63, 79, 80, 112, 120,
    153, 167, 168, 173, 174, 177, 179, 180,
    183, 204, 227, 228, 256, 257, 274, 275,
    325, 326, 344, 345, 349, 350
Burris, S., 147

Cantor, G., 71, 74–79, 126, 150, 185, 233
cardinality, 75, 185
Carnap, R., 24, 127, 202, 230
Cavaillès, J., 319, 348
Chevalley, C., 250
choice sequence, 179, 344, 350, 351
Church, A., 2–4, 65, 115, 150, 202,
    241, 242
Cohen, H., 94
Cohen, P., 76, 233
Cohn-Vossen, S., 60
combinatorics, 29, 38
completeness, 30, 157, 191, 198, 202, 211,
    225, 230, 234
computability, 2, 4, 28, 30, 79
  algorithmic, 50, 54, 115, 138
  mechanical, 47, 50, 241, 247
computers, 1, 4, 63
consistency, 73, 209, 213, 214, 219, 252, 280,
    319, 326, 339
constructivism, 26, 78, 140
continuum, 76, 185, 208, 233
Copeland, J., 4